Culture and Environment in Prehistoric Wales

Selected Essays

edited by

J. A. Taylor

BAR British Series 76
1980

B.A.R.

B.A.R., 122 Banbury Road, Oxford OX2 7BP, England

B. A. R. 76, 1980: "Culture and Environment in Prehistoric Wales".

ISBN 9780860540793 paperback

ISBN 9781407323480 e-book

DOI https://doi.org/10.30861/9780860540793

A catalogue record for this book is available from the British Library

This book is available at www.barpublishing.com

CONTENTS

INTRODUCTION

This book endeavours to achieve a dialogue between archaeologist and geographer and an integrated appraisal of both cultural and environmental changes which have taken place in Wales since the end of the last major cold stage, i.e. the last glaciation, the Devensian. It is initially based on a symposium held in Aberystwyth on September 2nd and 3rd, 1977 on the occasion of the second meeting of the Welsh Association, Y Gymdeithasfa Gymreig, the major objective of which is to promote interdisciplinary seminars on subjects relating to Wales.

Previous landmarks in the study of Welsh prehistory and its environmental background are four in number. First of all, in 1925, the late Sir Mortimer Wheeler's pioneer synthesis on Prehistoric and Roman Wales was published by the Clarendon Press of Oxford. This volume too was based on a series of public lectures—given at University College, Cardiff in 1922. In Sir Mortimer's own words:

> '...I offer a small scrap-book with the hope that it may soon be superseded by a work more commensurate with the material. The hills of Wales teem with the vestiges of early man; much of the evidence is unmapped, and of most of it we shall know nothing substantial without systematic excavation carried out, as it often must be, in remote and storm-ridden uplands with little expectation of more than the most meagre reward.'

This quotation anticipates my final plea made at the end of the last chapter of this volume for more prompt and continuing cooperation in the field between archaeologists and other scientists studying palaeoenvironments. A second quotation from Sir Mortimer, his final paragraph on p. 292, also foreshadows the interdisciplinary theme of this volume. He writes:

> 'In the name of science it is maintained that the fate of empires is determined by a few degrees of Fahrenheit, that the decline of the ancient Hellenic civilisation was due to the inroads of the material mosquito, and that King Arthur was possibly a sun-myth. All these suppositions may be true. But the clay is sometimes a little heavy on the boots.'

The second major landmark was that of Professor W. F. Grimes' Guide to the Collection illustrating the Prehistory of Wales, published in 1939 by the National Museum of Wales in Cardiff. The Museum's Department of Archaeology had encouraged field work specifically which is reflected in the range and quality of the Collection. The second edition of this book, re-named The Prehistory of Wales, appeared in 1951 and incorporated substantial extensions of the Collection and an up-to-date inventory of archaeological sites and excavations in Wales, compiled initially, by Dr. H. N. Savory. The Introduction begins:

'A full understanding of the prehistory of a country like Wales is impossible without some kind of understanding of the prehistory of Britain as a whole'.

This wide perspective still applies, as demonstrated in the essays in this volúme, especially those on the Upper Palaeolithic and the Mesolithic.

The third landmark was the publication in 1965 by Routledge and Kegan Paul of the volume entitled Prehistoric and Early Wales, edited jointly by Idris Foster and Glyn Daniel. This compilation also emanated from a symposium—that of the British Summer School of Archaeology which met at Bangor in August, 1959. The contributors included W. F. Grimes on the Neolithic in Wales and H. N. Savory on the Bronze Age in Wales. The volume was part of a series of 'Studies in Ancient History and Archaeology' under the general editorship of the late Dr. F. T. Wainwright, whom I was privileged to know as a friend and colleague, both of us having attended Prescot Grammar School, Lancashire, he several years earlier than myself. Glyn Daniel in the introduction to the book writes:

'In the lecture he gave in 1932 inaugurating the Reardon-Smith Memorial Lecture Theatre in the National Museum of Wales at Cardiff, Sir John Edward Lloyd... (sometime Professor of History at U. C. N. W., Bangor)... spoke of the role of the archaeologist and the historian sensu stricto in creating the early history of Wales. His lecture was called Wales and the Past: Two Voices, and these were the two voices he was talking about—archaeology and written history. This book discusses how these two voices now speak to tell us of the early part of Wales and the first appearance of man in the part of western Britain which by the sixth century A. D., when we end, was the country of Wales'.

The present volume embraces the last stages of the last cold period and the Holocene and ranges from the Upper Palaeolithic to the end of the Iron Age. But its two voices seek counterpoint between man on the one hand and environment on the other.

The fourth publication, again in 1965, was based on one of a series of annual climatological symposia, convened by the Editor at the University College of Wales, Aberystwyth, on the subject of Climatic change with special reference to Wales and its agriculture. This was published privately as Memorandum No. 8 under my editorship and went through two reprintings. It has long since been out-of-print but several copies have been lodged with the Huw Owen Library of the University College of Wales. Although it deals basically with agro-climatic relationships during the historic period, it is cited here because it constitutes a interdisciplinary look at man-environment relationships in the context of Wales. Hubert Lamb gave the first paper on the history of the Welsh climate followed by a joint team of historical geographers, historical and palaeobotanists. The Editor presented the final paper on recent and current climate trends and their impact on Welsh agriculture with the modern scene in mind but possibly with negative lessons for the prehistoric past, I wrote (p. 100):

'It is illogical to rank human motivations with environmental inducement. The former may or may not be conditioned by the latter. It will certainly predetermine the value and effect of the latter. It may accommodate it or surrender to it but it is always independent of it *sensu stricto*'.

The sense of this quotation would allow even the earliest of prehistoric colonisations a modicum of non-environmental stimulus in decision-making.

This volume then represents the fifth attempt to differentiate man and environment and their interactions within Wales and with special reference to the prehistoric period. Although the general objective is an integrated assessment, each essay represents an individual interpretation of either culture or environment for a specific period, and only in the final chapter (nine) does the editor attempt an overview. In Chapter One, Dr. David Bowen summarises, concisely and authoritatively, the progress of current research on the Late Pleistocene in Wales as a broad environmental framework for the Chapters which follow. Dr. Roger Jacobi offers a re-appraisal of the Welsh Upper Palaeolithic in Chapter Two in both British and European terms. He proposes that, subsequent to the gap in human occupation of these islands after the major glacial stage of the Devensian, man did not, for certain, return to these islands till around 11,500 B. P. In Chapter Three, the Editor discusses the environmental changes which took place during the Welsh Holocene, preparing the way for the ensuing five chapters on the successive archaeological periods from the early Mesolithic to the late Iron Age. Differential and sub-regional timings and impacts of climate changes are explained for the uplands, the western coastlands and the Borders of Wales. Ecosystematic models are presented for each culture period: they are designed to stimulate an integrated approach to both palaeoenvironment and archaeological reconstructions of past landscapes and colonisations, respectively. Dr. Roger Jacobi (Chapter Four) then examines the evidence for the Welsh Mesolithic settlement, followed by Dr. R. N. Savory (Chapter Five) on the Neolithic, Dr. Frances Lynch (Chapter Six) on Bronze Age monuments, Dr. C. B. Burgess (Chapter Seven) on the Bronze Age proper and finally Dr. Savory (Chapter Eight) again on the Iron Age. In the final chapter (nine), the Editor attempts a general perspective on the volume as a whole, focussing on man-environment relationships, models applicable thereto, and some remarkable and independently derived parallels which are emerging from both archaeological and palaeoenvironmental research taking place in Wales.

The Editor wishes to acknowledge the generous assistance of many who have helped with the initiation, preparation and publication of these essays. I thank my colleague Dr. David Bowen with whom I discussed the idea and structure of the volume from the very beginning. He provided much encouragement and support at the Symposium itself. Mr. Christopher Houlder also kindly advised on the archaeological content of the programme at the symposium. Dr. Roger Jacobi has provided continual valuable discussion and criticism on both archaeological and environmental points at many editorial stages. The book was initially inspired by the success of the Lancaster C. B. A. Symposium on the Highland Zone (Research Report No. 11, *The effect of man on the landscape: the Highland Zone*, edited by John Evans *et al.*, Council for British Archaeology, where a joint assessment, both environmental and archaeological, was attempted of Highland Zone changes in landscape and settlement in the prehistoric period. Both Geoffrey Dimbleby and Estyn Evans provided sympathetic promptings in the early stages of planning this volume.

Most of all, however, my gratitude is offered to the five other contributors who have made this book a reality. A composite work is rather like a wartime convoy of ships at sea. It must proceed at the speed of the slowest ship to keep the convoy intact and to maintain protective vigilance at all times. One or two ships in fact did not arrive and one or two additional ships did join the convoy, but no losses or casualties were suffered and no further bulletins on this subject will be issued. Suffice it to say that neither the Editor and certainly not B. A. R., our publishers, were responsible for the unprecedented delay of over two years in publication. We did not go to press till November, 1979 just after the last manuscript arrived in Aberystwyth.

My final appreciation is to those who made publication possible: British Archaeological Reports for efficiency and courtesy at all times; Professor C. Kidson for assistance with departmental cartographic and secretarial facilities; Mr. David Griffiths and his staff for photography; Mr. Michael Gelli Jones and his staff for cartography; Miss Gaynor Hamer, Miss Linda James and Miss Carol Parry for typing assistance; and Mr. Philip Seymour and Miss Janet Baxter for help in organising the symposium (see attendance list following this introduction). Philip Seymour also compiled and edited Appendix I which lists radiocarbon-dated sites in Wales.

Aberystwyth
February 10th, 1980

J. A. Taylor

Editorial Note

In chapters one, three and nine, the conventional radiocarbon dating years B.P. (i.e. before 1950 A.D.) has been used by D. Q. Bowen and J. A. Taylor. In the remainder of the chapters the archaeological convention of radiocarbon years b.c. has been used.

LIST OF FIGURES

Figure		Page

LIST OF PLATES

ALPHABETICAL LIST OF PARTICIPANTS AT THE CONFERENCE HELD IN ABERYSTWYTH IN SEPTEMBER 1977 ON THE SUBJECT OF 'CULTURE AND ENVIRONMENT IN PREHISTORIC WALES'.

Aalen, F. H. Mr.

Barnes, B. Mr.
Barnie, H. Mrs.
Benson, D. Mr.
Bowen, D. Q. Dr.
Bowen, E. G. Prof.
Bradley, R. I. Mr.
Brassil, K. S. Mr.
Burgess, C. Dr.

Caffell, G. Mrs.
Cain, T. R. Mr.
Case, D. Mr.
Chambers, F. Mr.
Charlton, P. E. Mrs.
Chitty, A. M. H. Mrs.
Clayton, D. Mr.
Cook, J. Mrs.
Cook, R. Mr.

David, G. C. Mr.
Davies, G. Miss
Davies, M. Mr.
Delaney, C. J. Mr.

Evans, A. H. Miss

Girling, M. Miss
Grimes, W. F. Prof.

Harris, C. Dr.
Houlder, C. H. Mr.

Jacobi, R. M. Dr.
James, T. Mr.
Jones, J. A. Mr.
Jones, J. A. Mrs.

Keepax, C. A. Mrs.
Kelly, R. Mr.

Green, H. S. Mr.
Green, M. J. Mrs.

Lawes, A. R. Mr.
Lynch, Frances

Miles, H. Mrs.
Moore, P. D. Dr.
Morgan, J. P. Miss
Mortingen, H. E. Miss

Newson, M. Dr.

Power, J. Mrs.

Ridge, T. S. Mr.
Roberts, L. J. Miss
Roese, H. E. Mr.
Rudeforth, C. C. Dr.

Savory, H. N. Dr.
Seymour, W. P. Mr.
Stone, A. Mrs.
Stone, T. A. Mr.

Taylor, J. A. Mr.
Tyler, A. Mr.

Vale, V. Miss
Vyner, B. E. Mr.

Walker, M. F. Mr.
White, F. Mr.
Williams, J. L. W. Dr.
Wiltshire, P. Mrs.

CHAPTER ONE

THE PLEISTOCENE SCENARIO OF PALAEOLITHIC WALES

D. Q. Bowen

Abstract

A conservative outline of the Pleistocene sequence of environmental change is sketched. It is based primarily on lithostratigraphy (rock-stratigraphic units), minimally assisted by some biostratigraphy, and inferences drawn therefrom. Other than for the Devensian Late-glacial, chronostratigraphy is based on the areal stratotype of Gower, South Wales.

Correspondence between the Palaeolithic archaeological and Pleistocene geological records is often ambiguous, and, save for the Devensian Late-glacial (Creswellian), ascription of artifacts to broadly defined geologic-climatic units (climatostratigraphy) is not always secure.

The sequence of events hinges upon the recognition of the Last Interglacial (deep-sea isotope stage 5e) marine transgression recorded by a raised beach unit. Aminostratigraphy, however, indicates that this event is not so precisely located as was hitherto thought. Antedating the Last Interglacial (sl) is a complex sequence of events not yet satisfactorily elucidated, including complete inundation by ice sheets. No geological evidence exists to support any theory of glacial transportation of the Preseli dolerites to Wiltshire.

Early and Middle Devensian time was cold with periglacial climates temporarily interrupted by brief interstadial ameliorations. No evidence for glaciation occurs until the Late Devensian, after 26 ka B.P., when, except for south-west Dyfed, much of Gower and the Vale of Glamorgan, Wales was extensively glaciated. After deglaciation, complete by ca. 14.5 ka B.P., and the climatic amelioration of a Late-glacial (Windermere) interstadial, severe periglacial conditions obtained between 11 and 10 ka B.P.

All disciplines with an historical interest in the Pleistocene Epoch are being subject to a phase of re-assessment in the light of the now established greater complexity of that span of time. In essence, older notions of it as having consisted of four or possibly six glacials, separated by interglacials,

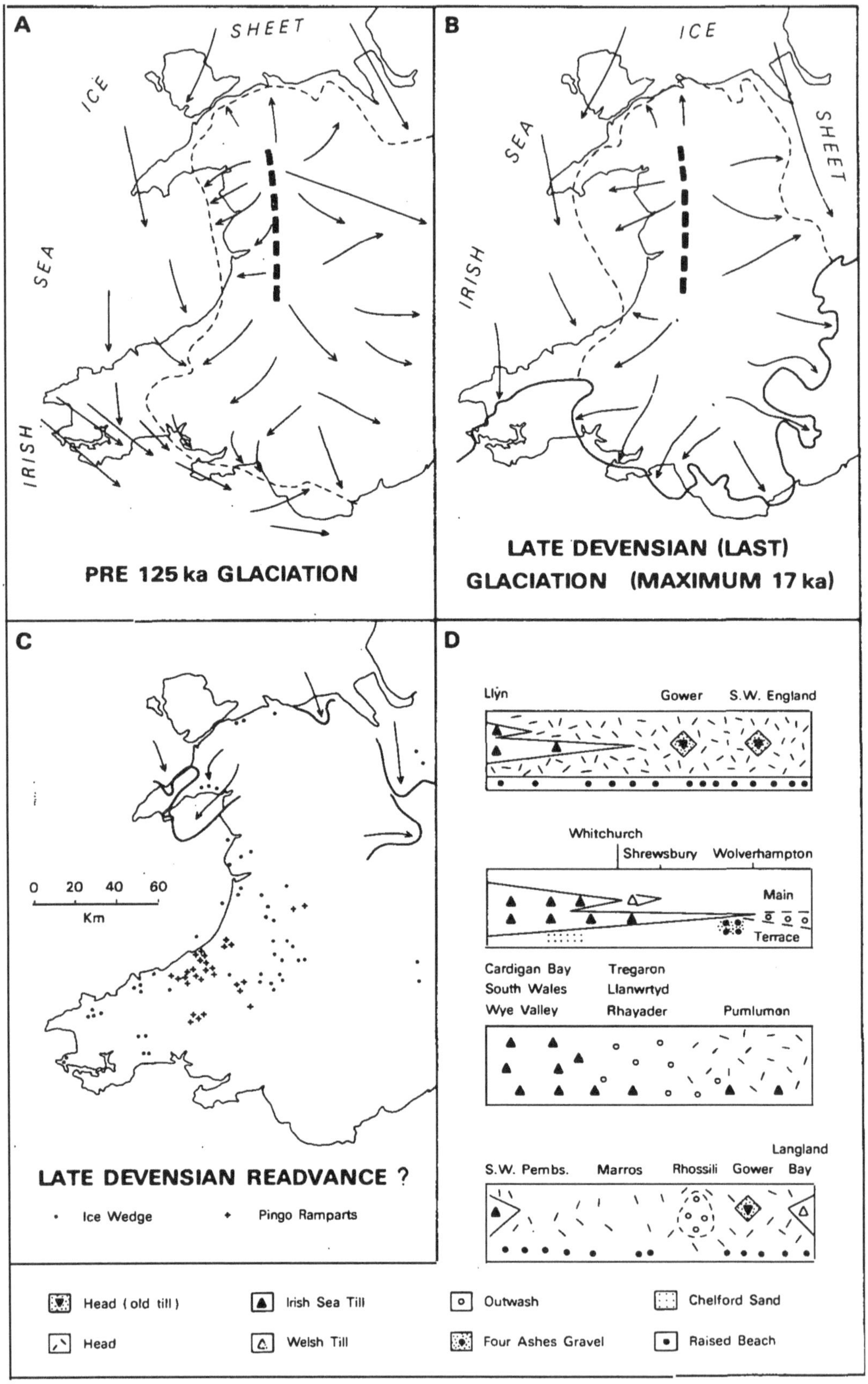
A
SHEET
ICE
SEA
IRISH
PRE 125 ka GLACIATION
B
ICE
SHEET
SEA
IRISH
LATE DEVENSIAN (LAST) GLACIATION (MAXIMUM 17 ka)
C
0 20 40 60
Km
LATE DEVENSIAN READVANCE ?
Ice Wedge
Pingo Ramparts
D
Llŷn
Gower
S.W. England
Whitchurch
Shrewsbury
Wolverhampton
Main
Terrace
Cardigan Bay
South Wales
Wye Valley
Tregaron
Llanwrtyd
Rhayader
Pumlumon
S.W. Pembs.
Marros
Rhossili
Gower
Langland Bay
Head (old till)
Head
Irish Sea Till
Welsh Till
Outwash
Four Ashes Gravel
Chelford Sand
Raised Beach

Figure 1.1. The Glaciation of Wales. A: previous to the *Patella* beach interglacial (isotope stage 5e). B: Late Devensian maximum ice extent. C: Extent of possible readvance in Gwynedd, Powys and Shropshire. This has been questioned in the Llŷn peninsula by Boulton (1977). Ice-wedge pseudomorphs and remains of pingos are all probably post-readvance and time-equivalent to the Younger Dryas. D: Diagrammatic and reconstructed sections illustrating the temporal-spatial relations of Welsh Pleistocene deposits. The raised (*Patella*) beach is Last Interglacial; Chelford Sand—Early Devensian; Four Ahses Gravel—Early and Middle Devensian. All Irish Sea and Welsh tills shown are Late-Devensian. Head deposits may be Early, Middle and Late Devensian. The Main Terrace of the River Severn is contemporaneous with the Late Devensian maximum advance, ca. 17 ka ago. The 'surrounding' of central Wales by Late-Devensian glacial deposits places the periglacial inheritance of that area in correct temporal-spatial perspective as Devensian Late-glacial in its fashioning—mostly time equivalent to the Younger Dryas.

are now redundant, for overwhelming evidence for a greater number of such events is to hand—and has been since Emiliani's (1955) pioneer work on the oxygen isotope stratigraphy of deep-sea cores. The development of this work and its present-day implications are described elsewhere (Bowen 1978). The task facing Quaternary scientists in most areas is formidable, being nothing less than a full examination of their data which was originally classified according to simpler models of Pleistocene history. In Wales such a task is considerably easier because its uplands functioned as gathering grounds for ice accumulation on several occasions and its margins were invaded by an ice-sheet with origins farther north in the Irish Sea (Fig. 1.1). This means that on each successive glacial occasion the combined effects of valley glaciers, ice-caps and ice-sheets would have effectively destroyed much, if not all, the evidence for antecedent events. In consequence it is only the last major glaciation that is represented in any widespread sense in Wales. Earlier events are represented principally along the Welsh borderlands (mostly in England), on the floor of the Irish Sea, and in South Wales beyond the maximum extent of the Late Devensian ice-sheet (Fig. 1.1). In South Wales sufficient evidence remains to reconstruct both the Last Interglacial (used in the sense that it refers to the deep-sea oxygen isotope stratigraphy Stage 5e, c. 125,000 years ago—but see below) as well as the events of Eary and Middle Devensian time prior to the development of the Late Devensian ice-sheet. At some highly localised sites for examples, the Gower caves, it is possible to find evidence for still earlier events, although this is not yet definitive (Bowen 1977a).

Viewed in a prehistoric and Palaeolithic context,the geologic record spans the phases to which the terms Mousterian, Aurignacian (Blattspitzen) and Creswellian apply. In some cases the Palaeolithic record is clearly related to and firmly within the context of the geological (Pleistocene) for example, the Creswellian site at Cathole, Gower. But in others it is not, as in the case of the Acheulian hand-axe found at Pen-y-lan, Cardiff, or the Levallois artifact allegedly from the raised beach at Goat's Hole, Paviland, Gower (McBurney 1965). In other cases radiocarbon assay places the archaeological context within such a geochronometrically determined time-scale (e.g. the 'Red Lady' of Paviland—Oakley 1968), but the geological context in lithostratigraphic terms is not precisely known. Thus the correspondence of the Pleistocene lithostratgraphic record with that of the Palaeolithic is not entirely unambivalent. It is through a combination of precisely established geological context, or failing that by the ascription of a particular Palaeolithic discovery within a broadly determined geologic-climatic unit (climatostratigraphy) that events can be characterised according to environmental parameters. The latter procedure, however, is less reliable, founded as it is on secondary inference, unlike that of precisely stratified objects in lithostratigraphic and possibly also biostratigraphic context. This distinction is one of some importance for climate and the floral and faunal make-up of an environment altered swiftly at certain times in the Upper Pleistocene (Last Interglacial and Last Glaciation), and ascription of a particular Palaeolithic event to a geologic-climate one may be radically in error. One thing seems clear. The geomorphological scenario for Palaeolithic man in Wales was basically similar to the present-day. This of course excludes some changes along the coastline as well as those in valleys where aggradation and incision of fluviatile gravels occurred. But the outline of plateaux surfaces, hillslopes, and even cliffed-coastline was already fashioned. In this context Palaeolithic Wales, at least as far as that record goes, and is presently understood, post-dated the considerable geomorphic changes which in landform terms occurred about the beginning of the Middle Pleistocene (c. 700 ka ago). The nearest indication is at Westbury-sub-Mendip in Somerset, where artifact Cromerian deposits (Cromerian III), are post-dated by more than 200 m of base-level fall (Bishop 1975).

Events antedating the Patella Beach Interglacial

At the present state of knowledge it seems reasonable to assume that the principal raised beach of Gower, the *Patella* beach of T. N. George (1932) is Last Interglacial in age. The term 'Ipswichian' should be avoided in Wales for even in England it is not clear what it means (Sutcliffe 1975) and it seems probable that two if not three such events are subsumed under that label (Bowen 1978). A suitable stratotype (type-site) has not yet been described in Wales and in that sense use of the term Pennardian by Bowen (1977b) is premature. It does seem likely, however, that the site on which he bases the name, Minchin Hole Cave, will emerge as the stratotype. Broad-based biostratigraphy aminostratigraphy, (Andrews, Bowen and Kidson, 1979), an uranium-thorium date from nearby Bacon Hole (C. Stringer - personal communication), as well as the lithostratigraphy along the adjacent coastline (Bowen 1973a), indicate that the *Patella* beach at Minchin Hole (outer beach) and along the coast, is Last Interglacial in age. Its importance lies in the use it allows of a widespread stratigraphic marker for, Gower, Carmarthen Bay, Pembrokeshire and parts of Gwynedd (Bowen 1977a).

At Minchin Hole at least two lithostratigraphic units antedate the *Patella* beach: the 'red cave-earth' at the base of an inner talus cone, and the 'inner beach', an interidal beach sand at c. 12 m O.D. (upper limit). It is possible that some of the beds immediately overlying the red cave-earth are also older but this is indeterminate at the moment. No artifacts have been recovered hitherto from any of the units at this site. Certain favoured situations then may well preserve a stratified record of pre-Last Interglacial events. The same is probably true for caves in the Carboniferous Limestone in South and North Wales.

At no locality outside the extent of Late Devensian glaciation in Gower does glacial material lie *in situ* in a position above the *Patella* beach. Derivative glacial materials do lie within the head overlying the beach, or occasionally, as on the north side of Port-Eynon Bay (Bowen 1971), occur as discrete lithostratigraphic units. These diamicton deposits, however, are technically periglacial head, and Devensian in age. Their original glacial sources lie *in situ* on the extensive plateaux surfaces of Gower and South Pembrokeshire, where a pre-*Patella* beach age is established from regional stratigraphic relations. They relate to an event of full and complete glaciation of Wales (Fig. 1. 1). Only limited evidence survives in North and mid-Wales, consisting of erratic pebbles incorporated into periglacial sequences of Devensian age. But in South Wales sufficient evidence remains to reconstruct the extent of glaciation and to indicate the directions of ice flow (Fig. 1.1). Irish Sea ice clearly overran the Pembrokeshire peninsula and penetrated at least as far as Pencoed and Ewenny in Glamorgan. It seems reasonable to correlate this advance with the one which reached north Devon and the Isles of Scilly. In comparison with South Pembrokeshire and Glamorgan, glacial deposits of pre-Patella beach age are unusually extensive in west Gower. They consist of an admixture of Irish Sea and Welsh materials (George 1933) and in their frequently undissected outcrop may be explained in at least two ways. First, that the Irish Sea ice-sheet was relatively debris-free south of Mynydd Preseli but that in west Gower, where it impinged upon the more locally generated Welsh ice-sheet, a more plentiful debris supply of Welsh derivation was available. Or second, that the contrast in dissection between west Gower and elsewhere reflects an age difference, the implication being that two separate glacial events are involved, that of West Gower being the younger, when Welsh ice incorporated Irish Sea materials at a time when that ice-sheet did not penetrate so far east (Bowen 1970). It is not possible to date the pre-Patella beach glaciation(s) and the use of labels such as Wolstonian is misleading and derives from the practice of forcing local evidence into preconceived pigeon holes of classification erected elsewhere by 'counting from the top'.

No artifacts have been discovered in these glacial deposits, but the archaeological implication of them is not inconsiderable because they relate to the ice-sheet glaciation alleged by some to have transported the bluestones of Mynydd Preseli across south Wales and the northernmost fringe of what is now the Bristol Channel into western England thence to the Salisbury Plain area where they were subsequently incorporated into the Stonehenge monument (Kellaway 1971). This thesis was subsequently reiterated and expanded (Kellaway et al. 1975) into one of universal glaciation of southern England and the English Channel outside the conventionally accepted extent of maximum glaciation (e.g. West 1977). The evidence on which such a reconstruction lies has been examined and found to be ambiguous to such an extent that in the strictest geological context the hypothesis cannot be reasonably entertained (Kidson and Bowen 1976). In the west, however, there can be no doubt that an Irish Sea ice-sheet did traverse the Bristol Channel (Fig. 1.1). The extent of its penetration beyond, however, is debatable, and in the absence of conventional evidence for glaciation, not proven. Its agency in transporting the bluestone rocks of Preseli, however, is still urged by some, B. S. John (1976), who argued that at one time there would have been bluestones in abundance no great distance away from Stonehenge.

The hypothesis is untenable for several reasons. Outcrops of the dolerite rock in question on Mynydd Preseli consist of narrow and linear features of limited areal dimensions. It is unlikely that glacial movement across them would have incorporated large masses of rock fragments into the ice. The Irish Sea ice-sheet did not effect appreciable erosion in this area and it is to be expected that other than for incorporation of tor-like outcrops (much as the outcrop is characterised today) of frost-shattered debris, bedrock proper would not have been greatly quarried and the elevation of the summits of the Preseli greatly reduced. This is borne out by the limited incidence of dolerite erratics south of the outcrops. J. C. Griffiths (1940) showed that an erratic indicator fan of this material may be traced only a short distance, to the Narbeth-Whitland district, after which no such rock fragments are known from the glacial deposits of South Wales. Moreover, it is pertinent to note that other indicator erratics from Pembrokeshire do occur: pyroxenic keratophyre from

New Inn has been discovered in Gower, and even in the outskirts of Cardiff (Griffiths 1940), while several others of known provenance lie in the ground south of Mynydd Preseli and in west Gower. The geological evidence is thus not consistent with the scenario of liberally scattered Preseli dolerite pebbles in western England awaiting prehistoric man's collection for use at Stonehenge as envisaged by some (John 1976).

The Patella beach Interglacial

As previously discussed, the Patella beach of Gower is dated as Last Interglacial (oxygen isotope stratigraphic stage 5e). Correlated with it on lithostratigraphic grounds are the similar raised beaches of west and north-west Wales (Bowen 1973a, 1974, 1977a). These include the raised beaches at: (1) Poppit, in the Teifi estuary, considered to be unfossiliferous by John (1965), but correlated biostratigraphically with marine interglacial deposits from a bore-hole in Cardigan Bay by D. L. Lear (personal communication); (2) at West Angle (although here some doubt attached to the interpretation of the alleged marine deposits) where pollen analytical investigations have revealed fully interglacial conditions (John 1965). There the presence of Hippophae pollen might encourage its dating as Hoxnian. But that label has no meaning in South-West Wales and is of limited value unless tied to a stratigraphic and geochronometrically dated sequence; (3) the deposits described by officers of the Geological Survey at Landshipping on the Daugleddau River, which are overlain by 'glacial' deposits lying outside the limit of the Late Devensian glaciation. These are poorly sorted and compare with similar ones upstream along the Western Cleddau which derive from catastrophic floods of meltwater directed via the Jordanstown and Esgyrn-Nant-y-bugail meltwater routeways which at one stage carried all the discharge from proglacial Lake Teifi as well as from the Irish Sea ice sheet in North Pembrokeshire, and are conceivably Devensian proglacial in age than earlier. There can be no doubt, however, that the coastline of Wales was occupied by more than one high sea-level as the inner beach at Minchin Hole shows. But until stratigraphic demonstration exists of other examples it is most reasonable to assume that they are the same age as the Patella beach. Recent work at Minchin Hole has demonstrated the reality of T. N. George's (1932) Neritoides beach, obscure in the excavations of the early 1970s, but now clearly separated from and younger than the Patella beach. Lithologically unusual as a beach sediment it is currently under investigation.

The significance of the Neritoides beach is unclear. It is, however, separated from the Patella beach by terrestrial deposits, but is unusual in that it contains a mixture of angular and rounded material within a predominantly shelly matrix. If the Patella beach is equivalent to O. I. Stage 5e then the Neritoides beach may be equivalent to O. I. Stage 5c. Estimates of world sea-level based on Uranium-Thorium dating has indicated that it stood only some 2 m below the present 105 ka ago (5c). It is, then, possible that a high sea-level obtained some 105,000 years ago when the Neritoides beach was deposited. It is, however, equally possible that the inner beach is 5e, Patella beach 5c (Bowen 1973c) and Neritoides 5a in age. Amino-acid determinations in progress may help clarify this question.

From the point of view of defining the Last Interglacial the chronostratigraphic evidence in continental Europe is unsatisfactory. At Grande Pile, in the Vosges Mountains, Woillard (1978, 1979) has shown that two discrete interglacials, defined by the appearance of mixed oak forest, occur after the classical Eemian (Last Interglacial). They could correspond to Oxygen Isotope Stages 5c and 5a. Thus, as palaeobotanically defined, three interglacials would occur in the Upper Pleistocene instead of, as currently accepted, one short one, or one long but complex one with sub-minima. To a large extent this is a matter of definition, but in the immediate context, serves to highlight the problems of characterisation with regard to environments. All the more difficult then is the ascription of artifacts to particular chronostratigraphic niches as is the case with the alleged Levallois find from the Paviland collection. It is by no means certain that Palaeolithic man was in Wales at this time, although his absence here, as in England where he was but scantily represented, is enigmatical at a time of generally optimal environmental conditions.

Notwithstanding the problems discussed above, palaeoecological reconstruction of the Patella beach interglacial from discoveries within the beach unit and immediately above it, as well as from the Bontnewydd Caves in Clwyd (Bowen 1974) show that climate was at least as warm as today and possibly warmer. Presumably the flora consisted of mixed oak deciduous forest as is hinted at from West Angle in Pembrokeshire, and shown by the pollen data of Campbell (1977) in Gower. Other data include terra fusca soil derivatives in the colluvium above the raised beaches in Gower, while terra rossa soils, possibly of the same age, have been recorded near Porthcawl (Crampton 1966). The stratigraphy of such palaeosols evident in pits prepared on the plateau surface at Hunts Bay, Gower, however, shows that the sequence is more complicated and has yet to be correlated satisfactorily with the coastal deposits (Clayden, in Bowen 1977b).

The Devensian Cold Stage

It is apparent from the foregoing that problems of defining the Last Interglacial are complementary to establishing the base of the Last (Devensian) Glacial Stage. Although the stratotype for the Devensian lies at Four Ashes, Wolverhampton (Morgan 1973),its base has not been satisfactorily defined. The Stage as a whole, however, is subdivided on the basis of radiocarbon years (base of Middle at 50 ka, and Late Devensian at 26 ka). Late-glacial time is thus part of the Late Devensian, although the two are frequently and erroneously used as synonyms. Such a practice has been argued by Mangerud et al. (1974) for Norden, and accepted as realistic for the British Isles by Pennington (1975). This, however, would place the maximum expansion of ice in the Middle Weichelian (Devensian), a procedure totally unrealistic in view of the onset and culmination of this event within globally recorded data: e.g. from isotopic analyses of ice-sheet cores or deep-sea cores. Every glacial has its Late-glacial and, in Europe, is recorded by the gradual amelioration of climate usually determined from the palaeobotony. With the stratotypic definition of a Windermere Interstadial (Late-glacial interstadial) in the Lake District (Coope 1977),is to be hoped that the matter is settled. In an archaeological context this would be desirably so, for Late-glacial time (of the Late Devensian) would more or less correspond with the time-span of the Creswellian culture.

Depending where the Last Interglacial-Devensian boundary is drawn, the Last Glaciation represents either 70,000 or 108,000 years. If the former then it includes four principal interstadials in the British Isles (Wretton, Chelford, Upton Warren and Windermere). But if the latter, it would include not only the above, but also hitherto unidentified interstadials known to occur at c. 105 and 82 ka ago in the oxygen isotope record. At the present time, and with the exception of the raised beach problem (above), and Campbell's (1977) pollen work in Gower, the identification of Early and Middle Devensian interstadials in Wales is still awaited.

Early and Middle Devensian time cannot be subdivided in Wales at present. These sub-stages are, however, represented by sequences of slope deposits displayed to advantage along the Welsh coastline (Bowen 1977a). In general they show that after initial disforestation at the end of the previous interglacial, soil erosion, sheet-wash erosion and colluviation occurred, followed by the effects of a frost climate on exposed bedrock slopes. Finer grained material, including loess-like silt, within some of these head deposits is under current investigation. Loess, for example, is more widespread than hitherto believed. Such head deposits also include lithofacies of several varieties, these being dependent on local geology, catchment characteristics, as well as the availability of materials for incorporation. In the latter category occur old glacial deposits which were subject to solifluction.

Samples of thermoclastic scree from excavations at Longhole and Cathole Caves, Gower, were subject to pollen-analytical investigation by Campbell (1977). He interpreted the data as representing the close of the Last Interglacial the entire Last Glaciation and Post-glacial. Moreover, at Longhole Cave the 'pollen sequence would appear to be the best thus far available for the whole of the British Last Glacial'. It shows two minor peaks of thermophilous woodland which he ascribes to the Last Interglacial and Chelford Interstadial. The latter shows a sharp increase in boreal forest, with Picea, Pinus and Betula, and is associated with two Middle Palaeolithic artifacts, presumably of Mousterian age. Higher up the sequence the discovery of Earlier Upper Palaeolithic artifacts ('Aurignacian') occurs where pollen values of Juniperus and Salix demonstrate a comparative abundance which he dates to the 'Middle Last Glacial' (presumably Upton Warren interstadial complex). At Cathole he interprets systematic changes in the flora to indicate Late-glacial Pollen Zones I, II and III, with Pollen Zone II being represented by a rise in Betula. Artifacts of Later Upper Palaeolithic aspect collected by Campbell (1977) are related by him to Pollen Zone I. Of some considerable interest is the appearance of thermophilous trees at this point in his pollen diagram, but this is not commented upon. Earlier, McBurney (1959) discovered Creswellian artifacts which Campbell considers to have come from a horizon time-equivalent to Pollen Zone III. Except for the event referred to the Chelford Interstadial by Campbell and the rise in Juniperus and Salix also recorded by him at Longhole Cave, no other indications of Early and Middle Devensian interstadial conditions are presently known in Wales. It must be added, however, that his pollen analysis was carried out on samples from limestone scree and sand, with a limited silt and clay fraction. These deposits lie on slopes and technically are colluvium and/or head. It follows that some of the pollen may be derived. This applies particularly to the basal layers, and the identification of a Chelford interstadial may rest on Last Interglacial pollen subject to incorporation into younger deposits. From another viewpoint, and if it is assumed that the pollen is not derived, the events recorded could equally be those of O. I. Stages 5e and 5c. This in fact is more likely because thermophilous trees are unknown from British interstadials yet they are recorded at the 'Chelford' horizon of Campbell's (1977).

Late Devensian glaciation in Wales involved both Welsh and Irish Sea ice-sheets (Fig. 1.1). The extent of glaciation has been one of considerable controversy in the past involving some bizarre reconstructions (Bowen in press), but the limit shown on Figure 1.1 is based primarily on the stratigraphic record and as such has been adopted by the Geological Survey on their Quaternary Map of Britain (1977). From the point of view of ice-free areas, only southern Pembrokeshire, western Carmarthenshire, west Gower, south Glamorgan and east Gwent were unglaciated. In these areas a severe climate obtained and loess deposition reached a maximum at this time. Following the principal deglaciation, it has been suggested that a readvance of the ice-sheet occurred in Gwynedd, Clwyd and the Shrewsbury district (the Welsh Readvance) (Bowen 1974). While the lithostratigraphic evidence seems sound enough in the Shrewsbury area, Boulton (1977) has reinterpreted the critical exposures in the Llŷn peninsula around Glanllynnau and concluded that no evidence for readvance occurs. The complex multi-till and outwash sequence can in his view be the product of supraglacial deposition. It may well be that this model can be successfully applied to other complex sequences.

The period between the disappearance of Late Devensian ice, about 15,000 years ago, and the opening of the Holocene Epoch, 10,000 years ago, broadly synonomous with Upper Palaeolithic or Creswellian, has been investigated throughout Wales, although there is a noticeable preponderance of sites in Gwynedd and a dearth in south-west Dyfed and south Wales generally. These Late-glacial sequences are commonly represented lithostratigraphically by a three-fold subdivision which include a single interstadial (Allerød), although it is claimed that a further interstadial (Bølling) may be represented at some sites. Pollen-analytical investigation has subdivided these sediments into conventional Late-glacial biozones I, II and III, although occasionally accomplished with some difficulty. It has also been assumed that they correspond to the continental sequence of Older Dryas Stadial, Allerød Interstadial and Younger Dryas Stadial: that is, a single climatic amelioration preceded and followed by colder conditions. Much depends on the criteria used to establish interstadial conditions for Coleopteran faunas indicate climatic amelioration antedates such indications by pollen (e.g. at Glanllynnau, Gwynedd, Coope and Brophy 1972). Throughout western Britain generally the pollen evidence indicates climatic amelioration at about 13,000 BP and rapid deterioration at about 11,000 BP, whereas the Coleopteran evidence would place climatic amelioration somewhat earlier, before 14,500 BP (Coope 1977).

Recently the Late-glacial was redefined at a stratotype in the Lake District and named the Winderemere Interstadial. It corresponds to the continental Bølling, Older Dryas and Allerød, but its lower boundary has not yet been agreed upon in the sense that it can be used as a standard scale (Coope 1977).

Hitherto no British stratotype exists for the post-Windermere climatic deterioration in climate, and the terms Younger Dryas or Loch Lomond Stadial remain in general use for this period.

The Welsh evidence was reviewed by Moore (1977) and although a general pattern is clearly discernible which is consistent with the rest of western Britain (Pennington 1975, 1977), there is a lack of radiocarbon dates for detailed correlation purposes. Climatic deterioration evident elsewhere at about 11, 000 BP is well represented in the pollen diagrams which, by virtue of their location inside and beyond cirque margins in Gwynedd and Powys (Brecon Beacons), allow the inference to be drawn that cirque glaciers and large snow patches existed at that time.

The geological and geomorphological evidence points to a rigorous climate and an extraordinarily rapid and effective operation of geological processes. Extensive scree formation occurred, solifluction took place when large volumes of material were moved downslope (virtually the whole of the Rhosili Bay, Gower, solifluction terrace was built up), and valley floors subject to gravel aggradation. Ice-wedge pseudomorphs and the remains of small pingos show that this event was accompanied by the existence of what was probably discontinuous permafrost. Elsewhere Coleopteran faunas characteristic of high arctic conditions occupied southern England so that there can be no doubt as to the severity of this episode, nor of the dramatic changes in landscape it produced (Bowen 1977c, Sissons 1979).

<u>Overview</u>

The geological record, as lithostratigraphy, is arguably complete insofar as Palaeolithic occupation of Wales is concerned. Subdivision of that record in terms of standard stages is difficult to achieve, a situation handicapped by biostratigraphic poverty and the lack of appropriated geochronometric dates. It is clear, however, that the record consists, at least in broad terms, of an important interglacial (that of the <u>Patella</u> beach), preceded by massive glaciation, and followed by the Devensian Stage. In greater detail, problems remain over the status and significance of the three marine units at Minchin Hole Cave, and the definition of the base of the Devensian Stage.

Biostratigraphic responses to fluctuations in the climatic conditions of Isotope Stage 5 and the Devensian Stage are sparsely recorded in Wales. It is true that the mammalian fauna of interglacial aspect as well as ofoold tundra biotype is known. But this is largely unrelated to the stratigraphic details of sites; hence its timing and environmental implication, not least for human occupation, is limited. Campbell's (1977) pollen investigations in Gower are lithostratigraphically secure, but chronostratigraphic dating is not definitive. This inability to characterise Early and Middle Devensian environments is frustrating in the light of what is known about the period from the only continuous pollen diagram for Northern Europe at Grand Pile, northeastern France (Woillard 1978, 1979), and deep-sea cores V19-29 from the East Equatorial Pacific, and core 12392-1 from the Northeastern Atlantic (Shackleton 1977), all of which display remarkably consistent correlation throughout the Last Glaciation and Isotope Stage 5, and reveal numerous climatic fluctuations. In general terms, however, it is encouraging to note that conventional stratigraphic placing of Mousterian, 'Aurignacian' and Creswellian cultures, coincide with times of climatic amelioration in what was otherwise an intensely cold, and moreover arid, succession of changing environments. Other than for the high sea-levels recorded by the raised beaches, sea-level remained below the present for all the period under consideration. It is possible that it rose to just under the present 60,000 years ago, but during the Upton Warren interstadial remained at least 40 m below, these estimates being based on the ^{230}Th dating of coral terraces in New Guinea (Bloom et al. 1974). During the Devensian Late-glacial it seems unlikely that sea-level rose about -40 m below present (Haynes et al. 1977). Ostensibly it would appear that the Devensian Late-glacial is a period with a comparative wealth of biostratigraphic evidence. It is true that the general vegetational succession has been established but precise palaeoecological characterisation is strictly limited. Not only may autecological evidence be highly misleading, but assemblage floras lack contemporary analogues, a persistent difficulty attending Pleistocene palaeoecology (Bowen 1979). Interestingly enough, the Late-glacial geomorphic scenario of Wales also lacks a precise environmental analogue today.

Acknowledgement

The author would like to acknowledge financial support for field work from the Sir David Hughes-Parry fund of the University College of Wales Aberystwyth.

REFERENCES

Andrews, J. T., Bowen, D. Q., Kidson, C., 1979. Amino-acid ratios and the correlation of raised beach deposits in south-west England and Wales. Nature, London 281, 556-558.

Bishop, M. J., 1975. Early records of man's presence in Britain. Nature, London 253, 95-97.

Bloom, A. L., Broecker, W. S., Chappell, J. S., Mathews, R. L., and Mesolella, K. J., 1974. Quaternary sea level fluctuations on a tectonic coast: New Th^{230}/U^{234} dates from the Huon Peninsula, New Guinea. Quaternary Research 4, 185-205.

Boulton, G. S., 1977. A multiple still sequence formed by a Late-Devensia Welsh ice-cap. Cambria 4, 10-31.

Bowen, D. Q., 1970. South east and central south Wales, in C. A. Lewis (ed.), The Glaciation of Wales and adjoining regions. London (Longman), 197-227.

Bowen, D. Q., 1971. The Quaternary succession of South Gower, in D. A. and M. G. Bassett (eds.), Geological Excursions in South Wales and the Forest of Dean, 135-142.

Bowen, D. Q., 1973a. The Pleistocene History of Wales and the borderland. Geological Journal 8, 207-224.

Bowen, D. Q., 1973b. The Pleistocene succession of the Irish Sea. Proceedings Geologist's Association, London 84, 249-272.

Bowen, D. Q., 1973c. The excavation at Minchin Hole. Gower 24, 12-18.

Bowen, D. Q., 1974. The Quaternary of Wales, in T. R. Owen (Ed.) The Upper Palaeozoic and post-Palaeozoic Rocks of Wales. Cardiff at the University Press. 373-426.

Bowen, D. Q., 1977a. The coast of Wales, in the Quaternary History of the Irish Sea. Geological Journal Special Issue No. 7, 223-256.

Bowen, D. Q., 1977b. Wales and the Cheshire-Shropshire Lowland. Guide-book for Excursions A8 and C8. X INQUA CONGRESS. 66pp.

Bowen, D. Q., 1977c. Late-Devensian Periglacial Environments in Upland Britain, X INQUA CONGRESS, BIRMINGHAM 1977 Abstracts, 49.

Bowen, D. Q., 1978. Quaternary Geology: A Stratigraphic Framework for Multidisciplinary Work. Pergamon Press. Oxford, New York, Toronto, Sydney, Paris, Frankfurt. 220pp.

Bowen, D. Q., 1979. Geographical Perspective on the Quaternary. Progress in Physical Geography. 3, 167-186.

Campbell, J. B., 1977. The Upper Palaeolithic of Britain (2 vols). Oxford.

Coope, G. R., 1977. Fossil Coleopteran assemblages as sensitive indicators of climatic changes during the Devensian (Last) cold stage. Phil. Trans. R. Soc. Lond., B, 313-340.

Coope, G. R. and Brophy, J. A., 1972. Late Glacial environmental changes indicated by a coleopteran succession in North Wales. Boreas 1, 97-142.

Crampton, C. B., 1966. Certain effects of glacial events in the Vale of Glamorgan, South Wales. J. Glaciol. 6, 261-266.

Emiliani, C., 1955. Pleistocene temperatures. J. Geology 63, 538-578.

George, T. N., 1932. The Quaternary Beaches of Gower. Proc. Geol. Ass. Lond., 43, 291-324:1933. The glacial deposits of Gower. Geol. Mag. 70, 208-232.

Griffiths, J. C., 1940. The glacial deposits west of the Taf. Unpublished Ph.D. thesis, Univ. of London.

Haynes, J. R., Kiteley, R. J., Whatley, R. C., Wilks, P. J., 1977. Microfaunas, microfloras and the environmental stratigraphy of the Late Glacial and Holocene in Cardigan Bay. Geol. J. 12, 129-158.

John, B. S., 1965. Pembrokeshire, in, C. A. Lewis (ed) The Glaciations of Wales. London, 229-265.

John, B. S., 1976. Pembrokeshire. Newton Abbot.

Kellaway, G. A., 1971. Glaciation and the stones of Stonehenge. Nature London 233, 30-35.

Kellaway, G. A., Redding, J. H., Shephard-Thorn, E. R. and Destombes, J. P., 1975. The Quaternary history of the English Channel. Phil. Trans. Roy. Soc. 279, 189-218.

Kidson, C. and Bowen, D. Q., 1976. Some comments on the history of the English Channel. Quaternary Newsletter 18, 8-10.

McBurney, C. B. M., 1959. Report on the First Season's Fieldwork on British Upper Palaeolithic Cave Deposits. Proc. Prehist. Soc. 25, 260-269.

McBurney, C. B. M., 1965. The Old Stone Age in Wales, in Daniel, G. and Foster, I. L. (eds) Prehistoric and Early Wales. London. 17-34.

Mangerud, J., Andersen, S. T., Berglund, B. E., Donner, J. J., 1974. Quaternary stratigraphy of Norden, a proposal for terminology and classification. Boreas 3, 110-126.

Moore, P. D., 1977. Vegetational History (of Wales) Cambria, 4, 73-83.

Morgan, A. V., 1973. The Pleistocene geology of the area north and west of Wolverhampton, Staffordshire, England. Phil. Trans. Roy. Soc. B 265, 233-297.

Oakley, K. P., 1968. The date of the 'Red Lady' of Paviland. Antiquity 42, 306-307.

Pennington, W., 1975. A chronostratigraphic comparison of Late-Weichselian and Late Devensian subdivision illustrated by two radiocarbon-dated profiles from western Britain. Boreas 4, 157-171.

Pennington, W., 1977. The Late Devensian flora and vegetation of Britain. Phil. Trans. Roy. Soc. Lond. B. 280, 247-271.

Shackleton, N. J., 1977. The oxygen isotope record of the Late Pleistocene. Phil. Trans. Roy. Soc. Lond. B. 280, 169-182.

Sissons, J. B., 1979. The Loch Lomond Stadial in the British Isles. Nature London 280, 199-203.

Sutcliffe, A. J., 1975. A hazard in the interpretation of glacial-interglacial sequences. Quaternary Newsletter 17, 1-3.

West, R. G., 1977. Pleistocene Geology and Biology. London.

Woillard, G., 1978. Grande Pile Peat Bog: A continuous pollen record for the Last 140,000 years. Quaternary Research 9, 1-21.

Woillard, G., 1979. The Last Interglacial—Glacial Cycle at Grande Pile in Northeastern France. Bull. Soc. Belge de Géologie 88, 51-69.

CHAPTER TWO

THE UPPER PALAEOLITHIC OF BRITAIN WITH SPECIAL REFERENCE TO WALES

R. M. Jacobi

Abstract

In this essay the radiocarbon and environmental evidence available for establishing the chronology of the British Upper Palaeolithic is reviewed and found to provide a framework not irreconcilable with the cultural sequences suggested for the continental mainland. The Earlier Upper Palaeolithic, with successive technologies equivalent to those of mainland "Blattspitzen assemblages", the Aurignacian and the Gravettian (Upper Perigordian) with "Font-Robert" points, is fitted between 36,000 and 25,000 radiocarbon years b.c. Doubts are expressed about the age suggested for the male burial from within Paviland of close to 16,000 b.c., and instead the burial is seen as an Aurignacian deposit. Resettlement of the island after the Devensian maximum can only certainly be documented from after about 9500 b.c. and correlates with the establishment of birch woodland during L-De II. The only evidence for any earlier settlement or exploitation, perhaps as early as the Oldest Dryas (or early Windermere), takes the form of shouldered and truncated artefacts resembling "Hamburgian shouldered points". Evidence is adduced for the continued exploitation of Britain during L-De III and an origin for our earliest Flandrian microlithic technologies is sought in the microliths associated with the Ahrensburgian stemmed points of L-De III.

Whilst we still have only the broadest outline of British Upper Palaeolithic sequences (Garrod, 1926; Campbell, 1977), it has been argued that the archaeological material, and with it the pattern of human settlement within Britain, can be divided into at least two chronologically discrete units—an "Earlier Upper Palaeolithic", taking in the "Aurignacian" and "Proto-Solutrean" of earlier writers, and a "Later Upper Palaeolithic", embracing those backed-blade industries termed "Cheddarian" and/or "Creswellian". Between the two is believed to fall a period when Britain was without human occupation, this gap corresponding to the maximum development of the Devensian ice sheets.

Both phases of settlement can in turn be argued to incorporate a number of chronologically discrete cultural entities, and it is the intention of this paper to look specifically at this archaeological evidence and, by so doing, to

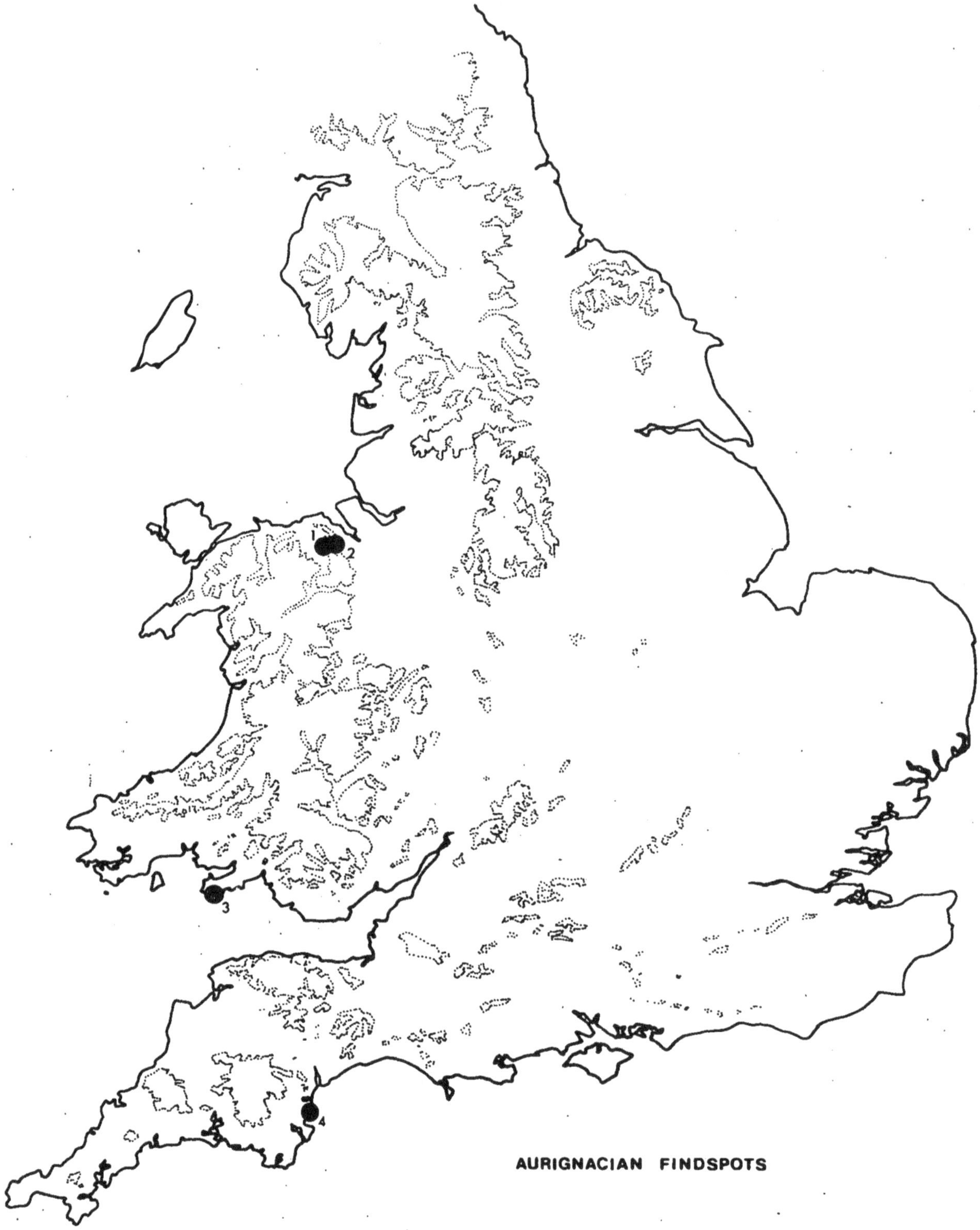

Fig. 2.1 Finds of "Aurignacian-type" artifacts. (1) Cae Gwyn Cave, (2) Ffynnon Beuno Cave, (3) Paviland, (4) Kent's Cavern

attempt both to define these cultural entities and to explore what, once defined, they have to contribute to a more refined understanding of the chronology of human activity in Britain, and more particularly in Wales, during the latter part of the Devensian. At the same time, a vigorous attempt must be made to counter a growing isolationism within British Upper Palaeolithic studies, and a deliberate effort will be made here to fit what evidence we have from Britain into the highly refined cultural framework that is rapidly emerging for north-western and central Europe.

"Leaf Points" and the Aurignacian

Material which, it would probably be agreed, belongs to the earliest stages of our Upper Palaeolithic must include, in the absence of any Châtelperronian this far north-west in Europe, the Aurignacian-type artifacts recovered from four British "cave sites" (Fig. 2.1) and, together with these, blades converted by ventral surface retouch or bifacial (often plano-convex) surface retouch into leaf-shaped spearheads. Such "spearheads" are certainly known from twenty-three findspots (Fig. 2.2) and are termed "leaf points". Marked by an open circle on Fig. 2.2 is the findspot of what is described as a "leaf or foliate point" of "szeletian" type at Chelsfield in Kent (Bouskill, 1979). This artifact has so far not been seen by the writer.

Otte (1977a), in a recent study of the Belgian Aurignacian, has distinguished among these assemblages four typo-technological groupings. In what is believed to be the oldest of these groupings "leaf points" (pointes épaisses de forme losangique) are suggested at both Spy and Goyet as having formed part of the same artifact assemblage as the more obviously Aurignacian tool forms. These "leaf points", with retouch largely confined to their ventral surface, resemble closely their counterparts in Britain.

Likewise three British cave sites have yielded both "leaf points" and specifically Aurignacian tool forms: these caves are Kent's Cavern, Paviland and Ffynnon Beuno, and in each case it might seem logical to assume the existence of the same polythetic assemblage as suspected for both Spy and Goyet. Confidence, however, in grouping this material is perhaps reduced when it is recalled that each of these sites, British and Belgian, was explored before the first world war. Thus, while the "leaf points" and Aurignacian-type tools may indeed have been excavated from broadly the same stratigraphic range within these caves—as appears to have been the case at Kent's Cavern—in no case were the excavations sufficiently rigorous that any sterile band of deposit would have been recorded which might have separated one group from the other and betrayed them as the products of different social groupings.

That Aurignacian tool forms were not present at every site from which "leaf points" have been recovered is shown by the absence of Aurignacian-type tools from surviving collections of finds from Soldier's Hole, Badger Hole, Hyaena Den, Pin Hole, Robin Hood's Cave, and Thakeham (near Pulborough, Sussex: Curwen 1949). At this last locality some fifty broken "leaf points" were recovered at the turn of the century from what must be presumed to have been a collapsed rock shelter. These were associated with blades (some with facetted platforms), utilized and retouched blades, burins developed on broken blades or sections of "leaf points", a "Corbiac burin", a chamfered blade", double platformed blade cores and cores developed on blade sections.

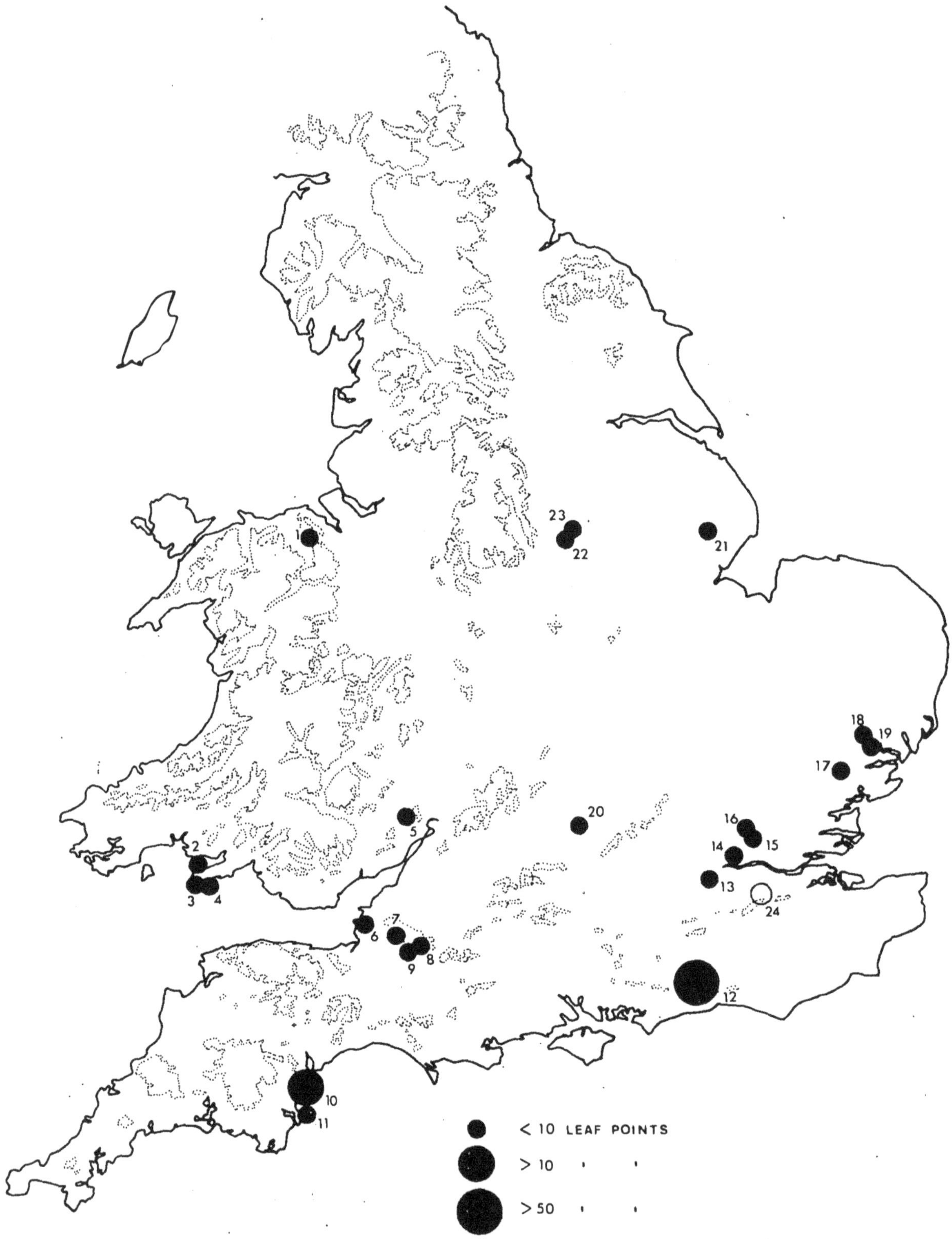
1
2
3
4
5
6
7
8
9
10
11
12
13
14
15
16
17
18
19
20
21
22
23
24
< 10 LEAF POINTS
> 10 " "
> 50 " "

Fig. 2.2 Finds of Leaf-shaped Spearheads from the British Isles.

Infilled Circles Key	1	Ffynnon Beuno Cave
	2	Nottle Tor
	3	Paviland
	4	Long Hole
	5	King Arthur's Cave
	6	Uphill Quarry
	7	Soldier's Hole
	8	Badger Hole
	9	Hyaena Den
	10	Kent's Cavern
	11	Bench Fissure
	12	Beedings, Thakenham
	13	Earl of Dysart's Pit, Ham
	14	Creffield Road, Acton
	15	Broxbourne
	16	Rikof's Pit, Broxbourne
	17	White Colne
	18	Bramford Road, Ipswich
	19	Barham
	20	Osney Lock, Oxford
	21	Salmonby
	22	Pin Hole Cave
	23	Robin Hood's Cave
Open Circle	24	Chelsfield (unconfirmed find)

There are thus in Britain a number of sites where "leaf points" have been found but where Aurignacian-type tool forms have not.

If we go on to consider only the Aurignacian component of the assemblages from Kent's Cavern and Paviland, a number of typo-technological features would seem to link these to their counterparts at Spy and Goyet and, to take the comparisons a stage further, the Aurignacian II of south-west France. These features are (1) that among the "Aurignacian scrapers" nosed forms (Grattoirs à museau") outnumber carinated forms (2) that burins busqués are present in all these assemblages (3) that burins (as a gross tool category) are less numerous than scrapers and (4) that "Aurignacian blades" and scrapers on "Aurignacian blades" are rare with the majority of tools developed instead on flakes rather than on blades.

If some form of correlation with the French Aurignacian sequence is believed meaningful and the possibilities of a close chronological identity can also be accepted, then immediately relevant to our British material must be dates from the Abri Pataud of between 29, 850 b.c. ± 310 (GrN 4531) and 30, 950 b.c. ± 700 (GrN 3116: Vogel and Waterbolk (1967) p. 114) and a date from Arcy-sur-Cure of 28, 850 b.c. ± 250 (GrN 1717: Vogel and Waterbolk (1963) p. 166). Rather younger dates from the Abri Pataud and La Rochette (Dordogne: Vogel and Waterbolk (1967) p. 113) are believed to reflect humic and rootlet infiltration, respectively.

If then we treat the Aurignacian-type artifacts from Kent's Cavern and Paviland as single populations—a procedure which it would be hard to justify—continental analogy might suggest an absolute age for this material of between 31, 000 and 28, 000 radiocarbon years b.c. If we go on to assume that, as in Belgium, the "leaf points" from these sites may be part of the same artifact population then, by definition, these would be of an equivalent age.

Recently, however, Hahn (1977) has shown that while Aurignacian occurrences in Czechoslovakia, Hungary, Rumania, Bulgaria and western Russia can regularly include rare (usually bifacially flaked) "leaf points", such points are absent from Aurignacian assemblages in Austria, Germany, Poland and France. This observation, and with it the proposition (now supported by Otte (1978) p. 134) that in westernmost Europe such "leaf points" are to be separated both culturally and chronologically from what can, I believe, be demonstrated to be a rather later Aurignacian technology, appears to be confirmed by the stratigraphic sequence within the Ilsenhöhle Cave below the Castle of Ranis in Thuringia (German Democratic Republic: Andree (1939) pp. 107-114, 362-4 and 406-412; Toepfer (1963) p. 13; Hahn (1970) p. 199; and Hülle (1977)).

Excavated from 1926 to 1939, the profile, 7 metres deep, disclosed archaeological material ranging from the Migration period to the Earliest Upper Palaeolithic or even Mousterian.

The lowest level, Ranis I (layer XI), is believed to be Middle Palaeolithic, but the presence of what could be interpreted as a "leaf point" ("Blattspitze") rather than a "Faustkeilblatt" (Hülle (1977) Taf. I) together with regular blades (ibid. Taf. nos. 3 and 4) may hint that this lowest layer contains instead an earliest Upper Palaeolithic assemblage with "leaf points". This could, extrapolating from earliest dates for Szeleta and Bohuniče, be as old as 40, 000 radiocarbon years b.c.

Ranis 2 (layers IX and X) is of particular relevance to us in that its stone industry includes a large series of "leaf points" developed on blades—the majority with retouch concentrated on their ventral faces. With these are some fully bifacially flaked pieces which possess markedly plano-convex profiles. All the "leaf points" are bi-points and the whole group bears an immediate resemblance—as recognized by Hülle (ibid. p. 121)—to those from the cave of Nietoperzowa, near Cracow, layer 6 of which is dated to 36,210 b.c. ±1,250 (GRO 2181: Chmielewski 1961), as well as both to the "leaf points" identified from the twenty-three findspots within Britain, and those from Spy and Goyet.

While the bulk of the "leaf points" come from layer X, a few were found also in the lower part of the overlying layer IX, and are separated only by the upper part of this layer from the rich stone industry of Ranis 3 (Layers VII and VIII). This latter industry is quite clearly Aurignacian and several writers (Andree (1939) p. 410 and Zotz (1951) p. 139) have convincingly compared it to the Aurignacian from Vogelherd V. Notably absent from this assemblage are "leaf points".

The observed succession, an Aurignacian without "leaf points" overlying a blade industry without Aurignacian tool forms but with (many) leaf-shaped spearheads, must be immediately relevant to the interpretation of such potentially confused assemblages as those from Spy and Goyet and from Kent's Cavern, Paviland and Ffynnon Benno. Thus, it might be reasonable to suspect that at each of these five sites, as at Ilsenhöhle, Aurignacian-type artifacts in fact occurred higher in the profile than an earliest Upper Palaeolithic assemblage with "leaf points" developed on blades. Similarly too the sterile band between the two assemblages may have been peculiarly thin. Hülle places Ranis 2 and 3 within the same interstadial (1977: Abb. 26). This, however, appears a doubtful interpretation if we are to accept the archaeological parallels which seem appropriate—Ranis 2 with Nietoperzowa and Ranis 3 with Vogelherd V. Layer 6 at Nietoperzowa is, as we have already commented, dated to close to 36,000 radiocarbon years b.c. and must hence fall within the Hengelo interstadial of the mainland. By contrast a series of radiocarbon determinations suggests a most probable position for Vogelherd V within either the 28th or 29th millennium b.c. (Hahn (1977) p. 168); that is within either an equivalent of the Arcy interstadial recognized further to the west in Europe or within the final part of the cold phase which immediately precedes it, and succeeds Hengelo.

Extrapolation from the results obtained at the Ilsenhöhle suggests exploitation of Britain, as far north as north Wales, Creswell, and south Lincolnshire at between 30 and 40,000 radiocarbon years b.c., and more probably at close to 36,000 radiocarbon years b.c., by groups who converted blades into leaf-shaped spearheads by means of surface flaking. That their technology did not incorporate specifically Aurignacian tool forms is suggested by the absence of such artifact classes from a number of British "leaf point" assemblages. A second period of settlement between about 30,000 and 28,000 b.c. by groups employing a technology equivalent to that used by mainland Aurignacian II populations can be inferred from assemblages recovered from within Kent's Cavern and Paviland and from isolated artifacts from Cae Gwyn and Ffynnon Beuno in north Wales. At Kent's Cavern, Paviland and Ffynnon Beuno, the two technologies may have existed in stratigraphical relationship one to another, but as at Ranis intervening deposits may have been peculiarly thin. Exploitation of Britain by such Aurignacian II groups could, if the technological analogy is accepted, be argued to have occurred within the Arcy interstadial.

Is it possible to test in any way this hypothesis against radiocarbon determinations which purport to date occurrences of either or both technologies within Britain? In attempting such an exercise it appears useful to consider first the determinations from Kent's Cavern and then, as a second group, the isolated dates obtained for Paviland, Cae Gwyn and Badger Hole. The determination of 26,550 b.c. $\pm$1600/-1300 (BM 602) from Robin Hood's Cave (Creswell) will be ignored since it is on an (effectively) unstratified (bear) bone and without relevant archaeological associations.

Kent's Cavern, Torquay, Devon

There are four dates from Kent's Cavern all on animal bone and all from the "loamy cave earth". Each is supposedly associated with an occurrence of Earlier Upper Palaeolithic artifacts. Of these, three group relatively closely, viz:

26,770 b.c. $\pm$ 450 (GrN 6202)
26,210 b.c. $\pm$ 435 (GrN 6201)
25,780 b.c. $\pm$ 350 (GrN 6325)

while one is rather older, viz:

36,320 b.c. $\pm^{1470}_{1240}$ (GrN 6324).

Each of the bones dated was collected by Pengelly during his excavation of 1865 to 1880. Pengelly excavated and recorded in a series of vertical spits with the result that a bone, even if from the same parallel, yard and level as a flint artifact whose age it is hoped to discover, could in reality have lain up to 30 cm above or below that artifact within the prism of soil being removed. In no case can the precise relationship now be determined. In a slowly accummulating deposit, such as the "loamy cave earth" at Kent's Cavern appears from its archaeological content to have been, considerable imprecision is necessarily introduced, since such 30 cm spits could incorporate items from more than one spread of cultural material. Pengelly, it should be remembered, was not attempting to identify, follow and record discrete layers of archaeological material within the cave earth. Thus, all we can assume is that a bone selected for dating, if found in the same spit, parallel and yard, was not found more than a foot vertically above or below an artifact whose age it is wished to investigate. The artifact could be younger, older or identical in age, but no way exists of determining which.

Each of the animal bones dated was unsplit, unworked and unburnt, and, in this context it must be remembered that beyond their use as shelters for human groups, caves will have been used both by hibernating bears and by carnivores. Since none of the bones is recorded as showing signs of human interference, it is possible that these arrived within the cave by mechanisms other than human introduction: thus GrN 6202 on the unburnt humerus of a bear may date a hibernating animal, while the dated portions of woolly rhinoceros

(GrN 6201: unburnt tibia), horse (GrN 6324: radius) and bison (GrN 6325: radius) could have been introduced into the cave system by carnivores. Indeed, significantly, some 228 hyena teeth are listed by Pengelly as deriving from this "loamy cave earth" (Campbell and Sampson (1971) p. 15), and this may well be a minimum estimate: (A. Turner, pers. comm.).

In using the dates from Kent's Cavern it is important to realize that the items dated are bones, which were not necessarily introduced into the cave by its human occupants. These, if Pengelly's recording is reliable, can only now be shown to have been recovered, not more than a foot either way vertically, of Earlier Upper Palaeolithic artifacts. The relationship of these bones to the population, or populations, of such artifacts cannot now be made more precise, and we can conclude only that into that part of the "loamy cave earth", into which Aurignacian-type artifacts and "leaf points" were being introduced, animal bones, which have given central dates of between 26,000 and 36,000 radiocarbon years b.c., were also being incorporated. This time range, however, does not deny the proposition that this "loamy cave earth" may once have contained both an Aurignacian tool assemblage, for which we have suggested an age of between 31 and 28,000 radiocarbon years b.c., as well as an earliest Upper Palaeolithic assemblage with "leaf points" potentially equivalent in age to Nietoperzowa and thus dating to about 36,000 radiocarbon years b.c.

Other determinations which it has been suggested relate to artifact assemblages containing either or both Aurignacian-type artifacts and "leaf points" are even less satisfactory from the archaeological viewpoint.

*Paviland, South Wales

Strictly without archaeological associations is a humerus of *Bos primigenius* deriving from Sollas's (1912) excavations at Paviland, and dated to 25,650 b.c. ± 1,300 (BM 1367: Molleson, 1978). If introduced into the cave by its human occupants, the age obtained suggests that the bone may be less likely to have belonged with the "leaf points" (Garrod (1926) fig. 8 No. 2; Sollas (1913) fig. 17, No. 118) or Aurignacian tool forms recovered than with a number of "Upper Perigordian" (Gravettian) flint, bone, and ivory tools found at the same time. For these a round age of about 26,000 radiocarbon years b.c. will be suggested (see below).

Cae Gwyn, Tremeirchion, Flintshire

The date from here of 16,050 b.c. $^{+1400}_{-1200}$ (Birm 146) is on a mammoth carpal (Rowlands (1971); Shotton and Williams (1971) p. 141) and there is no way of knowing at this distance from the original excavations—Cae Gwyn was dug in the 1880s (Hicks, 1886)—how closely (if at all) this was associated with the end scraper developed on an "Aurignacian blade" also found in the cave (Garrod (1926) fig. 14 no. 4; Campbell (1977) fig. 99 No. 2). This point is stressed in Nédervelde *et alii* (1973) p. 454). There is certainly no justification in quoting the date to support arguments that an Aurignacian-type technology continued this late within the British Isles.

It is true, however, that as late dates could be quoted for Aurignacian-type tool kits from Belgium (Trou du Renard: 22,580 b.c. ±470: LV 721), southern Germany (Bockstein Torle: 24,183 b.c. ±376 (H 4059-3386) and 18,450 b.c. ± 220 (H 4058-3355) Hahn (1977) p. 171) and lower Austria (Langmannersdorf: 18,630 b.c. ±170 (GrN 6659) and 18,310 b.c. ±200 (GrN 666): ibid. p. 170 , but in all cases the tool kits involved are highly distinctive with a heavy representation of burins—mostly notably "burins carénés"—together with retouched bladelets. Certainly no assemblage of this type can be recognized from Britain.

Badger Hole, Mendip, Somerset

A sample of burnt bone believed to have been associated originally with a "leaf point" assemblage from Badger Hole, Mendip, gave only a minimum age of 16,000 b.c. (Campbell (1977) p. 51).

Conclusion

It is possible to recognize from sites in Britain both "leaf points" and a range of Aurignacian-type tools. The evidence at our disposal is insufficient to confirm any combination of all these tool types into a single composite assemblage. It is possible, therefore, that as at Ilsenhöhle, an earliest Upper Palaeolithic technology, including leaf-shaped spearheads, and perhaps as early as 36,000 radiocarbon years b.c., preceded an Aurignacian-type technology. For the latter an age range of between 31 and 28,000 radiocarbon years b.c. is suggested. The spread of radiocarbon determinations from within Kent's Cavern would accommodate this hypothesis.

An alternative interpretation of these data could be that, as in east-central Europe, leaf-shaped spearheads remained "fashionable" over a considerable time-span (Hengelo into Post-Hengelo). Further that these may have existed within a broad range of technological associations of which the latest could be argued to incorporate Aurignacian-type stone tools. If this is so, assemblages with "leaf-points" at Pulborough, Soldier's Hole, Badger Hole and Robin Hood's Cave may precede temporally such composite assemblages as those from Paviland, Kent's Cavern and Ffynnon Beuno. The latter assemblages would, as in east-central Europe, be of Post-Hengelo age.

The Upper Perigordian (Gravettian)

At Spy (Namur: De Puydt and Lohest 1887), a technology, with stemmed and dorsally surface-retouched spearheads not dissimilar to the "Font Robert" stemmed points, well known from further south within Europe, occurred rather higher in the profile (the "Premier Niveau ossifère") than the "mixed" occurrence of Aurignacian-type tools and ventrally retouched "leaf-points" (the "Deuxième Niveau ossifère"). Equivalent stemmed and surface retouched spearheads from the open air site of Maisières Canal (Heinzelin, 1973), are believed to date close to 26,000 radiocarbon years b.c. (Otte (1976a) p. 335). A range of dates was, however, obtained for Maisières, using humic matter from the calcareous clay which contained the archaeological material, but there are clear discrepancies between many of these and the local stratigraphy, discrepancies which are attributed in large part to the difficulties in dating calcareous soil (Gilot (1971) pp. 47-48). However, considered acceptable by both de Heinzelin (1973 p. 45) and Otte (1976a p. 335) is the date of 26,015 b.c. ± 260 (GrN 5523) taken from immediately below the occupation horizon.

Artifacts recognized from Britain take the form of stemmed points ("Font Robert-type Points") and all except that from Pin Hole have flat dorsal surface retouch applied to their tips: in several cases these are loosely associated with other forms of surface retouched tools. All the examples of stemmed points recognized from Britain occurred as single examples and none is dated (Fig. 2.3).

From the south and east such points come from Peper Harow Park, near Godalming (Surrey: Winbolt 1929), from Mildenhall (Suffolk: Jacobi and Martingell, in preparation) and Bramford Road (Ipswich: Moire (1930) fig. 17). Near Gloucester at Barnwood (Clifford et alii (1954) pp. 180-183) a broken stemmed point occurred among a larger assemblage of flint tools hinting at an occupation site. Finds from caves or rock shelters comprise the much damaged mid-portion of a stemmed point from the South West Chamber of Kent's Cavern and the tang and shoulders of a stemmed point with extensive surface retouch from Paviland (Sollas (1913) fig. 17 No. 116). With this latter may have been associated the forepart of a "Pointe à face Plane" (Sollas (1913) fig. 17 No. 122). At Pin Hole Cave (Creswell: Armstrong 1929) a complete stemmed point is said to have been associated with the engraving, on a reindeer rib, of a masked figure and food debris which included ducks' eggs (ibid., p. 17). Rather more tentatively mapped as "Font Robert Points" are a pair of fragments from Cathole (Gower: Fig. 2.3 open circle and see below). In each case the pieces listed have precise parallels either or both at Maisières and further south at Cirque de la Patrie (Nemours: Schmider (1971) pp. 39-42).

Conclusion

From seven (or perhaps eight) findspots within Britain it is possible to identify stemmed points morphologically identical to the "Font Robert" points of Late Perigordian or Gravettian type from Belgium (Otte, 1977 pp. 256-258), north and northern-central France (Schmider (1971) and Larue, Combier and Roche 1956) and Germany (Hahn (1976) p. 109). These stemmed points, which could be argued to be equivalent in age to those from Maisières Canal, may imply the settlement, or at least the exploitation, of Britain as far north as Creswell during the climatic amelioration of Kesselt.

Absent from Britain—except perhaps from Barnwood—is any assemblage which combines these stemmed points with gravettes or micro-gravettes (Clifford et alii (1954) fig. 1 No. 7) and unknown from within Britain is any assemblage which combines also "éléments tronqués ou bitronqués" and "burins de Noailles". Dates of 25,110 b.c. ± 370 (GrN 4230) for a Périgordien à Noailles at the Abri Pataud and of 21,230 b.c. ± 1500 (GsY 69) for the same technology at the Abri du Facteur (for discussion see Delporte (1976) pp. 28-29) may not only suggest a broad age for part at least of these developments but may also indicate a date by which exploitation within this island had ceased The rather later exploitation of present-day Belgium is, of course, indicated by the date of 20,630 b.c. ± 470 (LV 721: Otte (1977) p. 253) from the Trou de Renard.

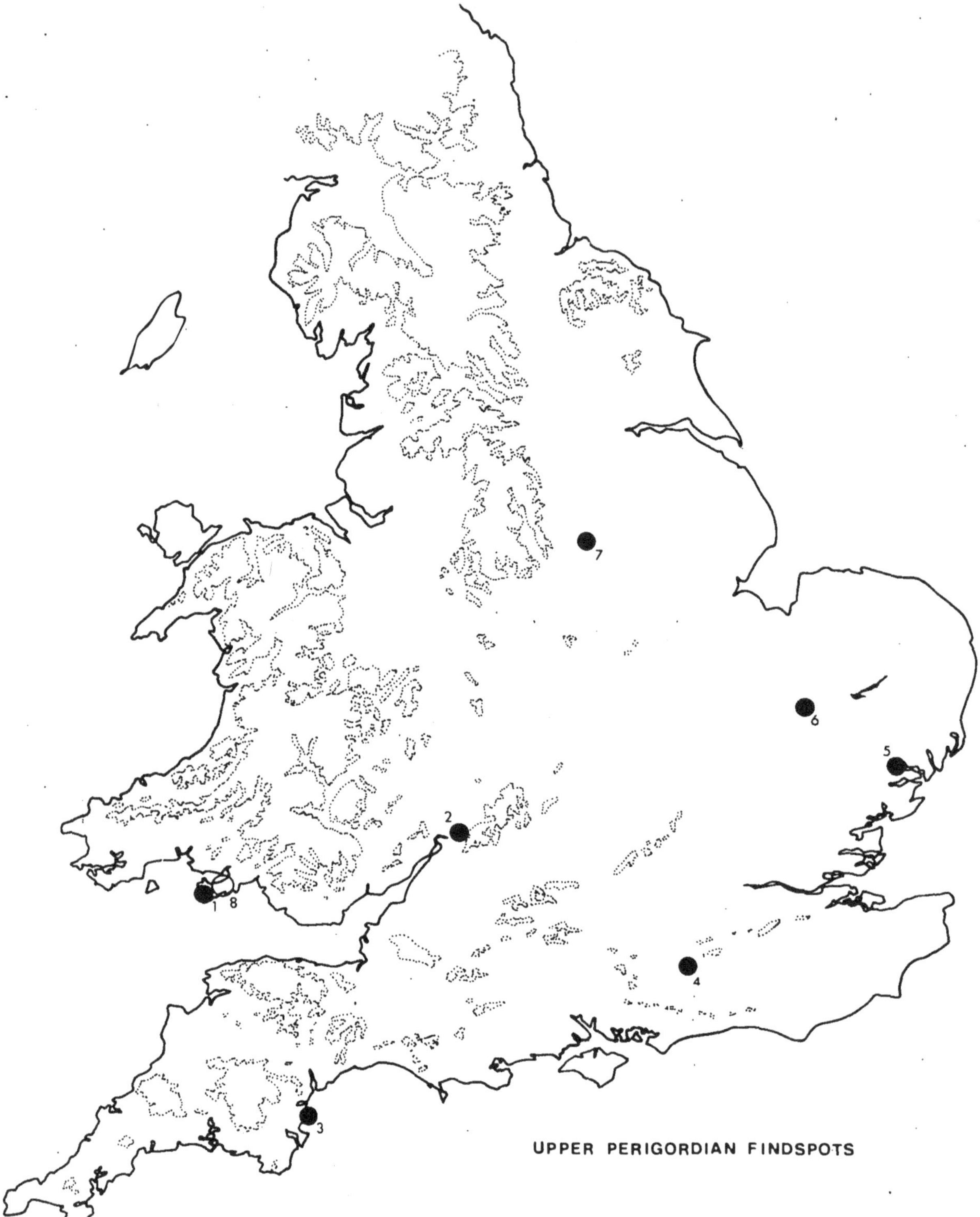

UPPER PERIGORDIAN FINDSPOTS

Fig. 2.3 (opposite) Upper Perigordian (or Gravettian) Find-spots in the British Isles.

Infilled Circles Key	1	Paviland
	2	Forty Acres Pit, Barnwood
	3	Kent's Cavern
	4	Peper Harow Park, Godalming
	5	Bramford Road, Ipswich
	6	Mildenhall
	7	Pin Hole Cave
Open Circle	8	Cathole Cave

The Gap in Settlement

It has been suggested (Bowen (1977) pp. 686-687; Wymer (1977) p. 104) that a human population was absent from Britain—and for that matter from Holland, Belgium and northern France—at the time corresponding to the maximum advance of the Devensian glaciers. By nearly 16,000 radiocarbon years b.c., these glaciers covered over half of England, most of Ireland and considerable areas of the adjacent sea beds. The British ice joined across the North Sea with the Scandinavian ice sheets which in turn covered the "flat lands" of Russia, Poland and Germany and also most of Denmark. Our next major problem is to define chronologically this gap in settlement.

We have already observed an absence from Britain of artifact types present in Belgium and believed innovated at between about 25,000 and 20,000 radiocarbon years b.c. We have also observed the absence of an Aurignacian technology equivalent to that of the Trou de Renard. In discussing the chronology of the occurrence of "leaf-points" in Britain it was suggested that determinations of BM 497 from Badger Hole,and Birm 146 from Cae Gwyn do not in themselves offer serious support for supposing the presence of human populations within the island close to 16,000 b.c. Two other dates, however, require also to be considered in this context since both relate to the period between 25,000 and 9,500 radiocarbon years b.c., when it will be suggested that any significant exploiting population may have been absent from Britain. These dates are 20,400 b.c. ± 620 (Birm 340: Shotton and Williams (1973) p. 458) from the cave of Ogof-yr-Ychen on Caldey and 16,510 b.c. ± 340 for the "Red Lady of Paviland" (BM 374: Barker et alii (1969) p. 289).

Ogof-Yr-Ychen: Caldey

At Ogof-yr-Ychen the dated scapula of woolly rhinoceros together with an adinole "scraper" and some flakes, formed part of a varied population, including also teeth of hyaena and rhinoceros and antlers of red-deer, washed, at a date unknown, into a swallet, roughly 2 metres by 1 metre across at its mouth and excavated to a depth of 6.4 metres (Nédervelde et alli (1973) fig. 1). The determination could obviously be taken as giving an age for the rhinoceros scapula, and the result obtained bears comparison with that for the mammoth carpal at Cae Gwyn (see above) and a date of 16,263 b.c. ± 310 on a musk ox from Clifford Hill, Northampton (BM 725: Burleigh et alii (1976) p. 30). This group of dates suggest the presence of a potentially exploitable megafauna within Britain close in time to the maximum ice advance. The determination, however, neither gives any indication of the date at which the scapula entered the swallet; nor allows the possibility of extrapolating the date to the human artifacts encountered also within the swallet. Indeed, the objects recovered could represent debris of widely differing ages and points of origin within the cave system, indiscriminately mixed within the pipe from which they were recovered, and (in part at least) accumulated therein at any date later than about 20,000 b.c. None of the artifacts is in any way culturally diagnostic.

The burial from Paviland

Infinitely more difficult of assessment in the date of 16,510 b.c. ± 340 (BM 374) on collagen from the skeleton of the "Red Lady of Paviland" (Barker et alii (1969) p. 289). Potentially, this result is of the greatest significance in being a date on human bone material, the skeleton of a 25-year-old male buried within the cave—if the date is accepted at face value—at a moment when, with lowered sea-level, access to the Goat's Hole must have been easier than today, and when, it has been suggested, glacier ice may have been only about 6 km to the north (Bowen, 1970: but see also John, 1971).

This determination on the "Red Lady" falls in age between a pair of dates on moss from silts overlain by Devensian Till at Dimlington (Penny et alii, 1969):

Dimlington	16,240 b.c. ± 250	(Birm 108)
Paviland (R.L.)	16,510 b.c. ± 340	(BM 374)
Dimlington	16,550 b.c. ± 400	(I 3372)

Coleoptera from these silts have been interpreted as indicating that the contemporary ice-front was only nearby at the time of their accumulation.

The determination for Paviland can thus be argued to suggest deposition of skeleton just before the Devensian ice sheets reached their maximum extent. Both the deposition of the skeleton and the maximum extension of these ice sheets could be argued to fall within the cold interval registered between the Lascaux and Laugerie interstadials of the mainland, while BM 375 could further be demonstrated to overlap considerably at a single standard deviation with a date of 16,310 ± 360 (LY 972: Evin et alii (1976) p. 81) for an initial Madgalenian horizon (level 20-18) in a cryoclastic scree within Laugerie Haute East and believed also to belong to the rather colder interval between Lascaux and Laugerie.

Objection has been made to BM 375 on the grounds that the result was obtained by using only 62 grammes of "bone powder", considerably less than the ideal minimum weight for material of such antiquity (Oakley (1968) p. 307). However, use of a small sample can be argued only to increase the standard deviation, not to alter the central date obtained (Burleigh (1979) pers. comm.) and the date cannot simply be dismissed on these ground alone.

The extended male burial from Paviland was a rich one. The skeleton or rather that part which remained undisturbed was "...enveloped by a coating of a kind of ruddle...the body... (being)...entirely surrounded or covered at the time of its interment with this substance. Close to that part of the thigh bone where the pocket is usually worn...surrounded also by ruddle (were) about two handsfull of small shells of the Nerita littoralis... At another part of the skeleton, viz in contact with the ribs (were) forty or fifty fragments of small ivory rods nearly cylindrical, and varying in diameter from a quarter to three quarters of an inch, and from one to four inches in length... (Also present were)...some small fragments of rings made of the same ivory and found with the rods... The rings when complete were probably four or five inches in diameter. Both rods and rings, as well as the Nerite shells, were stained superficially with red, and lay in the same red substance that enveloped the bones; they had evidently been buried at the same time with the woman..." (Buckland (1824) pp. 88-89).

The objects directly associated with the skeleton, and which could be taken as being dated by it, are the ivory "rods" and rings, the sea shells and the ochre. No flints were recorded as being associated, although the fact that many of the "Earlier Upper Palaeolithic" stone tools have traces of red ochre (ruddle) attaching to them, while none of the clearly Late-glacial artifacts exhibit such staining has been taken to suggest "... some sort of relationship..." between the burial and the earlier stone assemblage (Campbell (1977) p. 144). This, however, could be a most dangerous and potentially confusing assumption and the point has perhaps been overlooked that, if the burial and its associated ochre covering was inserted into a deposit containing Aurignacian and Perigordian artifacts, then some of these must inevitably come into contact with the staining, or become incorporated—like any other stones within the deposit—into the grave fill, and hence become stained. None of these archaeological data could then be used to contradict a hypothesis that the Goat's Hole may have been used as a burial site long after it had ceased to be used as an occupation site by Aurignacian or Perigordian groups (see also Molleson, 1976 and 1977).

Discussion

It appears valuable to offer some comments on the archaeological contents of the cave, although it seems impossible to reach any conclusion about the real age of the burial.

(1) (Four, and perhaps five, stone technologies appear to be represented):

Age ± 36,000 radiocarbon years b.c.	A Leaf Point Technology
31-28,000 radiocarbon year b.c.	An Aurignacian Technology
± 26,000 radiocarbon years b.c.	A Late Perigordian (Gravettian) Technology
± 9,500 radiocarbon years b.c.	A Federmesser Technology (Sollas (1913) fig. 14 Nos. 92-97; Garrod (1925) fig. 8 Nos. 7-11)
± 3,000 radiocarbon years b.c.	A Neolithic Technology (Sollas (1913) fig. 16 Nos. 112-115)

(2) The edges of the leaf points from Paviland (Sollas (1913) fig. 7 No. 28 and Fig. 17 Nos. 118 and 118A) are considerably notched and abraded (concassé) as a result of movement within the deposit during a phase of cryoturbation, as are many of the blades and flakes from the cave (cf. Sollas (1913) fig. 5 Nos. 1, 7, 9, 10, 18 and 19; fig. 7 Nos. 24-31; Fig. 10 No. 63; fig. 14 Nos. 103 and 104; also 106-108; and fig. 15 No. 2). Similar edge damage can be observed on artifacts from leaf point assemblages at Badger Hole, Soldier's Hole and Ffynnon Beuno. The Aurignacian-type tool forms and the Upper Perigordian (Gravettian) artifacts from Paviland retain their edges undamaged.

(3) With the Upper Perigordian (Gravettian) flint tools may originally have been associated a pair of particularly small chopped base points ("sagaies à base découpée"), one of ivory and one of bone (Sollas (1913) Pl. XXII No. 5), directly comparable in terms of both size and technology to an ivory point

from Maisières Canal (Heinzelin (1973) Pl. XLIII No. 5) and to examples from Goyet, Spy and the Trou Magrite (Otte (1977b) pp. 195-196), all sites with Upper Perigordian (Gravettian) occupation horizons. The points from both Paviland and the Belgian sites are markedly smaller than technologically comparable artifacts from Aurignacian contexts (Leroy Prost 1974).

(4) Flint working in the cave is suggested by an ivory punch (unpublished) and a bone "retouchoir" (Sollas (1913) p. 362 fig. 23 No. 1).

(5) The cave had clearly also been used as an ivory workshop. Buckland (1824; Pl. XXI fig. I and text p. 36), recovered the skull of a mammoth still retaining its tusks while among the surviving finds are broken and detached plates of ivory, at least one of which shows a series of cut marks (Sollas (1913), Pl. XXI No. 10). It was possible (Sollas (1913) pp. 362-3) to refit a perforated egg-shaped pendant (ibid. Pl. XXII No. 34) back into the cavity within a tusk fragment (recovered by Buckland) from which it had originally been extracted.

(6) Differences in cross-section suggest that the burial was accompanied by two bracelets, not just one as reconstructed in Sollas (1913, Pl. XXII No. 12). The shells found with the burial are perforated for suspension. There are 31 fragments of round-sectioned ivory rods which can be reduced to 23 sections after repair. The longest of these sections is 97 mm. Their diameters range from 8 to 14 mm: there are no pointed pieces and no traces of any bevel. Neither these "rods" nor the bracelets can be paralleled within the technology of the mainland Initial Magdalenian.

Instead, the "rods" resemble most closely the "Bâtonnets d'ivoire" from Aurignacian contexts at Spy and Marche les Dames (Grotte de la Princesse: Otte (1977a) fig. 5 No. 2) and interpreted as sticks of ivory awaiting division into bead blanks. A bracelet (perhaps unfinished) of ivory also comes from Spy so that the same Aurignacian workshop was producing both bracelets and batonnets, the very items found with the Paviland burial, while another bracelet comes from an Aurignacian context at the Trou Magrite (Otte (1976b) Pl. VI No. 4). Bracelets from a Gravettian context of the 24th millennium b.c. at the Magdalenahöhle (Weiss 1974) are ornamented with lines and dots, while the Paviland fragments lack ornamentation, and are therefore different.

Search of the surviving collections has further failed to reveal any raclettes, transverse burins on notch, or star-shaped perçoirs—all types known from the precisely contemporary mainland technologies of the Initial Magdalenian. Indeed, mapping of Initial Magdalenian occurences (Schmider (1971) pp. 63-110) reveals no evidence of settlement during the 17th and 16th millennia north of the Loire and west of the Essonne. If, therefore, we choose to overlook the technical problems surrounding the determination, ignore the Aurignacian parallels for the objects deposited with the body, and elect instead to accept that the burial is intrusive into older occupation deposits, it is necessary to explain the complete absence of artifacts and radiocarbon evidence for human settlement or exploitation of the area of southern Britain and northwestern Europe between Paviland and the distribution of Initial Magdalenian technologies—a distance, as the crow flies, of some 600 km.

1
2
3
4
5
6
7
8
9
10
11
12
13

Fig. 2.4 (opposite) Distribution of radiocarbon dates for Magdalenian Sites whose central points fall between 14,000 and 13,000 radiocarbon years bc (infilled triangles) or whose error expressed at a single standard deviation overlaps the years 14,000 to 13,000 bc (open triangles).

Key

1 Hallines
2 La Garenne
3 Lascaux
4 Le Flageolet
5 Combe Cullier
6 Grotte de St. Eulalie
7 Grotte Gazel
8 Le Rond du Barry
9 La Colombière
10 La Croze
11 Arlay (Grotte Grappin)
12 Grotte de Rigney
13 Munzingen

1
2
3
4
5
6
7
8
9
10
11
12
13
14
15
16
17

Fig. 2.5 (opposite) Distribution of radiocarbon dates for Magdalenian sites whose central points fall between 13,000 and 12,000 radiocarbon years bc (infilled triangles) or whose error expressed at a single standard deviation overlaps the years 13,000 to 12,000 bc (open triangles).

Key
1 Rocaux Sorciers
2 La Garenne (Abri Blanchard)
3 Lascaux
4 Laugerie Haute
5 Fongaban
6 Le Flageolet
7 Combe Cullier
8 Grotte de St. Eulalie
9 Duruthy
10 Canecaude
11 Grotte Gazel
12 Le Blot
13 La Colombière
14 La Croze
15 Arlay (Grotte Grappin)
16 Grotte de Rigney
17 Schussenquelle

The Late-Devensian Resettlement of Britain

In the second and longer part of this essay on the Upper Palaeolithic archaeology of the present-day main island of Britain, the first question to which we will have to attempt to find an adequate answer is just when, or rather how soon, after the maximum cold associated with the ice advances of the Middle Devensian, human groups recommenced exploitation of the island. Any answer to these questions can be regarded as totally satsifactory only if it is sympathetic to the cultural archaeology of mainland north-west Europe for the sub-evident reason that the initial colonizing population must be suspected as having originated within this mainland and most probably within its present-day north-western margins. Colonization of the present-day island may be suspected also to have taken place only after settlement of northern Belgium, the Netherlands and north-western Germany, precisely as the colonisation of Scotland and of Ireland succeeds settlement of England and Wales (Coles (1971); Mercer (1974); but see also Jacobi (1978) p. 328 Note I; Woodman (1978) pp. 199-205).

For this reason one must be suspicious of the proposition that human groups (using a technology whose characteristics cannot be paralleled within contemporary mainland Europe) were in the 13th millennium b.c. (GrN 6203) exploiting the area of the South-West Peninsula round Kent's Cavern. That is a millennium before the colonization of the Low Countries and north Germany can be shown to have been achieved and when, on the western margins of Europe, the northern limits of human settlement still lay, on radiocarbon evidence, perhaps as far south as the Ain (La Croze II: Desbrosse (1963); Evin et alii (1973a) p. 150 and La Colombière: Movius and Judson (1956); Haute Loire (Le Blot: Evin et alii (1973b) pp. 524-5) and Vienne and Indre (Roc aux Sorciers: Movius (1960) p. 370; Vogel and Waterbolk (1963) p. 169 and La Garenne, Abri Blanchard: Evin et alii (1975) p. 49) while cultural considerations might, at most, extend this settlement north only into Yonne (Grotte du Trilobite: Leroi-Gourhan et alii (1976) p. 1329). In the north-east of France, settlement during this millennium (Fig. 2.5) could be argued to have fallen slightly short of the limits achieved during the previous 14th millennium b.c., on the basis of radiocarbon dates from the Grotte Grappin (or Grotte St. Vincent) at Arlay (Jura: Evin et alii (1973b) p. 520, but see comments in Desbrosse and Evin (1971) p. 186). Similarities between specific items of bone and antler work (Desbross (1976) pp. 1350 and 1352) from Farincourt (Grotte II: Mouton and Joffroy (1956) and Rigney (Doubs: Glory 1961) and objects from Arlay could be used as the basis for arguments extending settlement during the 14th millennium b.c. as far north as Haute Marne (but see also Combier in Ferembach (1954) pp. 25-26 and Bourdier (1961) for discussion of chronology). This proposition of course if not diminished by the date of 12,990 b.c. ± 500 (LY 1191) on a vertebra of Coelodonta antiquitatis from Rigney: Evin et alii (1978) p. 49).

Certainly, within neither millennium (Figs. 2.4 and 2.5) can settlement or exploitation be shown to have reached as far north as the Pas de Calais, as could be argued to have happened at close to 14,000 radiocarbon years b.c. (Hallines: 14,050 ± 500 (Gif. 1712); Agache (1971) pp. 305-309 and Tuffreau (1976) pp. 1346-1347). Here the skeleton of a mammoth appears to have been visited on two separate occasions and it seems appropriate to speculate whether the find represents dismemberment of a fresh kill or rather the exploitation for raw materials of an older and perhaps freshly re-emerging collection of bones and ivory. If the latter interpretation is in fact correct, then this exploitation can only be assumed to have taken place at some date later than about 14,000 radiocarbon years b.c. (Gif 1712 is on a vertebra of mammoth), and there would be no certain evidence remaining for a human presence this far north-west within Europe during the Post-Lascaux and Pre-Angles (Arlette and A. Leroi-Gourhan (1964) p. 17) cold phase and indeed until that part of the Oldest Dryas which follows the Pre-Bølling oscillation.*

Incorporated into Figs. 2.4 and 2.5 are the radiocarbon determinations from Munzingen (Albrecht (1979) p. 79) and Schussenquelle (Lang 1962): see also Vogel and Waterbolk (1963) p. 168) but the possibility of reinterpreting the stratigraphies of both the Maszycka Cave near Cracow (Kozlowski (1966 and 1969) and of Nové Drátenické in Moravia (Klíma, 1949) as indicating use by Magdalenian groups within L De 1b or 1c instead of during L. De Ia, as suggested by Kozlowski (1971 pp. 56-59), reduces the scale of Magdalenian settlement and exploitation during both the 13th and 14th millennia b.c.

Indeed, an extension of Magdalenian settlement into Belgium and southern Germany can only certainly be demonstrated from within the 12th millennium b.c. whilst it is only potentially as early as the Meiendorf interval of L De 1a (Menke (1968) pp. 77-78; Schild (1973) pp. 49-50) that the Meiendorf and Poggenwisch variants of the Hamburgian (Tromnau 1975a and b) appear on the north European plain. Fig. 2.6 maps Magdalenian assemblages dated by radiocarbon to within this millennium. Represented by solid dots are sites attributed to the Meiendorf and Poggenwisch groups of the Hamburgian, and indicated by open circles are finds of Hamburgian material which, since they do not include "Havelte"-type points, cannot be shown to be younger in age than L De 1a and indeed, cannot be shown not to belong to either the Meiendorf or Poggenwisch groups.

While several commentators (Rust (1943) pp. 150-151 and Schwabedissen (1951a) pp. 67 and 65 Abb 2c) have pointed to artifacts from within England which appear morphologically identical to the shouldered and truncated ?speartips ("kerbspitzen") of these Hamburgian groups, there are no assemblages from within the island which combine only such speartips with long-nosed awls ("Zinken") and scrapers developed on elongated supports with marginal retouch. Although both shouldered and truncated pieces and "Zinken" can be demonstrated to be components of rather later technologies, mapped on Fig. 2.6 are finds of shouldered and truncated pieces and of "Zinken" from Britain which, since they are unaccompanied by other categories of backed artifact, cannot be shown not to have been discarded as early as the Meiendorf interval of L De 1a by Hamburgian groups exploiting at least the present-day eastern side of the island.

Fig. 2.6 attemps to map settlement during L De 1b and 1c—the Bølling and Older Dryas. It combines Magdalenian sites in Belgium and the Lower

---

* Significantly, Schmider (1977 p. 767) has likened the stone industry from Hallines to that from Habitation No. 1 at Pincevent dated (Gif 358) to the 10th millennium b.c.

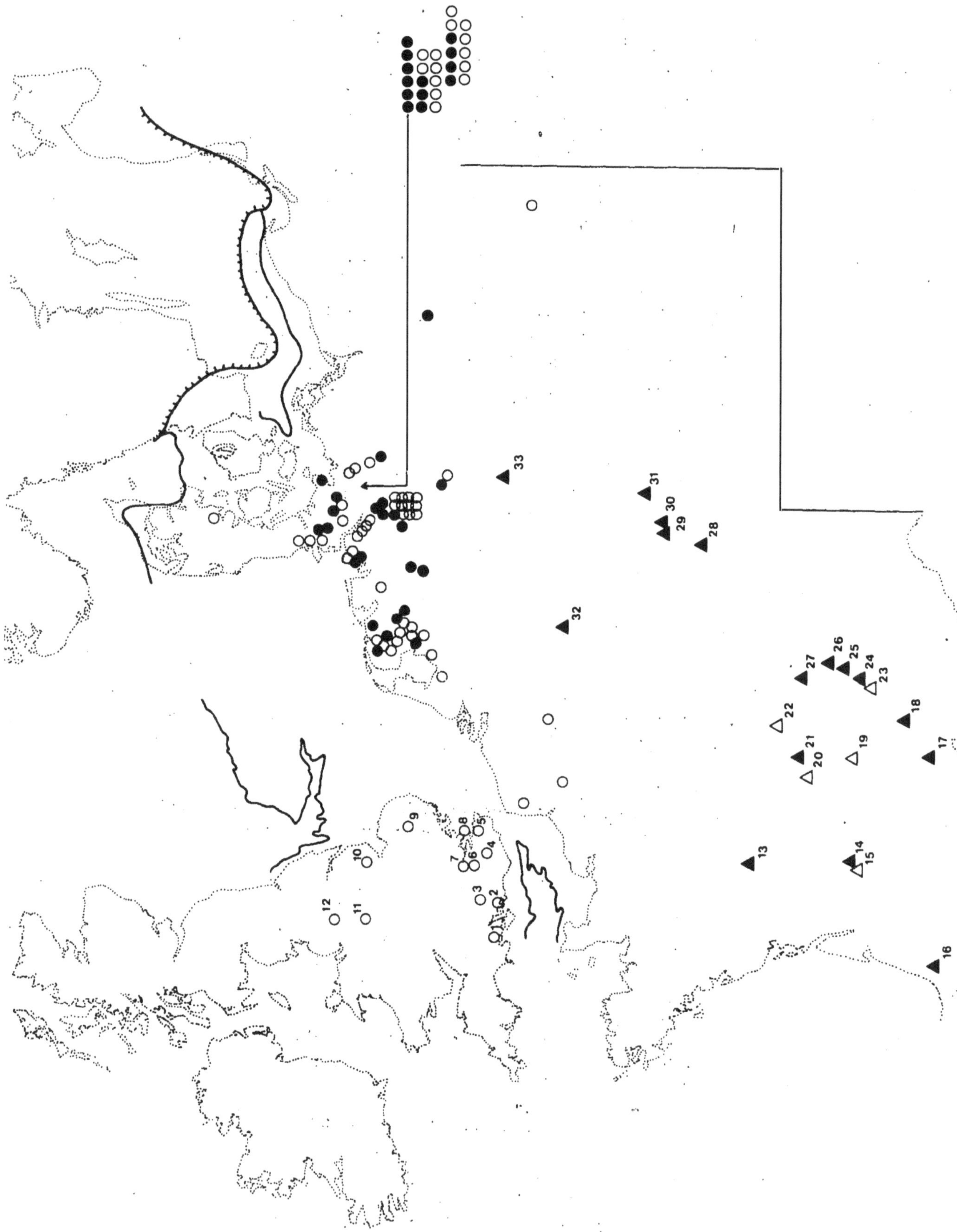

33
31
30
29
28
32
27
26
25
24
23
22
18
21
20
19
17
13
14
15
16
9
8
5
7
6
4
10
3
2
12
11

Fig. 2.6 (opposite) Distribution of radiocarbon dates for Magdalenian sites whose central points fall between 12,000 and 11,000 radiocarbon years bc (infilled triangles) or whose errors taken a at a single standard deviation fall at one extreme between 12,000 and 11,000 bc (open triangles). Also plotted are assemblages of the Poggenwisch and Meiendorf variants of the Hamburgian (infilled circles), and other mainland finds of Hamburgian material from which Havelte-type points are absent. (Open circles: data from Tromnau 1975a and b.) For Britain, Belgium and N. France are indicated finds of shouldered and shouldered and truncated pieces not associated with Federmesser-type backed artifacts.

Key (A) British Find-spots of shouldered and shouldered and truncated pieces not associated with Federmesser artifacts

1. Rush Corner, Bournemouth
2 Long Island
3 Fir Hill, Kingsley
4 Stonewall Park
5 Oare
6 Wallington
7 Wandsworth
8 Shoeburyness
9 Cranwich
10 Salmonby
11 Thor's Fissure Cave
12 Cupwith Hill

(B) Magdalenian Sites with radiocarbon dates falling between 12,000 and 11,000 radiocarbon years bc

13 Roc aux Sorciers
14 La Madeleine
15 Laugerie Haute
16 Duruthy
17 Grotte de Fontanet
18 Ebbou: Pont d'Arc
19 Le Blot
20 Longetraye
21 Vic le Comte: Enval
22 Viry: Grotte du Crest
23 Campalou
24 St Romans: Abri du Calvaire
25 St Thibaud de Couz
26 Grotte de Romains
27 La Colombière
28 Petersfels
29 Höhlensteinstädel
30 Kleinen Scheuer
31 Spitzbubenhöhle bei Eselsburg
32 Grotte du Verlaine
33 Bärenkeller bei Königsee Garsitz

5
4
3
2
1

Fig. 2.7 (opposite) This map of human settlement in N. Europe during the Bølling and Older Dryas can in view of the scarcity of solid archaeological information only represent a compromise between conflicting lines of evidence. Represented by infilled triangles are the Magdalenian sites of Pincevent, Etiolles, the Grotte du Coléoptère, Gönnersdorf and Andernach each independently dated either to the end of Bølling or early within Older Dryas. Infilled circles indicate sites with Havelte-type points which stratigraphies a at Luttenberg, Duurswoude II and Duurswoude IV show should be of Bølling and Older Dryas age. Half infilled circles represent assemblages of Meiendorf type since Lanting and Mook suggest (1977 p. 17) that the Betula peak in the pollen diagram for Meiendorf which overlaps the Hamburgian level may represent Bølling, a proposition not contradicted by the very unsatisfactory C14 evidence. If this suggestion is correct then any or all of the unassociated finds of shouldered and truncated pieces from Britain, Belgium and N. France could, if of Hamburgian origin, be of Bølling rather than of Older Dryas age. Represented by open triangles are finds of microlithic rectangles (data from Taute 1963) which a radiocarbon date from Olknitz in Thuringia (H830-840) places at the boundary between the Older Dryas and the Allerød. These microlithic rectangles are assumed to be of Magdalenian origin. Half infilled triangles represent Magdalenian sites within Poland. The absence of Magdalenian backed bladelet assemblages from Holland, West Belgium, Artois, the Somme Basin and N. Normandy, the areas in closest contact with Britain, and the presence in all these areas of stray finds of artifacts potentially of Hamburgian origin hint that if Britain was exploited prior to the Allerød that the technology of the exploiting group(s) will have been of Hamburgian type.

Key to N.W. European 'dated' Magdalenian sites

1 Etiolles
2 Pincevent
3 Grotte du Coléoptère
4 Gönnersdorf
5 Andernach bei Martinsberg

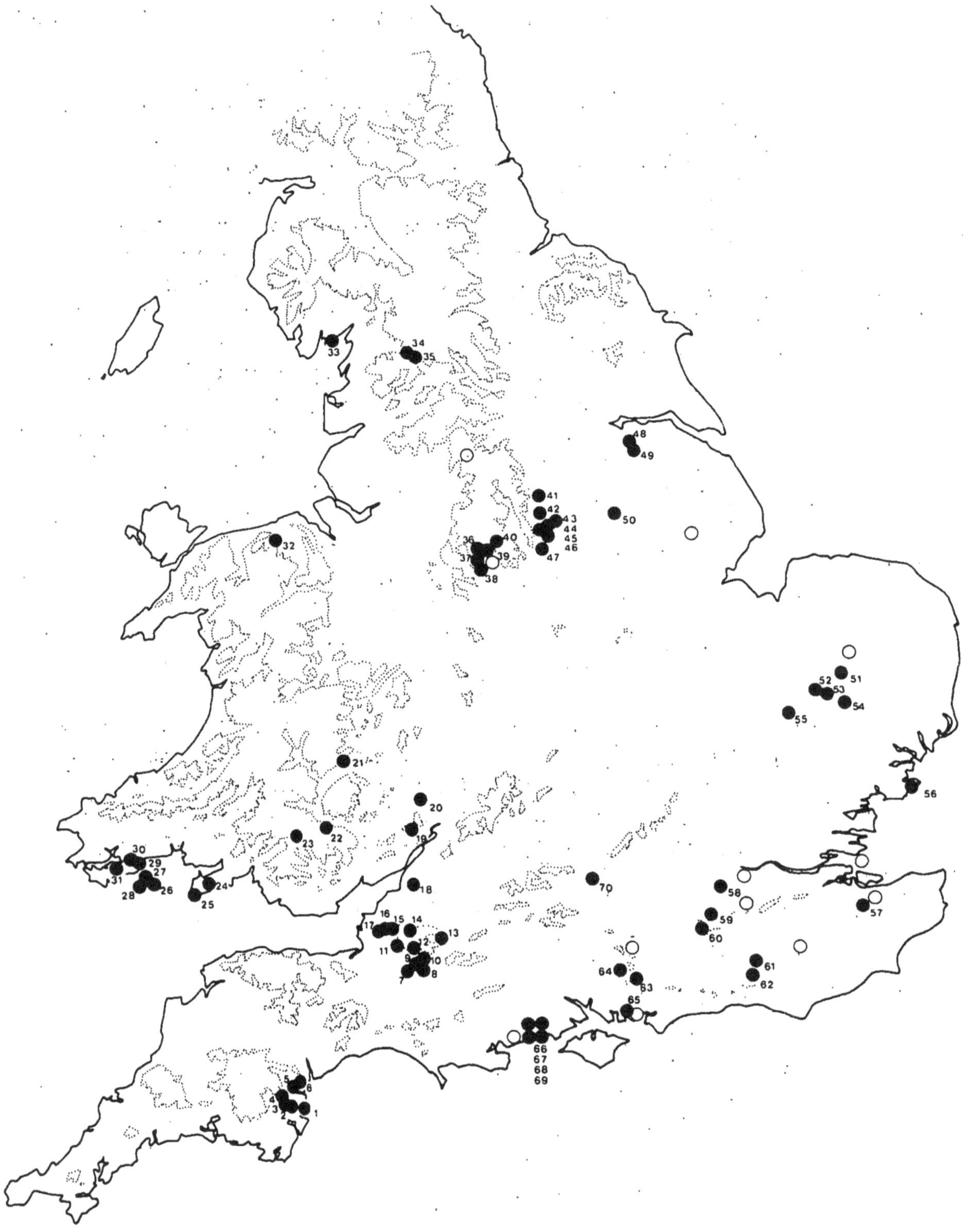
1
2
3
4
5
6
7
8
9
10
11
12
13
14
15
16
17
18
19
20
21
22
23
24
25
26
27
28
29
30
31
32
33
34
35
36
37
38
39
40
41
42
43
44
45
46
47
48
49
50
51
52
53
54
55
56
57
58
59
60
61
62
63
64
65
66
67
68
69
70

Fig. 2.8 (opposite) Infilled circles: finds of Federmesser-type artifacts in the British Isles. Continental evidence suggests that the earliest that any of these finds is likely to be is close to 9.800 radiocarbon years bc. Open circles: finds of unassociated shouldered and shouldered and truncated pieces. For key to these see the key to Fig. 2.6

Key

1	Kent's Cavern	36	Dowel Hall Cave
2	Tornewton	37	Ossom's Cave
3	Torbryan Caves	38	Elder Bush Cave
4	Three Holes Cave	39	Fox Hole Cave
5	Pixies' Hole	40	Stoney Low
6	Tramp's Cave	41	Anston Stones Cave
7	Flint Jack's Cave	42	Lob Wells Shelter
8	Great Oone's Hole	43	Mother Grundy's Parlour
9	Gough's (New) Cave	44	Robin Hood's Cave
10	Soldier's Hole	45	Pin Hole Cave
11	Callow Hill	46	Church Hole Cave
12	Sun Hole Cave	47	Langwith
13	Herriot's Bridge	48	Winterton
14	Aveline's Hole	49	Risby Warren
15	Banwell Camp	50	Morton
16	Banwell Caves	51	Great Fakenham
17	Hutton Cave	52	Undley
18	Bury Hill Camp	53	Mildenhall
19	King Arthur's Cave	54	London Bottom
20	Ledbury	55	Swaffham Prior
21	Arrow Court	56	Walton on Naze
22	Gwernvale	57	Bapchild
23	Rhigos	58	Syon Reach
24	Cathole Cave	59	Horsell
25	Paviland	60	Pyrford
26	Ogof Yr Ychen	61	Old Faygate
27	Nanna's Cave	62	Hawkin's Bridge Pond, Lower Beeding
28	Potter's Cave		
29	Little Hoyle	63	Sheet Common
30	Hoyle's Mouth	64	Butser Hill
31	Priory Farm Cave	65	Wiccor Shore
32	Plas-yn-Cefn	66	Hengistbury Head Site A
33	Kirkhead Cave	67	Hengistbury Head Site B
34	Kinsey Cave	68	Hengistbury Head Site C1
35	Victoria Cave	69	Hengistbury Head Site C2
		70	Crown Acres, Newbury

Rhine with the distribution of relatively localized and supposedly later "Hamburgian" assemblages with "Havelte"-type tanged points, artifacts shown stratigraphically at Duurswoude IV to be part of the technology of the Bølling and at Luttenberg and Duukswoude II to be part of that of the succeeding Older Dryas. So far such "Havelte"-type points are unrecognized from the British Isles. The find of a worked reindeer antler, dated to Older Dryas, on the bed of the Sound off Copenhagen (Clark (1975) pp. 85–86) indicates settlement, exploitation or exploration of the northern-most part of the central North European Plain during L De 1c, and exposes the gap within published archaeological sequences for north Germany (Schütrumpf (1958) p. 21 Tab 3; Clark (1975) p. 68 (Table 3). Finds which it is suggested may fill this gap are those of truncated microliths ("Rückenmesser mit retuschiertem Ende" and "Parallelseitige Messer mit retuschietem Ende") and microlithic rectangles ("Rechteckmesserchen" and "allseitig retuschierte Messerchen") mapped by Taute (1963 Karte 2) from northern Germany. These artifacts are suggested to derive from Magdalenian assemblages of the so-called "Ölknitz Group" and represent a settlement which in north-central Germany can be argued (Feustel (1974) pp. 190-199) to belong within L De 1c and earliest L De II.* If this equation is correct, it becomes possible to "people" the gap observed in published sequences for the area.

Failure to recognize from within Britain technologies with either "Halvelte"-type points or Magdalenian backed, or backed and truncated microlithic, argues against settlement or exploitation during L De 1b and 1c. To what then, do all suspected earliest British stone tool assemblages compare, and of what age are these?

In an attempt to answer both questions, discussion must be restricted to a specific part, albeit the major part, of the archaeological evidence available to us from within Britain. The discussion will concentrate upon those stone tool assemblages—amongst which I believe are those representing the earliest Late-Devensian re-settlement of Britain—which incorporate any or all of the following categories of backed artifact. These may be listed as:

(1) Angle-backed pieces, the equivalents of Bohmer's (1956) Creswell Points and the French "Pointe Azilienne à dos anguleux (Celerier (1977) p. 463

(2) Trapezoidal-backed pieces: the equivalents of Bohmer's (1956) and de Heinzelin's (1962) "Cheddar Points" in that "... la silhouette est trapezoidale: le bord abbatu dessin(ant) deux angles obtus, sensiblement égaux ..." (ibid. p. 34)

(3) "Shouldered and truncated pieces"- that is pieces which combine a distinct shoulder with an oblique distal truncation.

(4) "Convex-backed pieces"—a term intended to combine into a single grouping all forms of curved or "arch-backed" pieces whether or not these appear symmetrical in plan form or whether they come to a point at one or both extremities. Such "convex-backed pieces" may be elaborated by an additional oblique basal truncation, such pieces being traditionally termed by

---

* This suggestion is not denied by radiocarbon dates for assemblages with such rectangles from Olknitz and from Kůlna.

British prehistorians "penknife-points". These "convex-backed pieces" are equivalent to Azilian Points as defined by Celerier (1977).

To date it is possible to map seventy-one occurrences of assemblages which contain any or combine some or all of these categories of backed tool. Of these occurrences only forty-one (or 58%) come from caves or under rock-shelters—the remainder representing instead "open sites". Included into this map (Fig. 2.8), but not included in the above statistics, are finds made up only of shouldered and shouldered and truncated pieces. Since it cannot be demonstrated that such finds do not relate to a rather earlier Hamburgian exploitation of the island, their find-spots are marked by open rather than infilled symbols.

Both singly and, more significantly, in combination, these backed tool forms have, of course, precise analogues in the so-called "Federmesser" stone tool assemblages of northern and north-western Europe, where they are regularly combined, as in Britain, with convex-edged scrapers on either or both short or long supports, burins, awls (including the so-called "Zinken") and retouched pieces (Bohmers (1956) Pl. I and (1960) Pl. I). This "Federmesser" technology is believed within mainland Europe to belong at earliest, if not indeed in its entirety, within L De II, and clearly there will be a need within this essay to consider in some detail those occurrences in Britain where combinations of these various artifact categories would seem instead to have been attributed to various points within the preceding L De I.

It is necessary, however, to emphasize also that the total of directly relevant radiocarbon determinations for mainland Europe is still particularly small (Paddaya (1971) pp. 261-264; Lanting and Mook (1977) pp. 18-23), and in a large part the assumption that this technology belongs within L De II, and more specifically within its latter part, rests on the repeated association of Federmesser assemblages with the highly distinctive Usselo soil profile developed on top of the Younger Coversand I. A rather longer chronology, has, however, been argued from the recovery of Federmesser-type artifacts both from below this Usselo soil—as at Duurswoude-Oud Leger (Friesland)—and also from within the Younger Coversand II above this soil, as at Haule V, Siegerswoude II and Prikkedam (Newell (1973) p. 429 footnote 4). An alternative and seemingly valid interpretation of these observations would, of course, be to suggest that at Duurswoude the formation of the soil profile took place late within L De II rather than necessarily indicating a Pre-Allerød date for the stone tool assemblage from this site (Lanting and Mook (1977) p. 18), while a date of 9,010 b.c. ± 85 (GrN 1059) on charcoal from an Ahrensburgian culture layer, stratified within the Younger Coversand II at Geldrop I, (N. Brabant: Wouters 1957), may suggest that, locally at least, this coversand had begun to accumulate within L De II. Thus, the assemblages from Haule V, Siegerswoude II and Prikkedam may also belong within Le D II, albeit in all probability to a late stage within this interstadial.

Particular problems surround the interpretation of radiocarbon dates achieved on wood charcoal from these findspots, since it may be suspected in a number of cases—e.g. Horn Haelen (Limburg: GrN 497, 498 and 7297), Milheeze Ia (N. Brabant: GrN 2314) and Duurswoude-Oud Leger (Friesland GrN 607 and 4871)—that this charcoal originated not from artificial fire-spots associated with the occupation of these sites but from potentially later, natural forest conflagrations triggered by volcanic eruptions at the close of L De II.

In short, we are rapidly reduced to the very minmum of evidence from radiocarbon and pollen analysis. At Milheeze Ia fragments of worked wood from the peat bog adjacent to the settlement suggetsted its occupation during the Betula phase of L De II (Waterbolk (1962) p. 230)—that is between about 9,800 and 9,300 radiocarbon years b.c. A date of 9,490 b.c. ± 120 (GrN 1675) on charcoal from Budel II (N. Brabant) certainly appears too old for "Usselo" charcoal but in view of uncertainties surrounding the origin of this charcoal (Lanting and Mook (1977) p. 19) cannot certainly be regarded as a reliable date for the Federmesser-type artifacts from this site. Finally, at Een-Schipsloot, Federmesser-type artifacts can only be shown to be earlier than a localized peat band dated to 8,545 b.c. ± 60 (GrN 6341).

To summarize: three occurrences from Holland of Federmesser may by their positions in an undisturbed Younger Coversand II belong at earliest only to the final part of L De II. Within one of these assemblages, that from Siegerswoude II, angle-backed pieces far outnumber convex-backed pieces, while this site, together with Haule V, is incorporated by Houtsma (1975) into a small group of so-called "Creswellian" assemblages defined by a heavy representation of angle-backed pieces (Creswell points) and an equivalent rarity of convex-backed (Tjonger) points/pieces. Also included within this group is Zeijen, a site with an assemblage, the backed elements of which consist exclusively of angle-backed and trapeziform pieces (Bohmers (1956) PL. I). Two other sites would seem to date rather earlier within L De II: Budel II, which may be dated by GrN 1675 to close to 9,500 b.c.—that is to within the Betula phase of the interstadial, and Milheeze Ia, attributed on the basis of pollen evidence to this same Betula phase. Both sites are incorporated by Bohmers (1960) PL. I into his Tjongerian group of assemblages where convex-backed pieces out-number all other categories of backed artifacts—most usually angle-backed pieces, since both trapeziform pieces (as at Drunen III) and shouldered and truncated pieces (as at Fransman) are markedly rare. Whilst Bohmers believed it possible to identify a number of discrete and potentially socially significant groupings amongst these Dutch Federmesser assemblages, it has never been demonstrated that what variation is discernible cannot be explained in terms of a single developing technological sequence (for similar comments see Paddaya (1971) p. 266) and certainly the very limited chronological information would in no way deny such an interpretation. This observation is, I believe, an essential clue to understanding the variability visible between Federmesser assemblages within the present-day British Isles.

At the opposite (eastern) end of the European Plain, a technology with curved backed pieces—the Tarnovian—appears the equivalent to the Dutch Federmesser group of assemblages. This technology is believed dated, at two successive horizons within the Coversand complex at Calowanie just east of the Vistula, to 9,430 b.c. ± 95 (Calowanie level III; Chmielewski et alii (1975) p. 233; Shild (1973) Ryc. I) and 9,240 b.c. ± 65 (level IV: Vogel and Waterbolk (1972) p. 67. For artifacts see Chmielewski et alii (1975) Table LXIV-LXVII . At Calowanie this technology is replaced, within level V, believed dated to close to 8,800 radiocarbon years b.c. (8,870 b.c. ± 90 (GrN 5233) and 8,710 b.c. ± 100 (GrN 4966); Vogel and Waterbolk (1972) p. 67), by a technology with stemmed arrowheads (Chmielewski et alii (1973) Tabl. LXXXIV). However, at a second dune complex at Witów, 120 km to the west and between the Vistula and the Oder, an occupation horizon with curved-backed pieces (level III: ibid. Tabl. LXXI) is contained within a coversand and is higher in the profile than a "lower level" also with curved-backed pieces (ibid. Tabl. LXX) and dated to 8,865 b.c. ± 160 and 9,070 b.c. ± 170. Whilst this coversand at Witów does appear to be of L De III age (Wasylikowa (1964); Schild (1973) p. 61 and (1976) pp. 91-92), it is not possible to estimate how long into this zone a technology with curved-backed pieces persisted and just when it was complemented (Witów: level III) or replaced by a technology with stemmed pieces.

For the central part of the European Plain comparable chronological information is at a minimum. At Rissen, near Hamburg, an assemblage with curved-, straight- and angle-backed pieces (Rissen 14: Schwabedissen (1954) Taf. 12-20) was recovered from within a grey sand layer, "Schwarzgraue sandschicht" (ibid. Abb 20 and Taf. 10b: also Schwabedissen et alii (1957) Abb I), developed on dune sand. Above this grey layer is a coversand and, in a soil profile formed at its summit, a second stone industry, this time with Ahrensburgian stemmed microliths. This intervening coversand cannot, on cultural considerations, have accumulated later than at some time within either final L De II or within L De III. Of three age-determinations on burnt wood recovered from the grey sand layer, one of 7,330 b.c. ± 290 (Y 1573: Barendsen et alii (1957) p. 910), falling as it does within Fl. I, appears too young to fit the stratigraphic sequence suggested. Of two further dates—8,610 b.c. ± 200 (Y 157A: ibid. pp. 910-911) and 9,500 b.c. ± 300 (H 75/68: Münnich (1957) p. 197), neither conflicts with the suggested stratigraphy, although Y 157A would imply that deposition of the Younger Coversand did not begin before some point within L De III. The grey sand with the Federmesser-type artifacts is believed to be stratigraphically younger than, and to be separated by, the "dune sand" from a band of organic mud (Schwabedissen et alii (1957) Abb 3 and 5) dated to 9,950 ± 290 (H 18/11) and 9,600 b.c. ± 280 (H 21/18).

Tromnau (1975 a and b) has recently argued that Federmesser-type artifacts recovered, together with otherwise Hamburgian-type assemblages, from a series of findspots in north Germany should be regarded as either of L De 1b or 1c age, the total assemblage representing a separate (Teltwisch) group within a later Hamburgian (Tromnau (1975) pp. 35-42, 79-82 and Karte 5 (1975B) pp. 42-43). An alternative, however, would be to suggest these assemblages as being mixtures of artifacts of different ages discarded at particularly favoured hunting locations. At Teltwisch I these "Federmesser"-type artifacts were recovered from below a "Verbrodelte Bleichsandschicht" said to be equivalent to the Usselo soil further to the west, but it could be argued, as for Duurswoude-Oud Leger (see above), that the formation of this soil had taken place only late within L De II with the implication that these Federmesser cannot strictly be demonstrated to belong within an earlier pollen Zone.

The earliest central date from the North European Plain relevant to a Federmesser-type assemblage remains that from Westerkappeln (Westphalia: site C) of 9,850 b.c. ± 300 (KI 271: Günther 1973) but, taken at a single

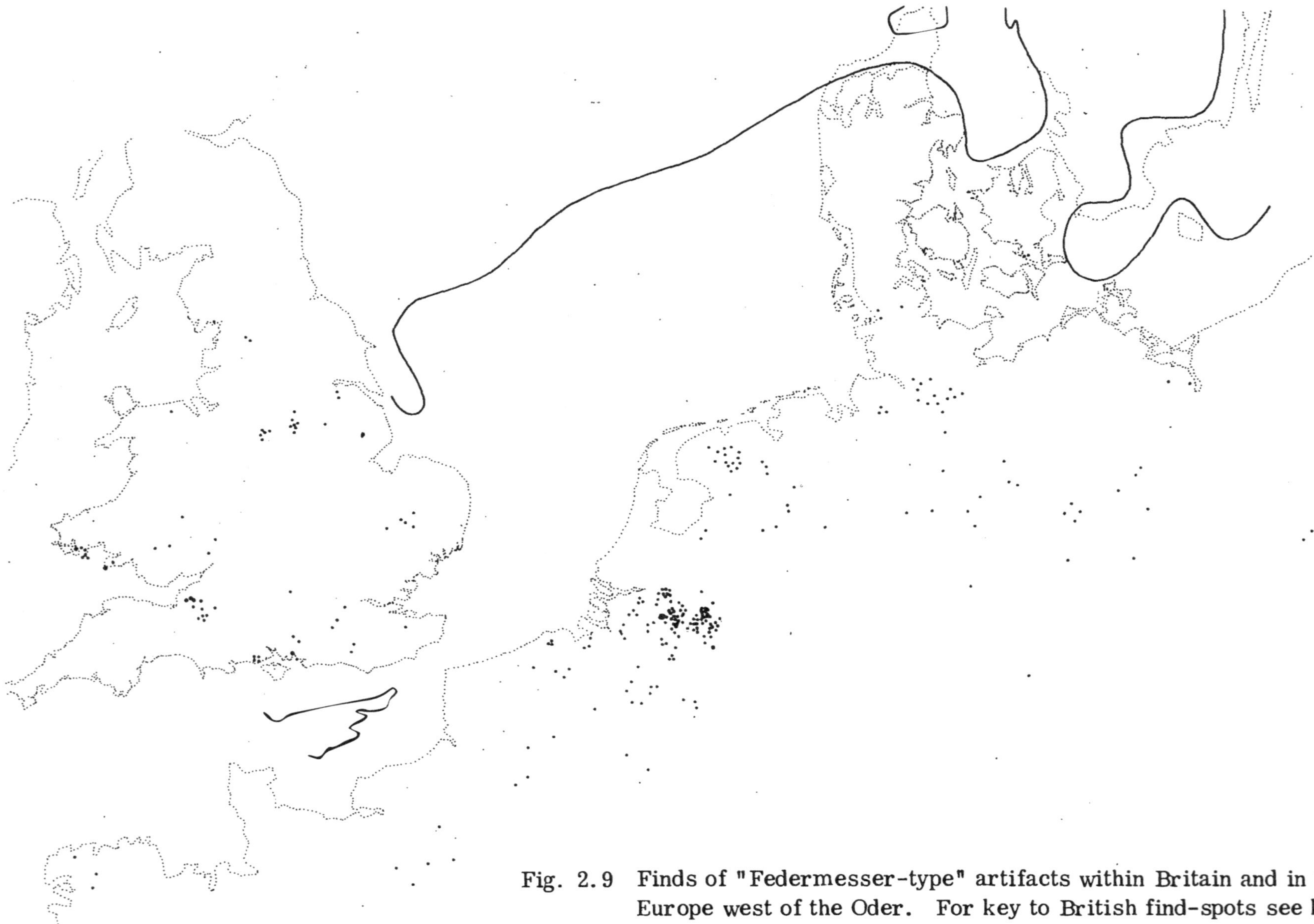

Fig. 2.9 Finds of "Federmesser-type" artifacts within Britain and in Northern Europe west of the Oder. For key to British find-spots see key to Fig. 2.8

standard deviation, it overlaps the oldest of the three determinations from Rissen (9,500 b.c. ± 300), the date from Budel II (9,490 b.c. ± 120) and the suggested age range for Milheeze 1a. It is only very slightly older than the oldest of the determinations from Calowanie (level III: 9,430 b.c. ± 95).

In no case then can a Federmesser technology be shown to be earlier than some point within the 10th millennium b.c. Schild (1976) p. 93 has linked the development of this technology to the local establishment of forest, and nowhere on the European plain can this technology be shown to be certainly older than the Betula phase of the Allerød.

This inability to demonstrate the existence of any Federmesser-type technology within the north-west part of the present European mainland earlier than the Betula phase of L De II must of necessity imply that it will only be at a date within or rather after this phase of the interstadial that the considerable expansion in the area of northern Europe settled, or at least exploited, occurred —an expansion (Fig. 2.9) which brought human groups as far west as the Priory Farm Cave (Grimes, 1933), the Hoyle (Savory, 1973) and Little Hoyle (McBurney (1959) pp. 168-169) in Pembrokeshire and to Nanna's Cave (Lacaille and Grimes (1955) pp. 96-120), Potter's Cave (Van Nédervelde and Davies 1975) and Ogof-yr-Ychen on Caldey (Fig. 2.10). To the north, contemporary groups exploited at least as far as the Victoria (Breuil (1922) pp. 275-276; Jackson (1945) and Lacaille (1946a) p. 76 and fig. 9) and Kinsey (Jackson and Mattinson 1932) caves of northern Yorkshire. Arguments will be presented below as to why this expansion cannot have taken place later than the end of the 10th millennium b.c. On the other side of the European Plain and within the same millennium, the first occupation sites appear on Zealand (Petersen 1973) and settlement could be argued to have been extended north to the very tip of Jutland as well as north-east into the area of present-day southern Sweden. On both sides of this plain this expansion of settlement—by over 400 kms on its western margin—appears to correlate closely with the local establishment of birch woodland (Fig. 2.11).

Defence, and it is to be hoped, confirmation of the thesis that Late-glacial resettlement of Britain was achieved only during this period of woodland development, requires a review of the chronological evidence relevant to each supposedly dated Late-glacial archaeological occurrence within the island. By this means it should be possible to discover whether in any case certain evidence can be discovered which need imply any earlier resettlement of the island. To impose some structure upon this discussion, sites will be considered in the order supposedly suggested by an initial acceptance of the radiocarbon and/or pollen evidence forthcoming from each. In every case we shall have particularly to assess either or both of (1) the likelihood of the association of the material(s) dated with the human artifacts recovered and (2) the validity of any stratigraphic interpretations made by previous commentators.

In structuring those parts of the discussion involving any assessment of the supposedly associated fauna(s) I have started from Stuart's recent proposition (1974 p. 257 and 1977a pp. 309-310) that Mammuthus primigenius, Coelodonta antiquitatis, Crocuta crocuta, Panthera leo, Bison priscus and Ovibos moschatus disappeared between the Middle and Late Devensian from the fauna to be found within these islands. While this observation requires to

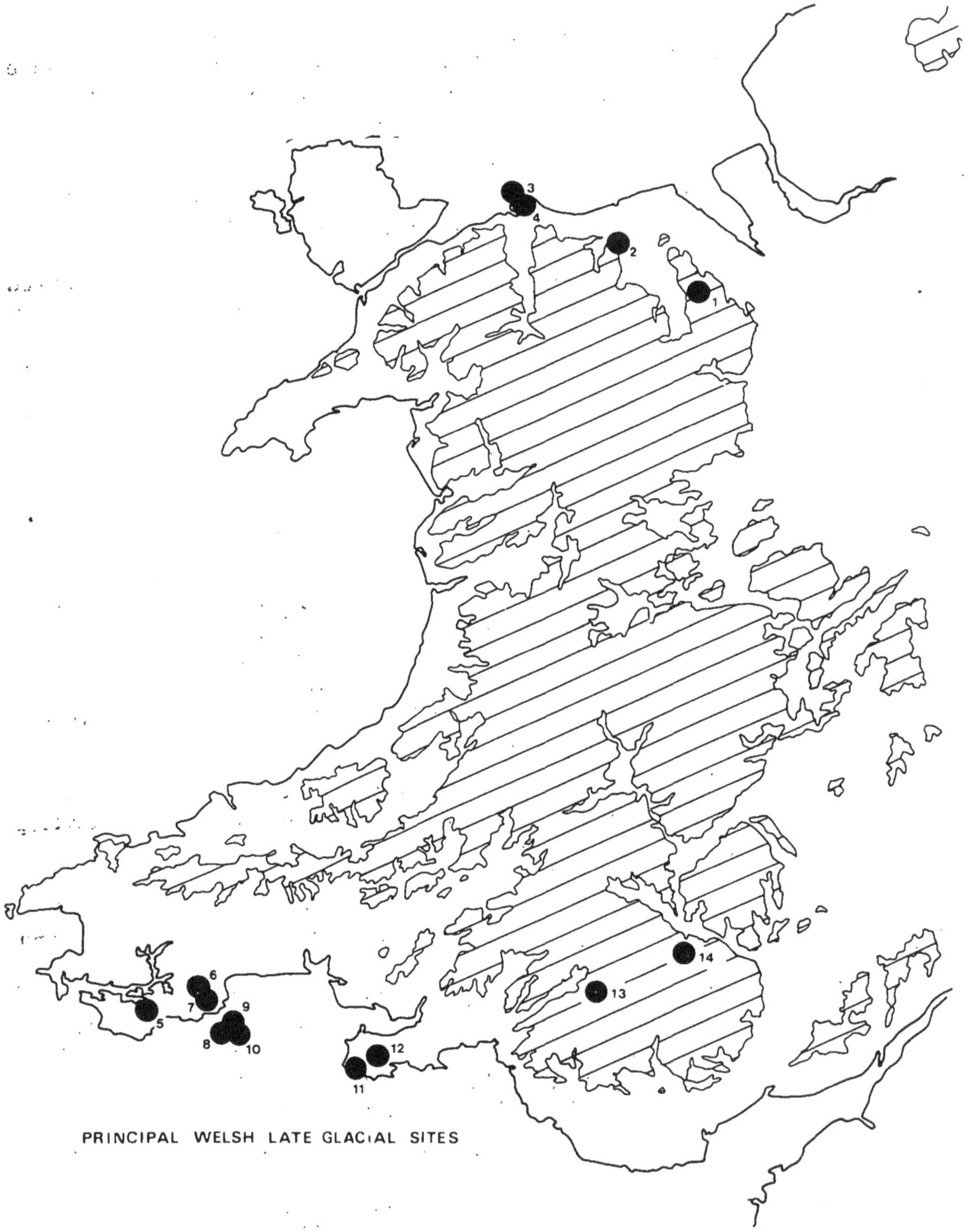

PRINCIPAL WELSH LATE GLACIAL SITES

Fig. 2.10 (opposite) Principal Welsh Late-glacial Sites

Key
1 Lynx Cave
2 Plas-yn-Cefn Cave
3 Ogof Tan-y-Bryn
4 Kendrick's (Upper) Cave
5 Priory Farm Cave
6 Hoyle's Mouth
7 Little Hoyle
8 Potter's Cave
9 Nanna's Cave
10 Ogof-yr-Ychen
11 Paviland
12 Cathole
13 Rhigos
14 Gwernvale

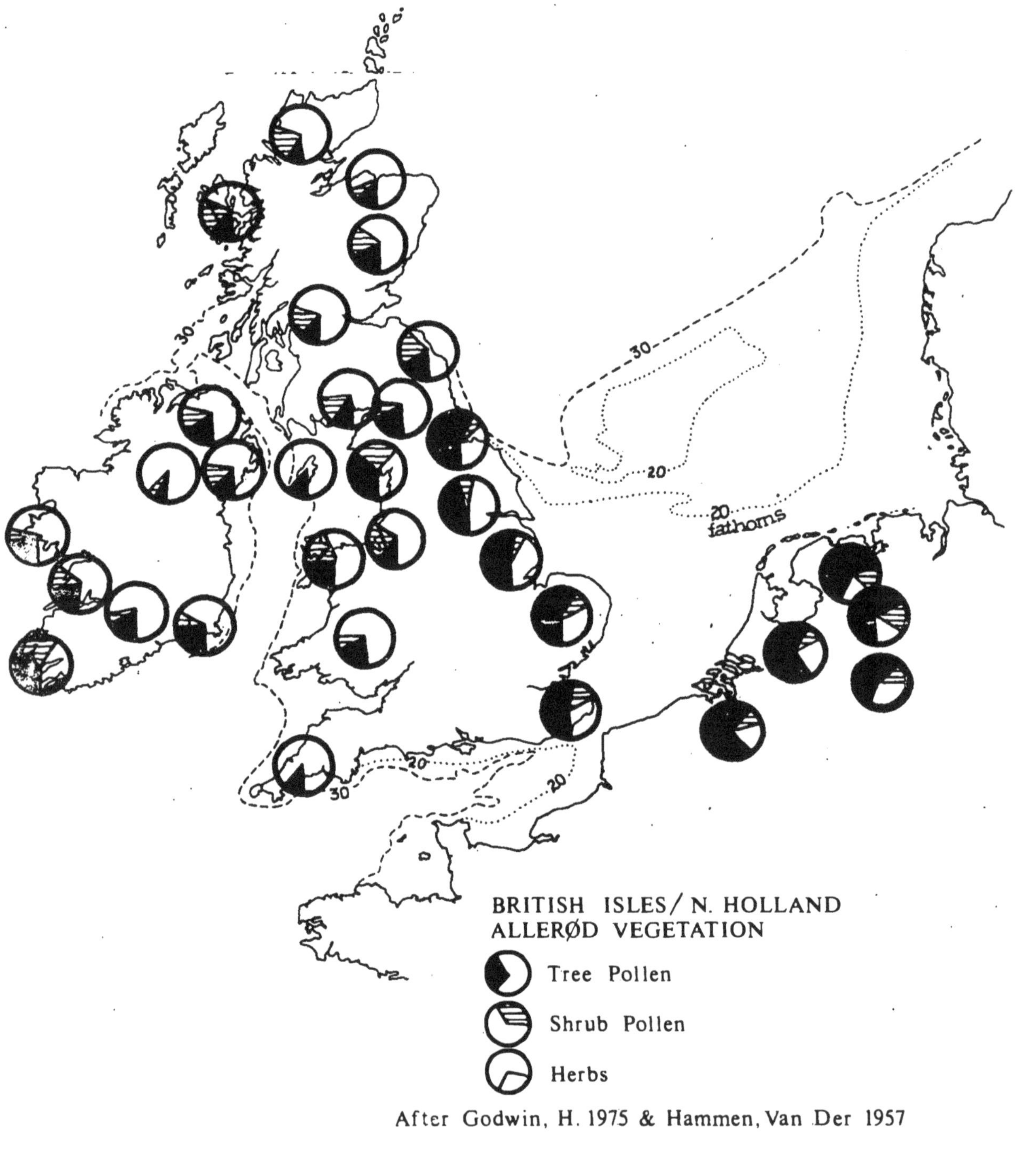

Fig. 2.11 Pattern of vegetation in Britain and N.W. Europe during the latter part of L-De II

be explicitly tested it is apparently confirmed by the small number of radiocarbon dates on reliably identified bones/antlers and by the sequences at both Soldier's Hole (Cheddar: Jackson in Parry (1931) pp. 58-62) and Elder Bush Cave (Bramwell 1964). Bones of all six animals are strikingly absent from L-De deposits accumulated in any cave or rock-shelter without older Early or Middle Devensian deposits. When we do have records of any or all of these animals from what associated cultural artifacts indicate to be L De occupation levels, they are all very noticeably from cave or rock-shelter locations which can in each case be demonstrated to contain, or to have contained, deposits with an earlier Devensian or even older fauna. In such cases it may be suspected either that the excavator was unable to make a sufficiently sympathetic separation of the relevant deposits—as at Kent's Cavern, Tornewton, King Arthur's Cave and Pin Hole—or that there is reason to suspect the incorporation of "older" fossil material into younger deposits. The latter process may, it is felt, explain the anomalies in the fauna from both Robin Hood's Cave and Mother Grundy's Parlour at Creswell. Again, apparent records for Sus scrofa, Capreolus capreolus and Castor fiber from L De deposits coincide with caves or rock-shelters known to have been used either as habitation sites or burial places (or both) during the Flandrian—as at Gough's Cave, Aveline's Hole and Chelm's Combe—and/or where excavation proceeded by reference to arbitrarily defined horizontal spits rather than natural stratigraphic units. (For an extreme example of this, see Balch 1927). If we can be strictly selective in our use of the mass of data potentially available and plot only those finds of ungulate material dated either (1) directly by radiocarbon or (2) recovered from within organic deposits and dated by pollen analysis, the following pattern of representation emerges for L De I-III and Fl. Ia. (Table 2.1).

THE CHRONOLOGICAL EVIDENCE: L-De Ia (The Oldest Dryas)

Kent's Cavern, Torquay, Devon

A single radiocarbon date of 12,325 b.c. ± 120 (GrN 6203) is available from the "vestibule" at Kent's Cavern. This date is on an unburnt tibia of Ursus arctos provenanced to the "Black Band", a three-inch lens of darker deposit with charcoal, bone fragments and artifacts sandwiched between a stony cave earth and an over-lying granular stalagmite. Fragments of charcoal, identified as Quercus sp., Ulmus sp. and Rhamnus catharticus (Campbell (1977) Appendix I pp. 203-207) from below this granular stalagmite, restrict the timing of formation of this latter deposit to the Flandrian. The stalagmite is said (Campbell (1977) p. 39) to contain Mesolithic and Neolithic artifacts and cannot therefore provide any useful terminus ante quem for the L-De occupation beneath.

Artifacts provenanced to the "Black Band" comprise angle-backed and trapezoidal-backed pieces, an artifact approaching a "Malaurie Point" (Campbell (1977) fig. 114 No. 6), an end scraper, a Magdalenian retouched blade with concave notch at its proximal end (de Sonneville Bordes and Deffarge 1975), an eyed needle of bone (Pengelly Find No. 1847) and a single, much damaged barbed antler spearhead (Pengelly No. 1970; Garrod (1926) fig. 4 No. 14; Campbell (1977) fig. 115 No. 4).

Table 2.I Pattern of Faunal Representation L-De I - Fl. Ia within Britain

	L De I	L De II	L De III	8,000bc	Fl. Ia
Perissodactyla					
Equus caballus L.			(RC) (RC) (RC*) (RC**) (P)		
Artiodactyla					
Sus scrofa L.					(P)
Megaloceros giganteus (Blumenbach)	Ia(RC) Ib/c(RC)		†(RC*)		
Cervus elaphus L.					(P) (P)
Alces alces (L.)		(P) (P)			(P)
Rangifer tarandus (L.)			(RC) (RC) (RC**)	(RC)	(RC) (RC***)
Capreolus capreolus (L.)					(P)
Bos primigenius Bojanus			(RC/COL)		(P) (P) (P)
Bos sp. and/or *Bison* sp.		(RC) (RC) (RC)			

(RC) Radiocarbon (P) Pollen (RC/COL) Radiocarbon combined with Coleopteran studies of enclosing peat lens
† uncertain identification * Antler fragment ? sp. combusted with metacarpal of *Equus*
** bones of *Rangifer* and *Equus* combined *** bone fragments including *Rangifer*

It is not possible to find a technologically equivalent group of backed blade material in mainland Europe of the 13th millennium b.c., the backed artifacts of what would generally be termed the Middle Magdalenian comprising instead only (effectively) microlithic backed bladelets, denticulated bladelets and frequently scalene triangles. Equally, barbed hunting spears are noticeably absent from sites of this date—appearing in the 12th millennium b.c. in the French part of their distribution (see, for example, Ly 922 for La Madeleine, Dordogne, layer 4: Evin *et alii* 1976)—the only exception being a spearhead from La Grotte Gazel (13,120 b.c. ± 270: Fif 2655) from a layer which has yielded also such potentially later Magdalenian artifacts as a sculptured spear-thrower and a contour découpé (Sacchi 1974), perhaps hinting at a later occupation.

Since the tibia of *Ursus arctos* dated was whole and unburnt there appears no reason necessarily to associate it with human activity within the cave complex. Equally, the record (Campbell and Sampson (1971) p. 15) of teeth of *Panthera leo, Coelodonta antiquitatis* and *Crocuta crocuta,* provenanced to the "Black Band", suggests mixing at the time of excavation with what must be regarded as elements from a rather colder, and in all, probability, Middle Devensian fauna, introduced into the cave by carnivore activity or earlier Upper Palaeolithic hunters. These teeth must be assumed to derive from the upper part of the stony cave earth which underlies the Black Band.

Summary

A date (GrN 6203) was obtained (1) on a bone which cannot be shown to have been introduced into the cave by, or interfered with in any way, by, its Late-Devensian inhabitants. It appears equally, if not more possible, that the bone may have derived from an hibernating animal. (2) The excavators appear to have collected as from the "Black Band" elements of an older fauna perhaps (in part) of Middle Devensian age. (3) The flint artifacts provenanced to the "Black Band" are of Federmesser type and compare exceptionally badly with what is known of the technologies of the 13th millennium b.c.. It could therefore be suspected that the tibia selected for dating was a rather older bone incorporated by the original investigators, along with the more obviously older elements listed by Campbell and Sampson. *Per se* the date cannot be taken as evidence for human settlement within Britain during Oldest Dryas.

L-De Ib: Sun Hole Cave, Somerset

An unburnt and unmodified humerus of *Ursus arctos,* recovered from the B 2-7 sharp scree of the Sun Hole Cave, is dated to 10,428 b.c. ± 150 (BM 524: Barker *et alii* 1971) and could, if associated with the artifacts from the cave, be taken to suggest settlement or exploitation as early as L-De Ib. Again, however, there is no specific argument for assuming this bone to have been introduced by the human occupants of the cave. Consideration of the very clear stratigraphic sections could be interpreted (Campbell (1977) fig. 13) as suggesting—and allowing also for the tilt of the deposits (*ibid.* fig. 14) —that the open thermoclastic scree *above* the humerus should join across a void created by previous excavators to link with either the scree of layer B 2-5 or with one of the lenses of scree within B 2-4, rather than that of B 2-8. On either reading of the stratigraphy, the sandy thermoclastic scree in which the humerus was found would be a part of B 2-4 (rather than B 2-7), a layer which incorporates pollen believed in part to be of L-De I age (*ibid.* p. 72).

There is thus considerable uncertainty in relating the dated humerus of Ursus arctos to that part of the profile into which artifacts have been projected. This uncertainty derives from the digging away of critical deposits during early campaigns within the cave.

It is clearly also very difficult to project artifacts accurately from these early campaigns (Tratman and Henderson (1928); Tratman (1955)—if the only stratigraphic information which we have about these is an arbitrary spit depth and a wide horizontal range—into the later, highly refined, stratigraphic sequence established by a subsequent excavator. The resulting variations in the stratigraphic attributions achieved for the same artifacts (and their recorded vertical ranges), when projected into the subunits of two such more recent profiles running at right angles to each other (Campbell (1977) fig. 12-14), confirm the very real ambiguities that can be introduced if any such attempt is made.

If our re-interpretation of the position of the humerus is correct and the sandy thermoclastic scree in which it rested is in reality a part of B 2-4, then, firstly, there is no artifact with a stratigraphic range, when plotted into either of the two profiles, that would fall below the level of the humerus. Secondly, there is no artifact with a stratigraphic range, when plotted, that falls entirely within layer B 2-4. Thirdly, any vertical range which overlaps with layer B 2-4 also overlaps with a higher, or a number of higher, layers. There is thus no artifact which cannot be shown to be potentially younger than layer B 2-4 and no artifact that can certainly be shown to be its contemporary.

Summary

A radiocarbon date of 10,428 b.c. ± 150 was obtained on a bone of Ursus arctos whose presence in the cave cannot be linked with human activity. Removal of critical parts of the deposits within the cave make it difficult to reconstruct the profile. It is possible to argue that the bone selected for dating derives from a deposit which cannot be certainly shown to have contained artifacts. Thus, there is no artifact which cannot be argued to be younger in age. Clearly, the stratigraphic evidence from this site appears ambiguous and it does not seem appropriate in view of this ambiguity to quote BM 524 in support of a re-settlement of Britain during L-De Ib.

L-De Ib/c: Kent's Cavern, Torquay, Devon

The second determination from this cave complex is that of 10,230 b.c. ± 100 (GrN 6204) on a metatarsal of Megaloceros giganteus from within a foot's vertical range, and up to a yard's horizontal distance, of a bilaterally barbed antler hunting spear (Pengelly find No. 2282; Garrod fig. 11 No. 15). Both are from within "...the second foot level of red loam (cave earth) having over it a stalagmite floor 18 inches thick..." (Pengelly, W. Kent's Cavern Monthly Report No. XXII.) As suggested for earlier determinations from the cave, it is not now possible to relate the position of the metatarsal to the positions of other items found within the same prism of deposit but the result would not be unreasonable for either this type of hunting spear or the convex-edged scraper found in the same prism. The date is thus particularly difficult to evaluate but it seems not unreasonable to comment that the bone cannot now

(1) be directly associated with any of the L-De artifacts from within the vestibule nor (2) can it be demonstrated to have been introduced by man. Contemporary mainland technologies at this time, L-De 1b, will have included in Holland either (or both) shouldered and truncated pieces and "Havelte"-type tanged points, the former being absent from with Kent's Cavern and the latter from Britain as a whole, while to the east and south of the Low Countries and even within their southern part, we would find a technology with numerous microlithic backed bladelets—a category which, together with the frequently associated "microperçoirs", is absent from Kent's Cavern as from all other British L-De sites.

<u>Summary</u>

A date of 10,230 b.c. ± 100 is not unreasonable for a bilaterally barbed spearhead of Late Magdalenian type. However, the bone dated cannot certainly be shown to have been closely related stratigraphically to the spear head, nor does it appear possible to demonstrate that the bone was introduced by human activity. Artifacts of a type to be expected, if settlement from the mainland had taken place during L-De 1b, are not to be found within the cave.

<u>L-De I (undivided): Cathole, Gower</u>

Evidence for occupation within Cathole during L-De I depends upon the interpretation of the pollen record for the site (Campbell (1977) fig. 72 and table 20).

It would not, however, be impossible to argue that, working down the profile, the pollen content of layer D is equivalent to the base of Flandrian II at Tregaron (Hibbert and Switsur, 1976) while that of layer C equates with either Zone Fl. Id at Tregaron or perhaps the top of Fl. Ic. This boundary is dated close to 6,250 radiocarbon years b.c. (Q 934 and Q 935). Artifacts from the interface between layers C and D appear to be Later Mesolithic—that is to say they could date to any time <u>after</u> about 6,800 radiocarbon years b.c. Artifacts from the interface B/C also appear to be Later Mesolithic.

Below this point it appears possible to interpret the pollen stratigraphy in two ways (1) and suggest, as does Campbell, that the lowered TP values for layer USB (11.4% deposit sample 9) represent L-De III, whilst the rather higher values for TP in UOB (30.6% total pollen) indicate L-De II. The reduced TP values in the rather lower deposits MSB, LOB and LSB would then all represent L-De I. An alternative reading would, however, be to suspect that the high values of <u>Cyperaceae</u> and <u>Gramineae</u> and other open-habitat plants, which are responsible for the reduced representation of tree pollen in sample 9, reflect temporary conditions local to the cave, and suggest that the pollen of deposit sample 8 (layer UOB), with 29.6% of the 30.6% TP made up of Birch, may belong within Fl. 1b while sample 7 from the midpoint of layer MSB, with its reduced TP (5%) and the highest values achieved by Juniper anywhere within the diagram (11%), represents Fl. Ia. This interpretation would not conflict with Evans's observations (Evans in Campbell (1977) pp. 205-209) of troglophile species of (late) Boreal Mollusca within the screes of deposits USB, UOB and MSB/LOB combined or recovery by McBurney (1960 fig. 1 Nos. 2, 9 and 16; Campbell fig. 131 Nos. 2, 3 and 7) from layer B (undivided) of a pair of microliths of early Flandrian type together with a microburin.

The pollen of layers LOB and LSB could be regarded as either of Earliest Flandrian, or of Latest Devensian age. The fact, however, the Late Boreal troglophile Mollusca were able to colonize air spaces within the scree network as deep as combined layer MSB/LOB (Evans pers. comm.) suggests the ease with which pollen could be down-washed into earlier sediments and later microfauna enter the deposits. It would not be impossible for (later) artifacts similarly to become displaced by slipping into or through such air spaces. Evidence for L-De occupation material within deposit B (undivided) is suggested by at least one backed artifact (McBurney (1960) fig. I No. 1; Campbell (1977) fig. 131 No. 1).

Summary

A single artifact from within deposit B appears to be of L-De type. The presence of Dicrostonyx torquatus, Lemmus lemmus and Microtus ratticeps (Sutcliffe and Kowalski, 1976) argues that layers USB, UOB, MSB, LOB and LSB may be of L-De age. The scree was, however, sufficiently open to allow colonization down to combined layer MSB/LOB by Late Boreal troglophile snails. With a scree sufficiently open to allow such colonization, we may suspect not just down-wash of pollen, but the intrusion of later artifacts particularly when small and of microfauna (cf. perhaps the record of Apodemus sylvaticus from layer UOB). Pollen spectra for layer B are interpreted as of L-De III and Fl. Ia and Fl. Ib type. The Late-glacial deposits of layer B and thesingle L-D-type artifact known to come from within this grouping of sub-layers, can only, on the suggested interpretation of the pollen, be argued to have an L-De III, or earlier, date. They cannot specifically be shown to be of L-De I age.

L-De I-III: Robin Hood's Cave, Derbyshire

At Robin Hood's Cave there appears, at first glance, to be a direct conflict between the evidence derived from pollen and small mammals and that of the radiocarbon dates. Federmesser-type artifacts were recovered from four contiguous layers, termed (Campbell (1977) pp. 64-69) from top to base, USB, OB, LSB and B/A.

A metacarpal of Equus together with an antler of ? Megaloceros from ?the middle part of layer OB was dated to 8,440 b.c. ± 90 (BM 603) while a humerus of Equus from the upper part (ibid. fig. 30) of layer LSB is dated to 8,640 b.c. ± 90 (BM 604: Burleigh et alii (1976) p. 22).

These determinations suggest that it was possible for objects, in this case animal bones, of L-De III age, to become incorporated into layers OB and the upper part of LSB. It cannot be demonstrated that flint artifacts, also of L-De III age, could not similarly become incorporated into layers OB and LSB. Acceptance of BM 603 and BM 604 implies both that the artifacts within layer OB and the upper part of LSB could be at least as late as L-De III and also that they cannot on present evidence be shown to be older (even though this may be suspected).

Artifacts from the lower part of layer LSB and from layer B/A below could be argued to be older than BM 604, that is older than some point within L-De III, but there is no precise way of estimating how much older these might be.

Within the pollen diagram presented for this site (Campbell (1977) fig. 77), there is a little above BM 603 and, within layer OB, a sharp rise in tree pollen, particularly of Betula, and it is suggested here that this represents the opening of Fl. Ia but in the open thermoclastic scree of layer OB some downwash of pollen may be suspected. If this Betula rise in the upper part of layer OB does indeed correspond to the opening of Flandrian Ia, then the lower TP values in the lower part of layer OB, and the very much lower TP values within layers LSB and B/A below, could be argued to correspond with L-De III.

On this reading of the evidence, there is no conflict between the radiocarbon dates and the pollen evidence from the cave while the artifacts from all three layers could be argued to be of L-De III age, if the pollen is not downwashed.

However, in the topmost of the four layers which we need to consider, layer USB, the pollen diagram shows sharply reduced values for TP. The same can be observed in the pollen count for deposit sample 7 at the top of layer OB. There is a single date for layer USB, taken on a broken humerus of Ursus arctos which was collected from the base of this layer and in contact with the point from which deposit sample 7 was collected (Campbell (1977) fig. 28). This was dated to 26,550 b.c. $\pm^{1,600}_{1,300}$ (BM 602). Clearly, this bone cannot be in its original position within the cave sequence, but no localized disturbance such as is indicated on Campbell (1977) fig. 28, was suspected at the time samples were submitted for radiocarbon dating (compare, for example, Campbell (1969) fig. 3 with Campbell (1977) fig. 28, and see comment in Burleigh et alii (1976) p. 22).

If no clear localized disturbance was visible, it appears logical to wonder how much of layer USB had not been disturbed by previous investigators of the cave. Alternatively, the "sandy thermoclastic scree", of which layer USB is composed, could represent the overlying of deposits of L-De III age by older deposits slipping down from a higher location within the cave, including sediment and fossils of Middle Devensian age. A trickle of older "fossil" material into these deposits, and presumably deriving from higher up within the cave, is suggested by the presence within layers USB, OB, LSB and B/A of teeth of Hyaena and Woolly Rhinoceros which must also be presumed to be of Middle Devensian date.

Summary

Radiocarbon dates from Robin Hood's Cave show that flint artifacts within layers OB and the upper part of LSB cannot be shown certainly to be earlier than L-De III. Artifacts within the lower part of LSB and in Layer B/A could be older but re-interpretation of the pollen diagram for the site suggests that these too may be of L-De III age. It is suggested that the topmost layer USB may not be in place and is made up of older deposits which have slipped from a higher position up the slope within the cave. Deposit sample 7, with a reduced representation of tree pollen, may incorporate some of this disturbed material which is poor in tree pollen.

L-De II: Poulton-le-Fylde, Lancashire

A findspot which, with initial acceptance of the radiocarbon dates, would seem to indicate human activity far north in the island, by at least the first half of the 10th millennium b.c., is Poulton-le-Fylde in Lancashire, where the skeleton of an elk (Alces alces) was recovered from open water deposits and was directly associated with a pair of barbed bone spearheads (Barnes et alii (1971); Hallam et alii (1973). The resemblance of these spearheads to examples (in antler) from the Proto-Maglemosian site of Star Carr has led to "backstage" speculation that these objects are really of Flandrian Ia/b rather than L-De II age. This speculation pointedly ignores the existence of a group of strikingly similar spearheads from a L-De II context at Dinslacken (West Germany: Stampfuss and Schütrumpf 1970). There appears to be no question that the 10 cm of shelly coarse detrital mud, in which the elk was discovered, is of Allerød age.

On the published section (Hallam et alii (1973) fig. 3), the elk remains are shown occupying the whole thickness of this 10 cm band of detrital mud—that is, as occupying deposits covering the whole of L-De II. Examination of colour transparencies of the profile (made possible by the kindness of Dr A. J. Stuart) indicates that, of the animal, at least the bones excavated under controlled conditions are covered by a part at least of the total thickness of detrital mud, seemingly hinting that, unless the bones had sunk through the deposit, the elk and the associated hunting spears must date to some point within the zone, and not to its very end. Equally, there would appear to be detrital mud visible below the bones, suggesting that the assemblage did not enter these deposits right at the opening of L-De II. There is thus no uncertainty over dating the assemblage to within L-De II but precisely where within the zone it belongs is not now possible to state with certainty.

Two dates relevant to this find have been published (Hallam et alii (1973) p. 110; Welin et alii (1974) p. 103). Both are on detrital mud—the first (if plotted onto Hallam et alii (1973) fig. 3) from the top 2 cm of this mud, and the second from its lowest 3 cm. They should, therefore, refer, respectively, to the closing and opening stages of L-De II.

9,715 b.c. ± 140 (IGS - C14/134 (ST 3836)
10,250 b.c. ± 160 (IGS - C14/135 (ST 3832)

Plotting of dates taken from profiles which (1) have continuous pollen diagrams covering both L-De II and III or (2) are from profiles where the zone boundary is marked also by a change in lithology from an organic-rich sediment to deposits which are primarily mineral in character, and (3) ignoring all isolated dates without backup information, indicate that the L-De II/III boundary should be placed between 9,000 and 8,800 radiocarbon years b.c. Equivalently, examination of all dates published for the opening of pollen Zone II revealed none that was older than 10,000 radiocarbon years which, even at the moment of original publication, was regarded as wholly satisfactory (for comments on early dates see Godwin, 1960). For both boundaries the determinations from Poulton seem rather too high and it seems a valid suspicion that this situation has arisen from selection of "detrital mud" as dating material. Such muds are, of course, built up largely or exclusively of the organic remains of submerged plants, and all dates from these muds are open to criticism because of possible incorporation of 14c-deficient carbon during the photosynthesis of these aquatic plants. In the case of the sediments at Poulton it is clear that the open water deposits were (and are) highly calcareous, while the deposits themselves had collected in a basin resting in till potentially rich in old carbonates. There is thus a strong probability that the apparent age of the samples has been increased due to hard water error, in so far as the water supply contains calcium carbonate and in so far as diffusing atmospheric CO_2 into the water will only partially have accounted for the origin of the assimilated carbon (Polach and Golson, 1966).

Summary

It is suggested that for the Poulton-le-Fylde site that the incorporation of 14c-deficient carbon by plants making up an Allerød detrital mud has caused samples of the mud to yield dates seemingly much too old for, respectively, the beginning and end of L-De II. Acceptance of this explanation suggests that this detrital mud may in reality have accummulated at an age equivalent to L-De II sediments elsewhere, that is, between about 10,000 and 8,800 radiocarbon years b.c.—as originally suspected by Barnes (Barnes et alii, 1971). Photographs suggest that the skeleton of Alces alces lay near the mid-point of the detrital mud, hinting at an age neither at the beginning nor the very end of L-De II.

Mother Grundy's Parlour, Derbyshire

The next group of material, the chronology of which requires investigation is the concentration of L-De artifacts excavated during 1924 by A. Leslie Armstrong and G. A. Garfitt from the left hand side of the platform in front of Mother Grundy's Parlour (Armstrong, 1925). For reasons that will become apparent I am deliberately restricting myself to the material from this excavation and will, since it appears irrelevant to the arguments of this paper, not attempt to marry up observations made by Armstrong with those of subsequent excavators, something which, it has been argued elsewhere, it is perfectly possible to do (Jacobi (1975) I Appendix II).

Artifacts were recovered by Armstrong from a profile 3 ft in thickness, the upper 30 ins of which was described as a "red sandy cave earth with stones". For the lowest six inches this changed to a yellow cave earth which contained items from a Middle Pleistocene fauna. Since clear lithological boundaries could not be recognised, except for this colour change, the deposit was divided into four arbitrary spits (ibid. fig. 4) described as the "Base, Lower Middle, Middle and Upper Middle zones". The "Base Zone", the thickest at 12 inches, incorporates both the 6 inches of the yellow cave earth and the lowest 6 inches of the "Red Sandy Cave Earth with stones".

Artifacts were attributed by Armstrong to these "zones". Preliminary analysis of the artifacts from his excavation suggests changes in technology between zones. The L-De artifacts of the "Upper Middle zone" are, however, mixed with one or more populations of tool-forms believed to be of Flandrian II age. The record of two very small, obliquely backed pieces (of Flandrian type)at the "top" of the "Middle zone", the attribution of, in two cases, conjoining portions of the same broken L-De backed artifact to the "Middle" and the "Upper Middle" zones, and the mixture of what are quite clearly Late- and Post-Devensian artifacts in the 6 inches of the "Upper Middle" zone all suggest either that at this point Armstrong's "zones" had failed conspicuously to correspond to a critical stratigraphical interface, or that there had been some natural (or artificial) disturbance of the upper part of the sequence—undetected by Armstrong—resulting in a genuine vertical mixing of artifact populations. If the latter is correct, our only clue as to the depth achieved by such mixing, at least in the part of the profile investigated by Armstrong, is the recorded observation that the deepest that a specifically Flandian-type artifact was recovered was the "top" of the "Middle" zone. Such disturbance could be argued to have obscured original differences in backed tool composition within the uppermost part of the profile.

While the sample of backed pieces initially recognised as of L-De age is small (a total of sixty-seven) a number of useful preliminary observations can, I believe, be made. Firstly, while convex-backed pieces are the most common backed artifact form in each of the lower three zones, there is a marked drop in their proportional representation between the "Lower Middle" and "Middle zones", whilst they are least important in the "Upper Middle zone". Shouldered and truncated pieces are present in only small numbers and decrease proportionally as one ascends from "zone" to "zone", the only fragment of such artifacts from the "Upper Middle zone" conjoining with its other half (Armstrong (1925) fig. 6 No. 18) in the "Middle zone" below. Trapezoidal-backed pieces are certainly present in the "Upper Middle zone", absent from the "Base" and "Lower Middle zones" and are represented in the "Middle zone" only by a fragment that conjoins with its other half provenanced to the "Upper Middle zone" (Armstrong (1925) fig. 13 No. 1). Angle-backed pieces appear to increase relative to other specifically L-De backed tool-forms from zone to zone and comprise almost half the backed pieces in the "Upper Middle zone", here outnumbering convex-backed pieces. There is thus a change from a dominance by convex-backed pieces to a dominance by angle-backed pieces from base to top.

Among other tool forms, "Magdalenian blades" (de Sonneville-Bordes and Deffarge (1974) pp. 98-99) are present in all four "zones" but are present in large numbers in only the lower two zones. In the "Base zone" burins outnumber scrapers while in the other zones scrapers outnumber burins. "Zinken" are present in the "Base" and "Lower Middle zones" (Armstrong (1925) fig. 8 No. 14 (upper part) and fig. 6 No. 13).

Three radiocarbon dates are now available which it is believed are immediately relevant to dating the lower part of this sequence—that is that portion of the sequence where convex-backed pieces are dominant among the backed tool forms and are apparently associated with "angle-backed pieces" and "shouldered and truncated pieces".

The first date of 9,370 ± 230 is on collagen from bone splinters (*Bos/Bison*) from the lowest 6 inches of the "Red Sandy Cave Earth and Stones" and it seems reasonable to suspect that this date, obtained on food debris, should apply to the flint artifacts and fragment of bevelled "assegai" from this horizon (Armstrong (1925) fig. 16 No. 4).

The second pair of dates is also on splinters of *Bos/Bison* bone but all marked as coming from the "hearth". This hearth occurred towards the centre of the spread of Late-glacial artifacts (*ibid.* (1925) fig. 3E) and was "at the top of the yellow cave-earth... roughly circular... (and)... constructed by scooping out a cup 9 inches deep, in the yellow cave-earth and lining the sides with tabular fragments of limestone... Black ashes, fragments of split and charred bone and stone pot boilers filled the hearth to a depth of 12 inches ..." (ibid. (1925) p. 152). The hearth and its infill would thus be contained in Armstrong's "Base zone".

The dates 9,335 b.c. ± 180 (Q 1483) Bone collagen
and 9,210 b.c. ± 170 (Q 1459) Bone collagen

agree closely with that from the upper part of the "Base zone", and the overlap is such at one standard deviation that there is a high probability that the same or a more or less contemporary event is being registered. These dates, it is believed, give an age to the earliest activity within this part of the shelter.

Summary

The determinations from Mother Grundy's Parlour appear significant in the following ways.

(1) There are three dates from a similar stratigraphic situation and these agree closely one with another. They would thus seem to provide our first "group" of dates for an occurrence of L-De artifacts.

(2) Each of the dates is on collagen from animal bones split to extract the marrow and thus representing food debris.

(3) Each group of fragments can be closely tied into Armstrong's excavation, either to a "Zone" (Q 1484) or to a built hearth (Q 1459 and Q 1483). In both situations only L-De type artifacts occurred.

(4) There is no evidence for any prior human occupation of the cave which could have produced the splinters, and none of the splinters combusted shows signs of carnivore or rodent interference.

(5) The dates are relevant to a particularly large but, horizontally, extremely localized artifact assemblage.

L-De III; Ossom's Cave, Staffordshire

Perhaps the determination which appears to present the fewest problems is that obtained on collagen from a large sample made up of splinters of reindeer bone, interpreted as food debris (and showing no trace of carnivore or rodent interference) and collected from within Ossom's Cave, near Grindon, Staffordshire, with an age of 8,640 b.c. ± 70 (GrN 7400) radiocarbon years. These splinters (the age of which is wrongly reported in Stuart (1977b) p. 77) were directly associated with a small group of struck flints including a complete "Penknife Point", an angle burin (Campbell (1977) fig. 155 No. 8), several retouched pieces and a collection of chippings, these last having been recovered from "around a block of stone used as an anvil...". A pitted bone of Bos/Bison is believed to have been used as a hammer (Bramwell pers. comm.). Associated food debris included reindeer, horse and Red deer—also Willow-Grouse, Ptarmigan and Arctic Hare (Bramwell pers. comm.; Bramwell (1955) and Stuart (1977) p. 77). The central date clearly falls within L-De III and, even when considered at three standard deviations, would still just lie within the range of determinations considered appropriate to this pollen zone. It also just overlaps the youngest of the Mother Grundy's Parlour determinations.

Summary

The determination of Ossom's Cave is (1) on food debris unequivocably associated (2) with flint artifacts, including a convex-backed piece, and (3), even when plotted at three standard deviations, still implies settlement within L-De III.

Broxbourne, Hertfordshire

"...A small flake-blade that was undoubtedly in situ..." was recovered by Warren from the extension south into Rikof's Pit at Broxbourne of a calcareous nekron mud (termed "Lake Marl D") of the Nazeing sequence (Warren in Allison, Godwin and Warren (1952) pp. 192-193). The artifact is said to have been "untrimmed" and was unaccompanied by any other evidence for human activity.

For the dating of this deposit we are dependent upon the interpretation of mollusc identifications made by A. S. Kennard (listed ibid. Table II and see also comments by Warren (p. 174) on sampling strategy), and Kerney (pers. comm. 1979) has suggested that "...bed D at Nazeing is of Devensian Late-glacial age. (That) the mollusc assemblage is that of zone Z" (Kerney (1977) p. 388)...but that "...a closer dating is really not possible on present evidence, though the high values of Trichia in the upper part of bed D are suggestive of a post-Allerød (i.e. pollen zone III) dating...". This interpretation in no way conflicts with the rather limited pollen evidence (Godwin in Allison, Godwin and Warren (1952) pp. 201-204), and confirms the suggestion made in the concluding Table in the Nazeing report that the calcareous nekron mud D is equivalent in age to the "Ahrensburg" culture of north-west Germany. That the accumulation of this nekron mud was followed by a phase of erosion and the deposition, still within L-De III, of a banded, calcareous mud (deposit M) within the new channel would suggest that deposit D cannot belong to the close of this zone.

Summary

An unretouched blade recovered from a calcareous nekron mud (deposit D) at Rikof's Pit, Broxbourne, can be argued to be of L-De III age, and would confirm exploitation of one part at least of the southern part of the island within this zone.

Flixton Site II, North Yorkshire

Bones representing the highly incomplete skeletons of at least three horses were recovered at 105-108 cms depth (Godwin in Clark (1954) fig. 24) from within fine detritus mud at Flixton Site II, and associated with a "small blade of flint" and a broken backed bladelet originally of triangular outline (Moore in Clark (1954) p. 192; also Clark (1954) Pl. XVI; Campbell (1977) fig. 141 No. 5). The bones can best be interpreted as non-meat bearing portions discarded after butchery. The fine detrital mud (100-108 cms) with twigs and Carex fragments, in which were the horse bones, was assigned by Godwin (in Clark (1954) p. 51) either "...to an early stage of Zone III or the preceding Zone II...". There is a radiocarbon date on organic mud from 100 cms (Godwin and Willis (1959) p. 66) of 8,463 b.c. ± 210 (Q 66).

Comparison of the relevant portion of the pollen diagram from Flixton Site II (97-115 cms: Godwin (1954) fig. 24) with the far longer and markedly more complete local diagrams for Flixton DB1 and Seamer A16 (ibid. figs 20a-20c, 21a and 21b also fig. 28) would suggest that the detrital muds, which contained the horse bones and associated artifacts, belong within L-De III with, from 105 cms (that is just above the horse bones), as at both Flixton DB1 and Seamer A6, the high Pinus values of the lower part of the diagram (105-115 cms) dropping steeply and being replaced by rising values for both Betula and Salix. At both Flixton DB1 and Seamer A6 this change is a feature of the latter part of L-De III. The curve for Ulmaria filipendula also begins to rise steeply above 105 cms, precisely as it does at the end of L-De III. A not unreasonable interpretation of the pollen diagram from Flixton Site II between 97 and 115 cms would be as follows.

(1) Sediments of L-De II cannot be recognized.

(2) The lower part of L-De III appears also to be absent.

(3) The profile from 97 to 115 cms covers the latter part of L-De III.

(4) The overlying gravelly sands from 79 to 97 cms represent solifluction during the final stages of L-De III.

(5) The horse bones, broken flint point and unretouched blade at 105-108 cms belong within L-De III and perhaps towards its mid-point.

(6) The 14c date of 8,463 b.c. ± 100 would seem to fit exceptionally well with this interpretation.

Messingham, Lincolnshire

A flint end scraper was recovered from within a lens of moss peat enclosed by coversands at Messingham, near Scunthorpe. An astragalus originally identified as of Megaloceros giganteus, but more recently as of Bos, was found in the same peat lens (May (1976) p. 28 and footnote 33). A sample of this moss peat, which also yielded a Coleopteran assemblage clearly indicative of a L-De III age (Buckland (1974) pp. 7-9), was dated to 8,300 b.c. ± 120 (Birm 349: Shotton and Williams (1973) p. 458). A second peat lens 1.5 metres higher within the coversands was dated to 8,600 b.c. ± 250 (Birm 707: Williams and Johnson (1976) p. 258). While the central dates of the two determinations are thus reversed, the two overlap considerably at a single standard deviation. The direct association of this assemblage of Coleoptera with a flint scraper and a bone of Bos demonstrates clearly both that man was capable of tolerating the environments to be found this far north during later L-De III and confirms the inference drawn from Robin Hood's Cave and Flixton that an ungulate fauna capable of exploitation existed within Britain at this time. The two determinations from Messingham fit well within the sequence of dates for coversand deposits in the Vale of York (Matthews (1970) and Gaunt et alii (1971) and with the group of dates from within the Younger Coversand II profile at Dennekamp (Vogel and Zagwijn (1967) p. 69) in Overijsel.

At Risby Warren, only eleven km NNE of Messingham, artifacts from a podsol, developed on top of the coversand, are of Maglemosian type (Site I: Jacobi (1975) fig. I.25; Riley (1978) fig. 2 Nos 2-9) and of Flandrian Ia to earliest Ic age. Similar assemblages from an identical stratigraphic position are known from Mell's Farm, (Messingham), Manton Warren and Cove Farm (Haxey),while at Risby Warren a group of thirteen "penknife points" was recovered from a thin band of eluviated clay below the coversands and above the limestone bed-rock (Armstrong (1932) p. 130; Lacaille (1946b) fig. 3; Dudley (1949) pp. 34-35 and fig. 13; May (1976) fig. 14). These penknife points have obvious analogues within the flint industry from Mother Grundy's Parlour, belonging (at least in part) to the second half of the 10th millennium b.c. In short, the radiocarbon and archaeological evidence confirm Catt's (1977 p. 125) suggestion that these coversands are the equivalent of the Younger Coversand II of the Netherlands.

Summary

At one point the north Lincolnshire coversands seal an occurrence of Federmesser (? of L-De II) age, whilst at least four points, artifacts of Earliest Flandrian type have been recovered from a soil profile developed upon these sands. At Messingham a flint scraper from a peat lens within these sands is dated by combined radiocarbon and environmental evidence to L-De III. The artifact can, of course, give no pointer as to the cultural connections of the locally exploiting group.

Anston Stone's Cave, South-East Yorkshire

The dating of, or rather the various interpretations that can be placed upon the several lines of evidence relating to the dating of the occupation within Anston Stone's Cave (also known as North Anston Cave and the Dead Man's Cave) have been discussed by several commentators (Mellars (1969) and (1974) pp. 74-75; Campbell (1977) pp. 103-104 and Jacobi (1978) pp. 298-300). For an account of the excavation see White (1970).

There is said to be a conflict (Mellars (1974) p. 275 footnote 126) between pollen spectra from deposit samples within the cave, which are interpreted as suggesting occupation within L-De II and L-De III (Campbell (1977) fig. 75), and three radiocarbon determinations obtained from the site (Barker *et alii* (1971) p. 167). These dates are on collagen separated from animal bone fragments and a reindeer antler "...in close association with the Creswellian artifacts..." (Mellars (1969) pp. 309-310). The samples were large, the specimens rich in collagen and there is no reason to suspect contamination of any sort (Burleigh (1979) *pers. comm.*). The dates obtained are:

			Quality of Association
7,900 b.c. ± 115	(BM 439)	Bone fragments including reindeer	***
7,990 b.c. ± 115	(BM 440A)	Bone fragments unidentified	**
7,800 b.c. ± 110	(BM 440B)	Reindeer antler	*

Jacobi (1978) pp. 298-300, has argued that it is not impossible to reconcile the pollen counts from inside the cave with counts obtained from peat bogs and of an age equivalent to that suggested by the radiocarbon dates from Anston Stone's Cave (for the discussion of theoretical problems in comparing pollen diagrams from such distinct environments, see Leroi-Gourhan (1964) pp. 7-8). A recent determination of 7,970 b.c. ± 130 (Q 1581) on collagen separated from a "massacred" female reindeer antler from the top of the Gough's Cave sequence at Cheddar, and which does overlap at a single standard deviation both

BM 439 and BM 440B, suggests that reindeer may still have been a part of our fauna (at least in north and west England) during the first centuries of the 8th millennium b.c. This new date (Q 1581) also makes the two determinations from Anston Stone's Cave on reindeer bone and reindeer antler, respectively, appear less "out of place".

Acceptance of the radiocarbon determinations and of the arguments presented in Jacobi (1978) would imply the use of Anston Stone's Cave by a "Federmesser"-type technology at a date close to the Late Devensian/Flandrian boundary. The alternative, an acceptance of the pollen zonation suggested by Campbell, with the implication that the radiocarbon dates are in some way too young, would place the use of this technology within L-De II and L-De III, a proposition clearly supported by evidence from Mother Grundy's Parlour, Ossom's Cave and perhaps also from Robin Hood's Cave. Neither interpretation, however, implies settlement earlier than some moment within L-De II.

Discussion

Rather than concluding this section by repeating much of the discussion, the salient points are presented on Table 2.II where I have attempted to assess the strength of association between the sources of chronological information—radiocarbon, pollen, Coleoptera—and the cultural material recovered from each locality. The strength of each association is scored: a ? if the materials dated cannot now be accurately related to this cultural material: a single star * if the date (or dates) can be shown only to be either from above or below an occurrence of archaeological material and do not therefore provide a direct date for this occurrence: a double starring ** implies that the sample dated (if bone, antler or charcoal) can clearly be demonstrated to be from the same stratigraphic unit as an occurrence of archaeological material but was not recovered in immediate association. In the case of pollen a double starring implies that this pollen could be suspected to have been downwashed into the layer from which the sample(s) counted were collected. Three stars *** imply that the sample (or samples) dated can be directly associated with the cultural occurrence whose age it is wished to establish—that is, in the case of bone fragments these represent food debris directly associated with artifacts (cf. Mother Grundy's Parlour and Ossom's Cave or that, in the case of organic deposits, these are either dated by pollen analysis, the sampling column for which crosses the deposits containing the occurrence of archaeological material (cf. Flixton Site II and Poulton-le-Fylde), or by a combination of radiocarbon and environmental studies (cf. Messingham).

The first implication of this scoring is, I believe, to suggest that the earliest that we can be certain of the presence of man within the island after the maximum ice advance of the Devensian is sometime within L-De II and not demonstrably within its earliest part. This is, of course, almost three millennia later than the date often canvassed for his return (Mellars (1974) p. 74; Evans (1976) p. 42 table 2 and p. 50; Campbell (1977) p. 199; Wymer (1977) p. 104). Such a revised chronology immediately explains the similarities discernible between the technology of our so-called "Creswellian" assemblage and that of mainland "Federmesser" assemblages, which have been argued also to be at the earliest of L-De II age. There appears no chronological or

Table 2.II

	Dating Methods Applied				Strength of Association			
	Radiocarbon	Pollen	Coleoptera	Mollusca	Radiocarbon	Pollen	Coleoptera	Mollusca
* Anston Stone's Cave	BM 440B: 7,800 bc ± 110	III ?			?	**		
	BM 439: 7,900 bc ± 115				?			
	BM 440A: 7,990 bc ± 115				?			
Messingham	Birm 349: 8,300 bc ± 120		III		***		***	
* Robin Hood's Cave	BM 603: 8,440 bc ± 90	III?			**	**		
	BM 604: 8,640 bc ± 90				**			
* Cathole Cave		III?				**		
Flixton Site II	Q66: 8,463 bc ± 100	III			* (above)	***		
Broxbourne				Mollusca Zone Z				**
Ossom's Cave	GrN 7400: 8,640 bc ± 70				***			
* Mother Grundy's Parlour	Q 1459: 9,210 bc ± 170				***			
	Q 1483: 9,335 bc ± 180				***			
	Q 1484: 9,370 bc ± 230				***			
Poulton-le-Fylde	† St 3836: 9,715 bc ± 140				* (above)			
	† St 3832: 10,250 bc ± 160	II			* (below)	***		
* Sun Hole Cave	BM 524: 10,428 bc ± 150				* (below ?)			
* Kent's Cavern	GrN 6204: 10,230 bc ± 100				?			
	GrN 6203: 10,325 bc ± 120				?			

† ? Dates too old because of "hard water error"
* Findspots of 'Federmesser' type material

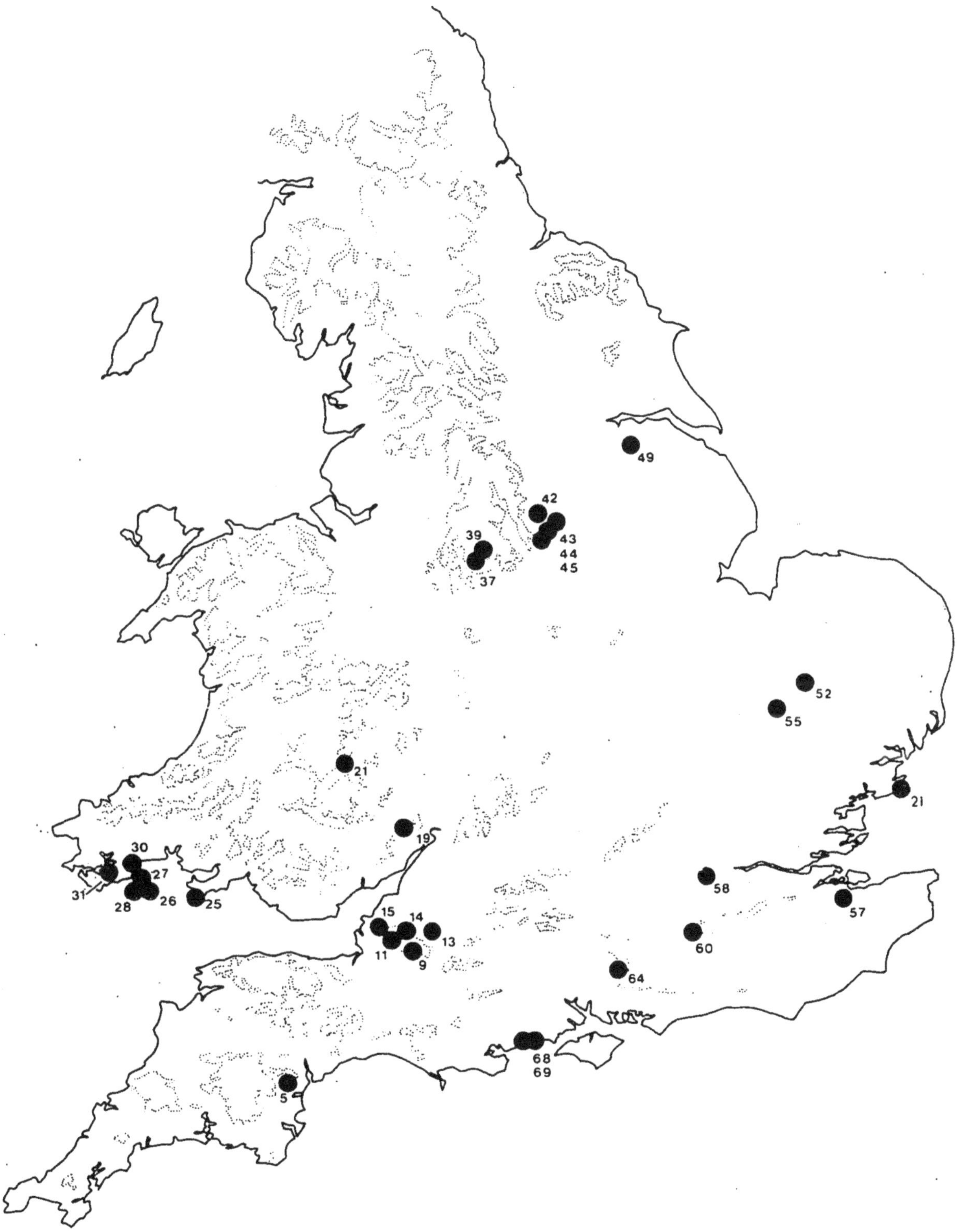

Fig. 2.12 Find-spots of Penknife Points in the British Isles
Key: for key to numbers see key to Fig. 2.8

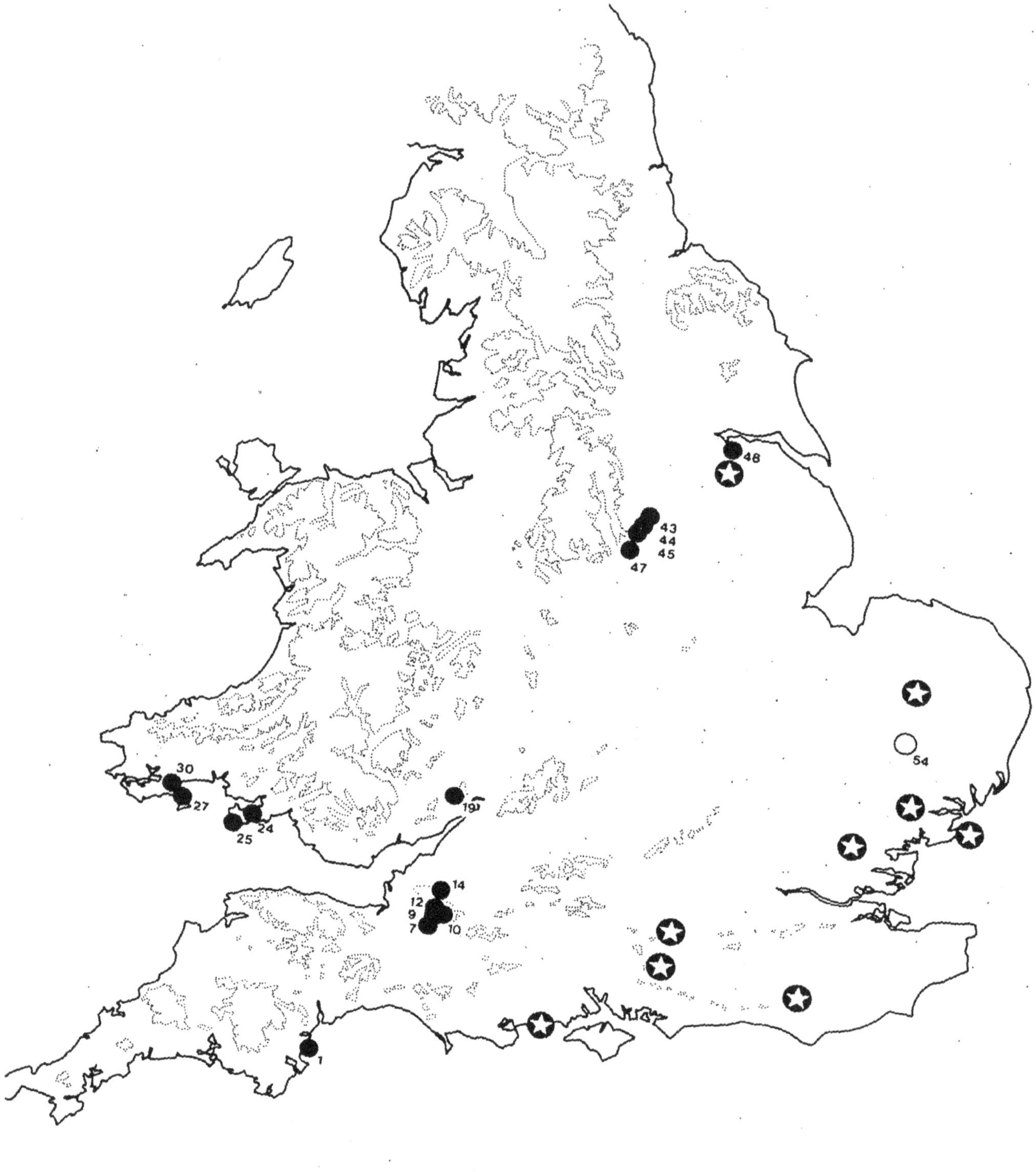

48
43
44
45
47
54
30
27
24
25
19
14
12
9
7
10
1

Fig. 2.13 (opposite) Find-spots of Trapeziform "Cheddar Points" (dots) in the British Isles. Solid dots represent certain find-spots and the single open circles an unverified occurrence. Also plotted (stars) are finds of "Bromme-type" stemmed pieces.

Key: for key to numbers see key to Fig. 2.8, and for key to Bromme-type stemmed pieces see Fig. 2.14 and key to 2.14

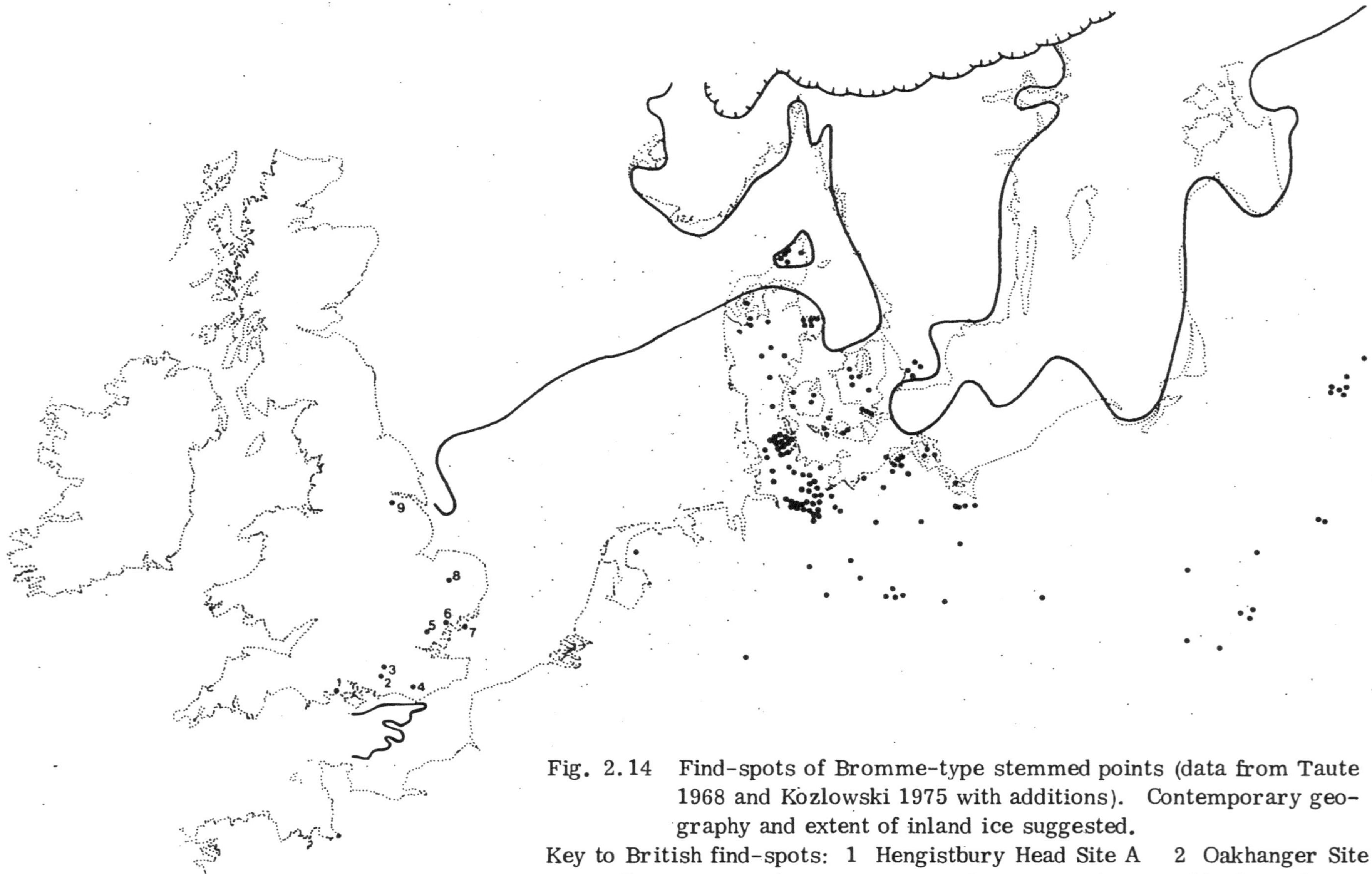

Fig. 2.14 Find-spots of Bromme-type stemmed points (data from Taute 1968 and Kozlowski 1975 with additions). Contemporary geography and extent of inland ice suggested.

Key to British find-spots: 1 Hengistbury Head Site A 2 Oakhanger Site VII 3 Headley 4 Newhouse Farm: High Hurstwood 5 Widford Brickpit 6 Near Manningtree 7 Stone Point 8 Cranwich 9 Brumby Wood

technological obstacle to associating exploitation and colonization of the island with groups from the area of the present-day mainland employing a "Federmesser"-type technology (but see also Schwabedissen (1954) Taf. 102b), with elements that could be argued to have prototypes in the latest Magdalenian technologies of the European plateaux (Kozlowski (1971) pp. 67-69). Indeed, while we would propose that backed blade assemblages within Britain which combine curve-backed pieces, angle- and trapeziform-backed pieces and shouldered and truncated pieces should be termed "Federmesser" assemblages, it is important to recognize that in reality these represent only the final stages in the settlement of the western edge of the Palaeolithic world by groups with a technology ultimately derived from the Magdalenian. The origins of the technology of groups who completed this settlement are much less clear (see below).

Arguable also as of Magdalenian derivation are the items of bone, antler and ivory recovered in association with this Federmesser technology from sites in western and northern England. This equipment includes the barbed heads of hunting spears made in bone (Poulton-le-Fylde), reindeer antler (Kent's Cavern: Garrod (1926) fig. 4 Nos. 13-15) and red deer antler (Aveline's Hole: Davies (1921) p. 69 and fig. 10 No. 1); assegais ("sagaies") with a single or a double bevel made of antler (Mother Grundy's Parlour: Armstrong (1925) fig. 17 Nos. 3 and 4; Fox Hole Cave: Bramwell (1971) p. 11 and fig. 4 No. 5; Victoria Cave: Breuil (1922) fig. 8 Nos. 1 and 2; Jackson (1945) Pl. XIa; Kinsey Cave: Jackson and Mattinson (1932) fig. 1 No. 13; King Arthur's Cave: Taylor (1928) fig. 2 No. 17); or fossil or traded ivory (Pin Hole: Armstrong (1925) pp. 172-173 and fig. 14 No. 1, also Pl. XXII fig. 5); eyed bone needles (Kent's Cavern: Garrod (1926) fig. 4 No. 16; Cathole McBurney (1959) p. 268; Church Hole Cave: Garrod (1926) fig. 31 No. 3 and ?King Arthur's Cave; Taylor (1928) p. 64 and fig. 2 No. 18. See also Stordeur-Yedid 1979) and shaft-straighteners ("bâtons de commandement") of reindeer antler or bone (Syon Reach, London, and Gough's (New) Cave: St. George Gray in Parry (1929) pp. 111-113 and Pl. XIX Nos. 1-2, also Parry (1931) p. 49). Other items of equipment include thonging needles of antler (Foxhole Cave and Church Hole Cave: Garrod (1926) fig. 31 No. 6) and tallies, calendars, calculators, pocket rulers or gaming aids formed by regularly indented lengths of bone (Gough's (New) Cave: Parry (1931) fig. 2 No. 2; Hawkes <u>et alii</u> (1970) and Tratman (1976), and Chelm's Combe: Harrison (1976) fig. 32). Objects of adornment comprise perforated teeth of <u>Vulpes</u> (Gough's (New) Cave: Parry (1929) Pl. XX No. 2) while sea shells <u>Lævicardium crassum</u> and <u>Buccinum undatum</u> were brought inland from the coast (Gough's (New) Cave: Tratman (1950) p. 225 footnote). Artistic manifestations are restricted to the horse's head engraved on a section of rib bone from Robin Hood's Cave (Boyd-Dawkins (1877) p. 592), if this is in fact a genuine British find; and to scratched pebbles from Gough's (New) Cave (Parry (1931) pp. 47-49 and fig. 2 No. 1). An engraving of a horse's head, also on a rib bone, and claimed to have been found near Sherborne (Dorset: Smith-Woodward 1914) appears to be a modern production and to be modelled on the engraving from Robin Hood's Cave (for discussion and further references see Halstead (1978) p. 12).

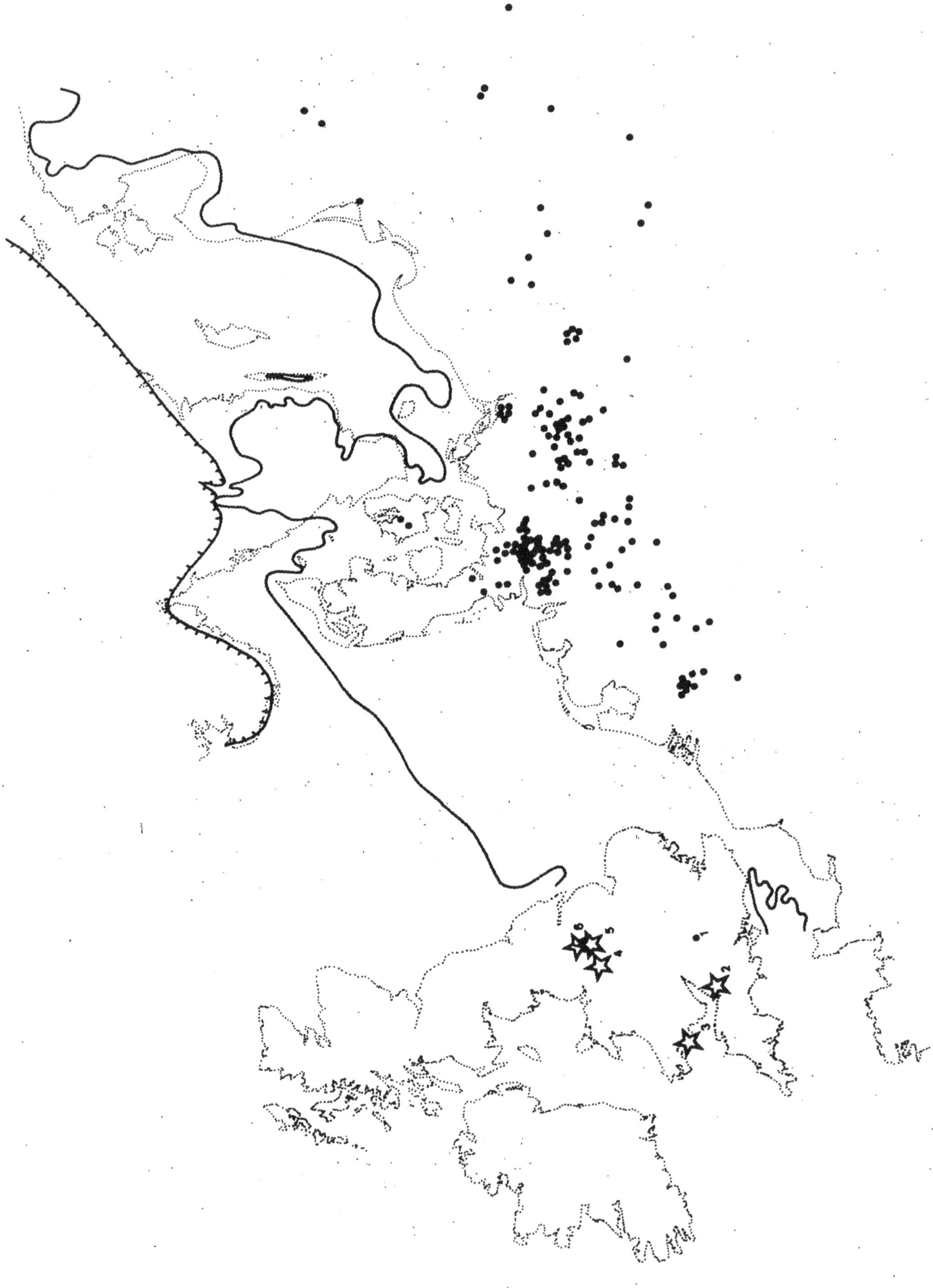

Fig. 2.15 (opposite) Find-spots (solid circles) of Ahrensburgian-type stemmed points (data from Taute 1968; Becker 1971; and De Laet 1979 with additions). Also indicated (open stars) are British "Federmesser" assemblages which can be argued to be contemporary with the Ahrensburgian (Sun Hole is included amongst these on the basis of unpublished radio-carbon dates communicated by C. F. Hawkes). Also suggested is contemporary geography and the extent of mainland inland ice.

Key:	Infilled circle	1 Avington VI
	Open stars	2 Sun Hole Cave
		3 Cathole Cave
		4 Ossom's Cave
		5 Robin Hood's Cave
		6 Anston Stones Cave

A second implication of Table II would seem to be that an exploiting population was present within both northern and southern England and perhaps Wales (Cathole) during L- De III—a population which Table I suggests will have had to adapt to the primary exploitation of Equus caballus, Rangifer tarandus and Bos/Bison (?Bison bonasus L.: Degerbøl, 1964) instead of, as during the wooded second part of L-De II, Alces alces and Bos/Bison. By contrast, the very sparse distribution of stemmed Ahrensburgian arrowheads north of Latitude 55° (Schild (1976) p. 94) compared to the large number of finds of the rather earlier (Fischer 1978) Bromme-type tanged spearheads (compare figs. 2.14 and 2.15) suggests that, on the other side of the northern European Plain, groups chose during L-De III to leave the tundra relatively underexploited and selected preferentially to exploit the tundra/taiga ecotone where large kills of reindeer could be made in autumn and the meat frozen and stored for winter consumption.

The third implication of Table II is that not only was a Federmesser technology innovated during L-De II but it continued to be employed at least in the north of England and south-west Wales (?) into some part of L-De III. This appears to be the implication of GrN 7400 from Ossom's Cave and of the chronological data, difficult as these are to interpret, from Robin Hood's Cave, Anston Stone's Cave and Cathole.

"Dated" assemblages of Federmesser within Britain are too few and too small to order into any typo-technological sequence. However, there is nothing within the evidence as outlined here that does not support the contention that angle-backed pieces are most numerous in assemblages able to be shown to belong later in the local sequence, as at Anston Stone's Cave, Robin Hood's Cave and the "Upper Middle Zone" at Mother Grundy's Parlour. By contrast, convex-backed pieces, notably "Penknife Points", are most numerous in what are argued to be the initial stages of the British Federmesser technology, as in the "Middle", "Lower Middle" and "Base Zones" at Mother Grundy's Parlour. The trends detectable at Mother Grundy's Parlour (see above) appear to be continued as one ascends Parry's spits at Gough's New Cave (statistics presented by Campbell 1977).

Further, trapezoidal-backed pieces ("Cheddar Points") occur in only the upper part of the Mother Grundy's Parlour sequence, increase slightly within the Gough's Cave sequence, and are recorded from Robin Hood's Cave and among early finds from Cathole, both sites where occupation has been suggested as belonging within L-De III.

The distribution of "Penknife-Points" (Fig. 2.12) and "Cheddar Points" (Fig. 2.13) within Britain is markedly different, and the absence from southern and eastern England of a backed-tool type, which is hypothesized as being most common late in the sequence of our Federmesser assemblages, may suggest an absent or rarity of such assemblages in this area. Is there any archaeological material which might be argued to fill this apparent gap?

Steilspitzen Gruppen in Britain?

A number of stray finds have been made in England (and it has been suggested in Wales: Campbell (1977) fig. 133 No. 3*) of tanged artifacts which in terms of both size and overall morphology fulfil precisely definitions of Late-glacial Bromme points. Bromme points have been described as possessing a "...pedoncle court et large obtenu par retouches abruptes eventuellement alternes, dessinant des crans. Pointe peu ou même non retouchée. C'est aussi la pointe de Teyjat ou pointe à soie magdalénienne..." (Heinzelin (1962) p. 36). This definition could, perhaps, be strengthened by adding (1) that the tang is always developed on the bulbar end of the blank, (2) that it is formed by retouch executed from the ventral surface and (3) that the short, relatively broad tang is always parallel sided with its margins trimmed so as to run parallel to the main axis of the blank. Finally (4) there is an absence of flat surface retouch at any point on the artifact and what retouch (if any) there may be at the tip takes the form of vertical chipping.

The number of tanged artifacts from Britain which fulfil this definition is small: isolated (rather doubtful) artifacts from Lincolnshire (Brumby Wood, near Scunthorpe) and Norfolk (Cranwich: Wymer 1971): three find spots in central and northern Essex (Jacobi 1979), a single find from Sussex (Woodcock (1978) p. 13 and fig. 5 No. 5) and three findspots again in Hampshire (Oakhanger VII (Box L-12 of the excavation), Headley, and Hengistbury Head Site A: Mace (1959) fig. 9 Nos. 75 and 76). Deliberately omitted from this list and from Fig. 2.14 are Hengistbury Head Site C2, (Campbell (1977) fig. 161 no. 9), Portland Site I and Verne Ditch, also on the Isle of Portland (Palmer (1970) fig. 2 and (1971) fig. 2), since the relevant artifacts do not appear to fulfil the definition outlined. It was not possible to trace a tanged artifact from Haslemere figured at an unknown scale in "Bygone Haslemere" (Swanton and Woods (1914) p. 3 fig. 4) and so this find too has had to be omitted.

Acceptance of these British finds as "Bromme points" would imply a total east to west distribution of this spearhead type (Becker (1971) pp. 135-137) of over 2,000 kms from the Neimen Basin to the Hampshire Avon and with a spread north to south of rather under 400 kms (Taute (1968) Karte 3 and

---

* Delporte (1976) p. 40 suggests that this artifact may be a "Font Robert Point", a proposition strongly supported by the presence among the same collection of fifty-three flints excavated by Colonel Wood in September-October 1864 from the Upper Cave at Cathole of what is quite clearly the triangular sectioned stem of just such a point (B.M., Natural History). This piece closely resembles the stem of the Broken Font Robert Point identified from Paviland (Sollas (1913) fig. 17 No. 116; Garrod (1926) fig. 8 No. 1; Campbell (1977) fig. 95 No. 5). Fauna recovered by Wood and listed by Garrod (1926 p. 65) includes such Middle Devensian forms as Elephas primigenius, Coelodonta antiquitatis, Panthera leo and Crocuta crocuta. It is therefore quite possible for the cave to have originally contained evidence of an occupation of Later Perigordian (Gravettian) date as well as Federmesser-type artifacts (Garrod (1925) fig. 9 No. 3; Campbell (1977) fig. 131 No. 1 and fig. 133 No. 1).

Kozlowski (1975) Carte I). At no point does this immense distribution (Fig. 2.14) overlap that of the chronologically rather earlier Teyjat points of the French Later Magdalenian (de Sonneville Bordes 1969 and 1970) while the identification of such points from the northern parts of Jutland and from southern Sweden (Salomonsson 1964) suggests a real extension of settlement or at least of exploitation at the time these "Weaponheads" were "fashionable", north and north-east beyond the limits achieved by groups employing a "Federmesser"-type technology.

Since none of the British finds is from an excavated context it is impossible to know whether they were once elements within more polythetic assemblages which included a range of backed tool forms, or whether there existed in Britain, as in Denmark and southern Sweden (Mathiassen, 1946; Salomonsson, 1964, and Andersen, 1972) a technology wherein Bromme type "points" were the only category of knife or weapon head to be represented.

Restricted chronological data indicate that innovation of these tanged pieces certainly took place at some moment within L-De II (Iversen, 1946) and perhaps during the last centuries of the 10th millennium b.c. (Chmielewski et alii (1975) p. 267 and Kozlowski (1975) pp. 140-141). Evidence for continued use into L-De III (Andersen (1970) p. 97 and (1972) pp. 46-47) is more ambiguous and it may indeed be possible to reinterpret drawings of the profile of Nørre Lyngby (Jessen and Nordmann (1915); Iversen 1936 and 1942) as well as the published spectra as suggesting that the tanged piece from here need not be as late as L-De III in age.

In northern Germany between the Weser and the Oder there is a total overlap (Taute Karte 3, 4 and 9) in the distributions of (1) "Federmesser"-type assemblages (2) finds of Bromme-type tanged pieces ("Lynbgy spitzen") and (3) a technology which combines stemmed points greatly reduced in size and weight ("Ahrensburgian Poinsts": Bohmers (1956) p. 9) with obliquely backed microliths ("Zonhoven points": ibid. p. 29), microlithic triangles and bitruncated bladelets of rhombic outline ("Vierecke"). That this third technology is to be associated with a local uptake of archery is suggested by finds of stone arrow-shaft straighteners (Taute (1968) Taf. 17, 38 and 82 and Rust (1943) Taf. 88) as well as at Stellmoor by the recovery of some hundred pinewood arrowshafts (ibid. Taf. 91-96) and two pieces of pinewood interpreted as bow ends (ibid. pp. 192-193 and Taf. 97-98: but see also Clarke (1963) p. 62 footnote 2). In north Germnay (Stellmoor: Schütrumpf 1943) and in western Poland (Calowanie level V: Chmielewski et alii (1975) Table LXXXIV) this third technology clearly belongs within L-De III, but a date from Geldrop I (N. Brabant) of 9,010 bc ± 85 (GrN 1059) suggests a local innovation within L-De II, in an area where the heavier Bromme-type tanged pieces are not to be found.

Recent work by Froom in the Kennet valley has resulted in the recovery at Avington VI (Froom, 1972) stratified below a Later Mesolithic assemblage of a flint industry combining a stemmed microlith (or "Ahrensburg Point") with an obliquely backed microlith, a triangular microlithic, scrapers, cores and long blades (Froom (1978) pers. comm.). Whilst this is the first certain record of Ahrensburgian material from within Britain, it is not unlikely that previous finds of microlithic stemmed points will have been "typologized away" as artifacts of later date (see, for example West (1973) fig. 11 No. 1). So far, however, from Britain there are no axes, adzes or sockets of reindeer antler (Lyngby-beile: Schwantes 1923) or escutcheon-based spearheads, the former recorded from as far west as the Scheldt (Ellewoutsdijke: Zeeland) and the Maas (Rekem, Belgium Limburg: De Laet (1979) p. 115 and fig. 38) and the latter from as far west as Roermond in Dutch Limbourg (Wouters (1955) fig. I Nos. 13 and 25) and both part of the L-De III technology of the mainland (Clark (1975) Map 6).

These finds, an assemblage with an Ahrensburgian stemmed point (Fig. 2.15), stray finds of Bromme-type tanged points (Fig. 2.14) and a rather greater number of discoveries of "Federmesser"-type artifacts (Fig. 2.8), suggest that in south-east England three distinct technologies may perhaps be represented precisely as in north Germany and southern Denmark. It may further be suggested that in both areas these form a chronological sequence through the latter part of L-De II and into L-De III. Tanged artifacts may, at least in northern Germany (Paddaya 1973) and Poland (Chmielewski <u>et alii</u> (1975) p. 267) appear (first) in a Federmesser context but many of these apparent (surface) associations could perhaps better be interpreted as simply the rather later losses of weapons tipped with Bromme-type points at favoured hunting sites (for a list of findspots see Paddaya (1973) pp. 186-187). These tanged artifacts, interpreted as heavy spearheads (Becker (1971) pp. 135-137) can further be argued to have become replaced through time by archery equipment tipped with the smaller and lighter Ahrensburgian stemmed points (Brinch-Petersen 1970; Fischer (1978); but for size overlap of these pieces with Bromme-type tanged points see Clark (1975 fig. 6), and this change can certainly be placed within L-De III. It remains speculative to which technology the broken triangular microlith from Flixton Site II and the end scraper from Messingham—both of which can be argued to be L-De III date—actually belong. The microlith from Flixton could, however, be regarded as lying outside the range of Federmesser equipment recovered from British sites and to resemble rather more closely triangles from Ahrensburgian sites.

If these hints as to technological developments in south-east Britain are correctly interpreted, and if in northern England and south-west Wales a "Federmesser" technology does indeed continue into (or even through, as would be implied by acceptance of the 14C dates from Anston Stone's Cave) L-De III, it then becomes necessary to recognize moments within later L-De II and within L-De III when there may have been, precisely as has been suggested for Holland (Taute (1968) p. 265) , a pair of contemporary technologies present within the island. These in turn must imply the activities of a pair of social groupings and clearly a major priority of research into the British Upper Palaeolithic must be to investigate the archaeology of L-De III.

<u>The origins of the earliest Flandrian microlithic technology</u>

Within England and Wales there is no convincing evidence (<u>pace</u> Campbell (1977) pp. 186-190 and 195-197) for the evolution of our earliest Flandrian microlithic technology from any of the L-De Federmesser assemblages which we have discussed. There is certainly no pattern among either group of assemblages within which Federmesser forms become rarer with time, and such forms as obliquely backed microliths are innovated and become more numerous.

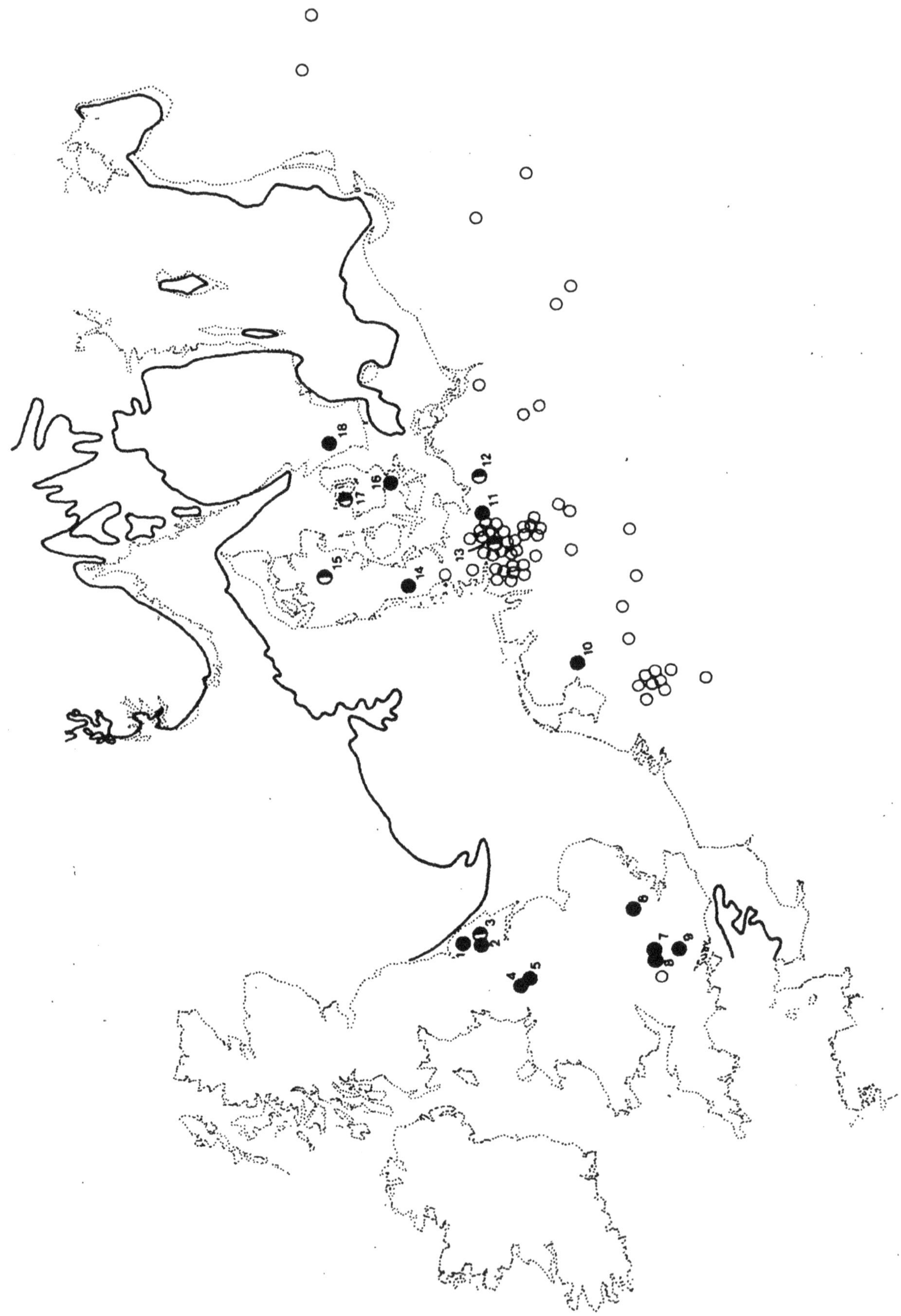
1
2
3
4
5
6
7
8
9
10
11
12
13
14
15
16
17
18

Fig. 2.16 (opposite) Early Mesolithic assemblages dominated by obliquely backed pieces which (solid circles) are dated by radiocarbon to between, or whose ages overlap at a single standard deviation, the years 8,000 to 7,000 bc or are dated by pollen analysis (half infilled circles) to zones IV or V of the Flandrian. These finds are plotted against find-spots of microliths from Late-glacial stemmed point contexts (open circles).

Key (infilled and half infilled circles)

1	Money Howe	10	Gramsbergen
2	Star Carr	11	Duvensee Wohnplatz I
3	Flixton Site I	12	Hohen Viecheln (Lower Settlement)
4	The Lominot	13	Pinnberg KI
5	Warwock Hill (south)	14	Draved Syd
6	Broxbourne Site 106	15	Klosterlund
7	Thatcham	16	Barmose I
8	Marsh Benham	17	Vig
9	Oakhanger Site VII (2)	18	Bare Mosse IV

Equally, there is no trend among our earliest Flandrian flint industries for microliths which bear the closest resemblances to "Penknife Points" to be most common in the earliest of these assemblages. In fact, the converse appears to be the case with such forms relatively more numerous towards 7,000 radiocarbon years b.c. than towards 8,000 b.c. Angle-backed, trapezoidal-backed and shouldered and truncated pieces are all absent from our earliest Flandrian assemblages while the large scalene triangles, which do certainly resemble small Angle-backed pieces, can be argued to belong to the middle, rather than the beginning, of the 8th millennium b.c. All these observations would tend to emphasize the differences between the technology of our Federmesser assemblages and that of our earliest Flandrian ones.

If we cannot demonstrate in Britain any logical progression from a Federmesser technology towards the microlithic technology of the earliest Flandrian, dominated as the latter is by obliquely backed pieces and associated with bi-truncated rhombic and trapezoidal microlithis, also isosceles triangles, it may be useful to look elsewhere either inside or outside Britain for evidence of such an evolution.

The technology of L-De III age whose flint industries <u>can</u> contain any or all of the microlithic shapes found in the earliest Flandrian technologies not just of Britain but also of Denmark, Germany, Holland and north Belgium is that of the Ahrensburgian, and the Ahrensburgian, it has been suggested, may not be entirely absent from Britain. Indeed, it may be most economical to visualise the microlithic technologies of the earliest northern European Mesolithic as being the end product of a process of selection against the relatively "elaborate" stemmed points and in favour of rather simpler forms of tip piece and/or side barb—that is towards obliquely backed pieces, triangles, rhombic pieces and trapezoids (so-called Star Carr trapezes: Brinch Petersen 1970). Suggested as belonging to this earliest (post-stemmed point) microlithic technology would be Barmosen on Zealand (Tauber (1970) p. 130 and Johansson (1971); Klosterlund in north Jutland (Mathiassen 1937; Tauber (1970) p. 130) sites round Draved Bog, south Jutland (Tauber (1973) p. 87), Pinnberg 1b (Rust (1953) Taf. 13), sites of the "Halterner Gruppe" of the Lower Rhine (Arora 1976), Gramsbergen in Overijsel (dated 7,370 b.c. ± 60 (GrN 7793): Stapert <u>in press</u>) and such Belgian sites as Zonhoven (Jacobi 1976 p. 62). For distribution of these sites see Fig. 2.16.

In Britain the earliest radiocarbon dates for an Earliest Flandrian microlithic technology derive from south-east England, the earliest individual date being from Thatcham Site IIIA, where a group of artifacts from the south-west corner of the site, including obliquely backed and trapezoidal microliths and an adze resharpending flake, is associated with charcoal dated to 8,415 b.c. ± 170 (Q 659), that, is, to very close to the L-De/Flandrian Ia boundary. Immediately associated with Q659 is also a fauna of Red deer and <u>Bos</u>, identification of <u>Bos</u> tying in well with the date of 8,400 b.c. ± 200 IG 2707) for <u>Bos</u> at Kildale Hall (Simmons, 1975) as well as with the determination from Messingham (see above). If these microlithic assemblages are indeed to be associated with a use of archery equipment, it is open to speculation whether the appearance of hafted core adzes in Britain and southern Scandinavia reflects a regular selection of trunk wood, rather than branches, for working down into bow staves and arrowshafts. Indeed, as a corollary, the extreme rarity

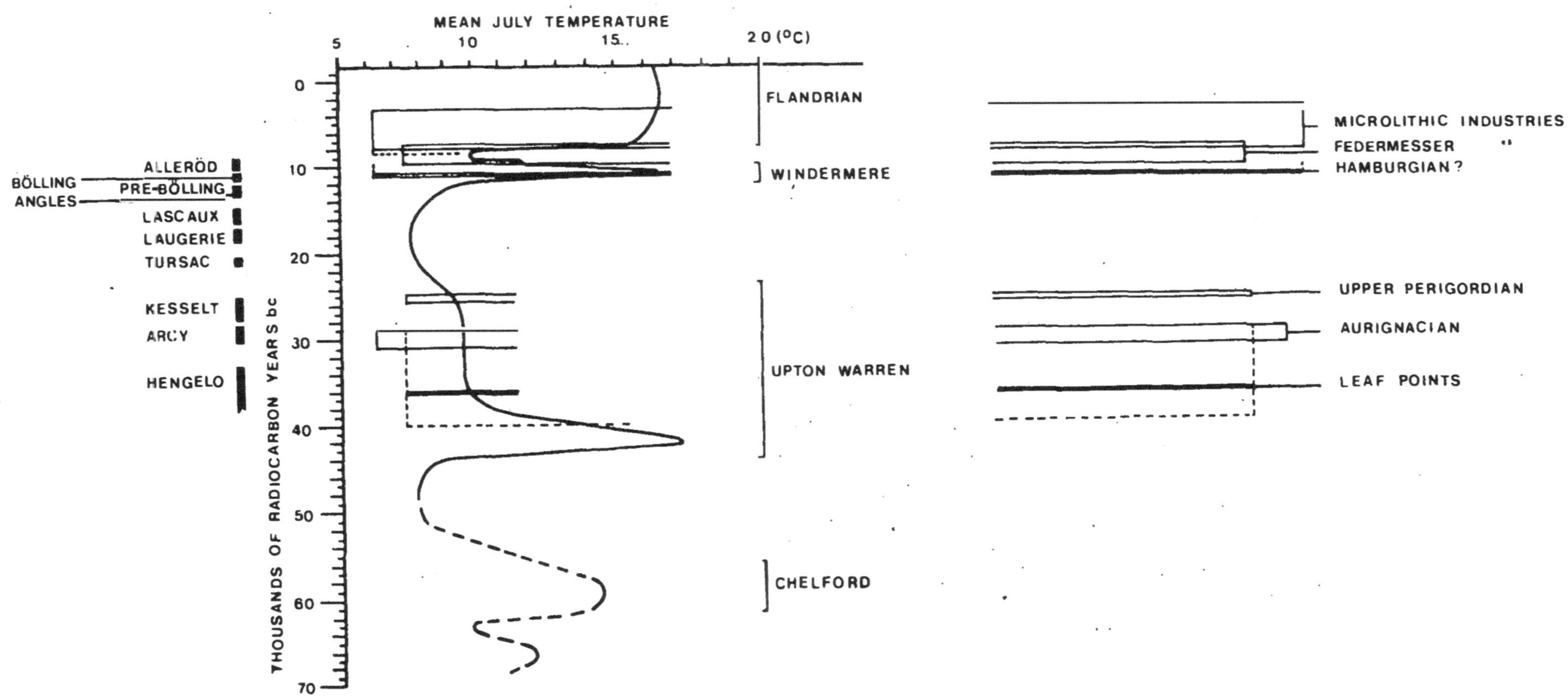

Fig. 2.17 Cultural evidence for Upper Palaeolithic activity in Britain plotted against (1) mean July temperature as derived from studies of fossil Coleoptera and (2) plotted against interstadials recognized from mainland Western Europe on the basis of pollen analyses. While a most probable date of about 36, 000 bc is suggested for the manufacture of leaf points, dotted lines suggest that some might be as early as 40, 000 bc when the first such spearheads appear in Central Europe, while the latest may have been associated with Aurignacian-type artifacts and hence as late as 28, 000 radiocarbon years bc. Space did not allow most probable dates for theBromme-type stemmed points found in Britain or for the Ahrensburgian assemblage from Avington VI to be indicated. This information is recoverable from the text. On the basis of the latest date from Anston Stones Cave and the earliest of the determinations from Thatcham Site IIIA an overlap between the latest occurrence(s) of a Federmesser technology and the earliest technology with microlithic equipment (but without stemmed points) is allowed for, with a dotted line indicating the oldest of the Thatcham determinations (Q 659).

of adzes in Mesolithic assemblages from high ground may reflect the poor development of suitable straight timber in such windswept environments.

Earliest dates for this microlithic technology across the island suggest its uptake later as one moves north and west with the earliest dates from north Wales clustering close to 6, 800 radiocarbon years b.c., and for Scotland closer to 6, 000 b.c. (Coles 1971; Mercer, 1974).

Conclusion

Settlement or exploitation of Britain after about 40, 000 radiocarbon years b.c. and up until the opening of the Flandrian can be shown to have been episodic. The cultural sequence prior to the maximum advance of the Devensian ice sheets appears to duplicate that to be found in Belgium except that a human population can be argued to have ceased to exploit Britain rather earlier than it did Belgium. After the maximum advance of the Devensian ice sheets, the pattern of exploitation detected resembles, in the south-east of the island, that established for the area of northern Germany between the Weser and the Oder, and, in the north and west, the cultural sequence for western Belgium, where no certain trace of a Hamburgian has been identified and where no assemblages with stemmed points intervene between a "Federmesser"-type technology and assemblages of the "South Dutch Boreal Mesolithic".

On Fig. 2.17 are plotted suggested dates for the human exploitation of the present-day island. These are plotted against a curve of July temperatures for lowland central Britain as suggested by Coleopteran evidence (Coope (1977) fig. 5.1), while on the same diagram are indicated major interstadials as recognized for mainland Europe (Leroi Gourhan and Renault-Miskovsky, 1977). Assemblages with leaf points, it is suggested, could have appeared during the Hengelo interstadial and may date to close to 36, 000 radiocarbon years b.c. At a number of sites these may once have been stratified below an Aurignacian comparable to the Aurignacian II of the mainland, which is in turn associated with the Arcy interstadial. Radiocarbon dates suggest a maximum time range for this technology of between about 31, 000 and 28, 000 radiocarbon years b.c. Completely separate from either of these two technologies is an occupation close to 26, 000 radiocarbon years b.c. which left the Font Robert Points recovered from at least eight findspots within the island. These Font Robert Points represent the most distinctive elements from within an Upper Perigordian or Gravettian technology. An absence of Éléments tronqués ou bitronqués as well as of Noailles burins hints that the island may have ceased to be exploited close to 25, 000 radiocarbon years b.c. This phase of exploitation falls within Kesselt.

Each of these phases of exploitation fits within the Upton Warren interstadial.

Acceptance of the proposition that the adult male skeleton recovered from within the Goat's Hole Cave (Paviland) was deposited close to 16, 500 b.c. (that is just before the ice sheets of the Devensian reached their furthest extent and within the cold phase between Lascaux and Laugerie) would require some explanation for the total absence of settlement traces in the 600 kms between Paviland and the Paris basin. It appears justified to speculate as to whether the burial belongs instead to one of the rather earlier phases of exploitation of the island.

A number of shouldered and truncated pieces from eastern England cannot be shown not to be as old as L-De Ia and to be attributable to either the Meiendorf or Poggenwish variants of the Hamburgian. If they are indeed equivalent in age then this would hint at an exploitation of Britain during the warmest part of the Windermere interstadial. Equally, however, they cannot be shown not to be components of a Federmesser technology, innovated into Britain during the 10th millennium b.c.—that is, during L-De II. As on the mainland, there is no evidence to suggest that in Britain this technology belongs to any earlier moment in time. During L-De II Britain can be shown to have been exploited as far west as central southern Devon and westernmost Wales while to the north the island was exploited at least to the middle Pennines and to the southern Lake District. In the north and west of the island it could be argued that this technology continues into L-De III while in the south at least it may have been superseded by technologies with firstly Bromme-type tanged spearheads of a type found as far east as Lithuania, followed during L-De III by a technology where newly acquired archery equipment was tipped with Ahrensburgian stemmed points. At Avington VI and at many sites on the mainland these stemmed points were accompanied by a range of microlithic equipment identical to that found in the earliest Flandrian. This Flandrian microlith equipment is found earliest in south-east England and only rather later in Wales and northern England, and it is again only during the Flandrian that exploitation and settlement of Ireland and Scotland finally took place.

Acknowledgements

I must thank all the museums and private individuals who allowed me to study material in their possession, and also for their quite exceptional hospitality. Dr. D. A. Roe read and commented on an early draft of this paper, as did Mr. S. N. Collcutt. Professor D. T. Donovan, Dr. A. J. Stuart and Dr. D. A. Bramwell all offered valuable advice on fauna and stratigraphy. Dr. V. R. Switsur undertook the dating of the samples of bone from Mother Grundy's Parlour, which were made available by the British Museum (Natural History). Dr. M. P. Kerney commented on the Mollusca from Broxbourne, while Miss Theya Molleson and Dr. R. Burleigh both discussed with the writer the dating of the "Red Lady" of Paviland. Finally I have to thank Mr. F. R. Froom who communicated details of his finds of Avington VI, and Mrs. J. Atkins who at very short notice undertook to type the script.

BIBLIOGRAPHY

Agache, R., 1971. Informations archéologiques: Circonscription du Nord et Picardie. Gallia préhistoire Vol. XIV, pt 2, pp. 271-310.

Albrecht, G., 1979. Magdalénien - Inventare vom Petersfels: Seidlungsarchäologische Ergebnisse der Ausgrabungen 1974-1976. Tübinger Monographien zur Urgeschichte Band 6, pp. 83.

Allain, J., 1974. Rapport du Directuer: Circonscription du Centre. Gallia Préhistoire XVII, fasc. 2, pp. 446-469.

Allison, J., Godwin, H. and Warren, S. H., 1952. Late-Glacial deposits at Nazeing in the Lea Valley, North London. Royal Society Phil. Trans B, Vol. 236, pp. 169-240.

Andersen, S. H., 1970. Senglaciale bopladser ved Bro. Fynske Minder, pp. 85-100.

Andersen, S. H., 1972. Bro: en senglacial boplads på Fyn. Kuml pp. 6-60.

Andree, J., 1939. Der eiszeitliche Mensch in Deutschland und seine Kulturen. Ferdinand Enke Verlag. Stuttgart, pp. I-IX and 758.

Armstrong, A. L., 1925. Excavations at Mother Grundy's Parlour, Creswell Crags, Derbyshire 1924. J.R.A.I., Vol. IV, pp. 146-175.

Armstrong, A. L., 1926. Excavations at Creswell Crags, Derbyshire, 1924-1926. Trans. Hunter Arch. Soc., Vol. III, No. 2, pp. 116-122.

Armstrong, A. L., 1929. Pin Hole Cave Excavations, Creswell Crags, Derbyshire. Discovery of an engraved drawing of a masked Human Figure. P.P.S.E.A., Vol. VI, pt. I, pp. 27-29.

Armstrong, A. L., 1932. Upper Palaeolithic and Mesolithic Stations in North Lincolnshire. P.P.S.E.A., Vol. VII, pt. I, pp. 130-131.

Arora, S. K., 1976. Die Mittlere Steinzeit im Westlichen Deutschland und in den Nachbargebieten. Beiträge zur Urgeschichte des Rheinlandes II. Rheinsiche Ausgrabungen Band 17, pp. 1-65.

Balch, H. E., 1927. Chelm's Combe Shelter. Proc. Somerset Arch. and Nat. Hist. Soc., Vol. LXXII (1926) pp. 97-123.

Barendsen, G. W., Deevey, E. S. and Gralenski, L. J., 1957. Yale Natural Radiocarbon Measurements III. Science, Vol. 126, pp. 908-919.

Barker, H., Burleigh, R. and Meeks, N., 1969. British Museum Natural Radiocarbon Measurements VI. Radiocarbon, Vol. 11, No. 2, pp. 278-294.

Barker, H., Burleigh, R. and Meeks, N., 1971. British Museum Natural Radiocarbon Measurements VII. Radiocarbon, Vol. 13, No. 2, pp. 157-188.

Barnes, B., Edwards, B. J. N., Hallam, J. S. and Stuart, A. J., 1971. Skeleton of a Late Glacial Elk associated with Barbed Points from Poulton-le-Fyle, Lancashire. Nature, Vol. 232, pp. 488-489.

Becker, C. J., 1971. Late Palaeolithic Finds from Denmark. Proc. Prehist. Soc., Vol. XXXVII, pt. 2, pp. 131-139.

Bohmers, A., 1947. Jong-Palaeolithicum en Vroeg-Mesolithicum. In Gerder et alii (eds.), Oudheidkundig Bodemonderzoek in Nederland. J. A. Boom & Zoon, uitgevers, Meppel, vide esp. pp. 129-199.

Bohmers, A., 1956. Statistics and Graphs in the Study of Flint assemblages. II. A Preliminary Report on the Statistical Analysis of the Younger Palaeolithic in Northwestern Europe. Palaeohistoria, Vol. V, pp. 7-26.

Bohmers, A., 1960. Statistiques et Graphiques dans l'étude des industries lithiques préhistoriques. V. Considérations générales au sujet du

Hambourgien, du Tjongérien, du Magdalénien et de l'Azilien. Palaeohistoria, Vol. VIII, pp. 15-38.

Bourdier, F., 1961. Le Bassin du Rhône au Quaternaire: Géologie et Préhistoire. Paris Editions du C.N.R.S. 2 vols.

Bouskill, I. W. C., 1979. London Borough of Bromley: Investigations and Excavations during the Year. Arch. Cantiana, Vol. XCIV, 1978, pp. 282-283.

Bowen, D. Q., 1970. The Palaeoenvironment of the "Red Lady" of Paviland. Antiquity, Vol. XLIV, pp. 134-136.

Bowen, D. Q., 1977. Hot and cold climates in prehistoric Britain. Geographical Magazine, pp. 685-698.

Boyd Dawkins, W., 1877. On the mammal Fauna of the Caves of Creswell Crags. Q.J.G.S., Vol. XXXIII, pp. 589-612.

Bramwell, D., 1955. Third Report on Excavations at Ossom's Cave. Peakland Arch. Soc. Newsletter, No. 13, pp. 7-8.

Bramwell, D., 1964. The Excavations at Elder Bush Cave, Wetton, Staffs. North Staffs Journal of Field Studies, Vol. 4, pp. 46-60.

Bramwell, D., 1971. Excavations at Fox Hole Cave, High Wheeldon, 1961-1970. Derbyshire Arch. J., Vol. XCL, pp. 1-19.

Breuil, H., 1922. Observations on the Pre-Neolithic Industries of Scotland. Proc. Soc. Antiq. Scot., Vol. LVI (1921-22), pp. 261-281.

Brinch-Petersen, E., 1970. Le Brommëen et le cycle de Lyngby. Quartär, Bd 21, pp. 93-98.

Buckland, P. C., 1974. Environment and archaeology in the Vale of York. South Yorkshire Studies in Archaeology and Natural History I, pp. 6-18.

Buckland, P. C., 1976. The use of insect remains in the interpretation of archaeological environments. In D. A. Davidson and M. L. Shackley (eds.), Geoarchaeology: Earth Science and the Past. Duckworth, pp. 1-X and 408. Vide esp. pp. 369-396.

Buckland, W., 1824. Reliquiae Diluvianae or Observations on the Organic Remains contained in Caves, Fissures, and Diluvial Gravel, and on other Geological Phenomena attesting to the action of an Universal Deluge. London. John Murray, Albermarle Street, pp. 1-VII and 303.

Burleigh, R., Hewson, A. and Meeks, N., 1976. British Museum Natural Radiocarbon Measurements VIII. Radiocarbon, Vol. 18, No. 1, pp. 16-42.

Campbell, J. B., 1969. Excavations at Creswell Crags: Preliminary Report. Derbyshire Arch. J., Vol. 89, pp. 47-58.

Campbell, J. B., 1970. The Upper Palaeolithic Period. In J. B. Campbell, D. Elkington, P. Fowler and L. Grinsell The Mendip Hills in Prehistoric and Roman Times. Bristol, pp. 5-11.

Campbell, J. B., 1977. The Upper Palaeolithic of Britain: A Study of Man and Nature in the Late Ice Age. Clarendon Press, Oxford. 2 vols.

Campbell, J. B. and Sampson, C. Garth. 1971. A New Analysis of Kent's Cavern, Devonshire, England. University of Oregon Anthropological Papers, No. 3, pp. 1-VII and 40.

Catt, J. A., 1977. Loess and Coversands. In F. W. Shotton (ed.), British Quaternary Studies: Recent Advances. Clarendon Press, Oxford. Vide esp. pp. 221-230.

Celerier, G., 1977. Inventaire morphologique des Pointes Aziliennes en Périgord: un projet de rationalisation. In La Fin des Temps glaciaires en Europe. Colloques Internationaux du C.N.R.S., No. 271, pp. 461-466. pp. 227-238.

Chmielewski, W., 1961. Civilization de Jerzmanowice. Instytut Historii Kultury Materialnej Polskiej Akademii Nauk. pp. 92.

Chmielewski, W., Schild, R. and Więckowska, H., 1975. Prahistoria ziem Polskich I. Paleolit i Mezolit. Polska Akademia Nauk: Instytut Historii Kultury Materialnej, pp. 457.

Clark, J. G. D., 1954. Excavations at Star Carr. C.U.P., pp. 1-XXIII and 200.

Clark, J. G. D., 1963. Neolithic Bows from Somerset, England, and the Prehistory of Archery in North-western Europe. P.P.S., Vol. XXIX, pp. 50-98.

Clark, J. G. D., 1975. The Earlier Stone Age Settlement of Scandinavia. C.U.P., pp. 1-XXV and 282.

Clifford, E. M., Garrod, D. A. E. and Gracie, H. S., 1954. Flint Implements from Gloucestershire'. Ant. J., Vol. XXXIV, pp. 178-187.

Coles, J. M., 1971. The Early Settlement of Scotland: Excavations at Morton Fife. P.P.S., Vol. XXXVII, pt. 2, pp. 284-366.

Coope, G. R., 1975. Climatic Fluctuations in North West Europe since the Last Interglacial, indicated by Fossil Assemblages of Coleoptera. In A. E. Wright and F. Moseley (eds.), (1975) Ice Ages Ancient and Modern. Geological Journal, special issue No. 6, pp. 153-168.

Coope, G. R., 1977. Quaternary Coleoptera as aids in the interpretation of Environmental History. In F. W. Shotton (ed.) British Quaternary Studies: Recent Advances. Clarendon Press, Oxford, pp. 1-XII and 298.

Curwen, E., 1949. A flint dagger factory near Pulborough, Sussex. Ant. J., Vol. XXIX, pp. 192.

Davies, J. A., 1921. Aveline's Hole, Burrington Coombe: an Upper Palaeolithic station. Proc. Univ. Bristol Spel. Soc., Vol. I, No. 2 (1920-21), pp. 61-73.

Davies, J. A., 1923. Aveline's Hole, Burrington Coombe: an Upper Palaeolithic station. Proc. Somerset Arch. and Nat. Hist. Soc., Vol. LXVIII (1922), pp. 21-26.

Davies, M., 1975. Ogof Tudno—A New Archaeological Cave on the Great Orme, Llandudno. Privately circulated, pp. 2.

Degerbøl, M., 1964. Some remarks on late and post-glacial vertebrate fauna and its ecological relations in northern Europe. J. Anim. Ecol., Vol. 33 (Supplement), pp. 71-85.

Delibrias, G., Guillier, M. T. and Labeyrie, J., 1964. Saclay Natural Radiocarbon Measurements I. Radiocarbon, Vol. 6, pp. 233-250.

Delporte, H., 1976. L'organisation du Périgordien supérieur en France et ses rapports avec le Périgordien d'Europe occidentale. U.I.S.P.P. IXe Congrès: Colloque XV, pp. 7-51.

Desbrosse, R., 1965. Les sagaies magdaléniennes de la Croze (Ain). Revue archéologique du Centre, No. 15-16, pp. 327-334.

Desbrosse, R., 1976. Les civilisations du Paléolithique supérieur dans le Jura et en Franche-Comté. In H. de Lumley (ed.) La Préhistoire française, Vol. I, pt 2, pp. 1348-1357.

Desbrosse, R. and Evin, J., 1971. Datations au C14 de Gisements magdaléniens du Jura et des Préalpes du Nord. Actes de VIIIe Congrès International de sciences Préhistoriques et Protohistoriques (Belgrade), Vol. 2, pp. 179-187.

Dudley, H. E., 1949. Early Days in North-West Lincolnshire. Scunthorpe: W. H. and C. H. Caldicott, pp. 1-XVI and 255.

Evans, J. G., 1976. The Environment of Early Man in the British Isles. Paul Elek: London, pp. 1-XIV and 216.

Evin, J. et alii., 1973a. Lyon Natural Radiocarbon Measurements III. Radiocarbon, Vol. 15, No. 1, pp. 135-155.

Evin, J. et alii., 1973b. Lyon Natural Radiocarbon Measurements IV. Radiocarbon, Vol. 15, No. 3, pp. 514-533.

Evin, J., Marien, G. and Pachiaudi, Ch., 1976. Lyon Natural Radiocarbon Measurements VI. Radiocarbon, Vol. 18, No. 1, pp. 60-88.

Evin, J., Marien, G. and Pachiaudi, Ch., 1978. Lyon Natural Radiocarbon Measurements VII. Radiocarbon, Vol. 20, No. 1, pp. 19-57.

Ferembach, D., 1954. Note sur une mandibule presumée du Magdalénien III avec une introduction sur la stratigraphie et la datation du gisement par J. Combier. Bull. Soc. Anthrop. Paris, Vol. 5 (10th series), pp. 25-34.

Feustel, R., 1974. Die Kneigrotte: eine Magdalénien-Station in Thüringen. Weimar: Hermann Böhlaus Nachfolger, p. 224.

Fischer, A., 1978. På sporet af overgangen mellem palaeolithicum og mesolithicum i Sydskandinavien. Hikuin, No. 4, pp. 27-50.

Froom, F. R., 1972. Some Mesolithic Sites in South-West Berkshire. Berks Arch. J., Vol. 66, pp. 11-22.

Garrod, D. A. E., 1926. The Upper Palaeolithic Age in Britain. Oxford: Clarendon Press, pp. 211.

Gaunt, G. D., Jarvis, R. A. and Matthews, B., 1971. The Late Weichselian Sequence in the Vale of York. Proc. Yorks. Geol. Soc., Vol. 38, pt 2, No. 13, pp. 281-284.

Gilot, E., 1971. Louvain Natural Radiocarbon Measurements X. Radiocarbon, Vol. 13, No. 1, pp. 45-51.

Glory, A., 1961. La Grotte de Rigney (Doubs). Anciennes fouilles de M. Jacques Collot. B.S.P.F., Vol. 58, pp. 389-400.

Godwin, H., 1954. Lake-Stratigraphy, Pollen Analysis and Vegetational History. In J. G. D. Clark (1954) Star Carr. Vide esp. pp. 25-69.

Godwin, H., 1960. Radiocarbon dating and Quaternary history in Britain. Proc. Royal Soc. B, Vol. 153, pp. 287-320.

Godwin, H. and Willis, E. H., 1959. Cambridge University Natural Radiocarbon Measurements I. Radiocarbon, Vol. I, pp. 63-75.

Godwin, H. and Willis, E. H., 1962. Cambridge University Natural Radiocarbon Measurements V. Radiocarbon Vol. 4, pp. 57-70.

Grimes, W. F., 1933. Priory Farm Cave, Monkton, Pembrokeshire. Arch. Cambrensis, Vol. LXXXVII, pp. 88-100.

Grimes, W. F., 1951. Prehistory of Wales. National Museum of Wales, pp. 1-XVII and 288.

Gross, H., 1951. Die Moorgeologische Datierung des jungeren Magdalénien in Deutschland. Eiszeitalter und Gegenwart, Vol. I, pp. 166-171.

Gunther, C., 1973. Der Federmesser-fundplatz von Westerkappeln, Kr Tecklenburg. Bodenaltentümer Westfalens, Bd XIII, pp. 5-76.

Hahn, J., 1970. Recherches sur L'Aurignacien en Europe centrale et orientale. L'Anthropologie, Vol. 74, pp. 195-220.

Hahn, J., 1976. Das Gravettien im westlichen mitteleuropa. U.I.S.P.P., IXe Congrès Colloque XV, pp. 100-120.

Hahn, J., 1977. Aurignacien: Das ältere jungpaläolithicum in mittel- und osteuropa. Fundamenta, Reihe A Band 9, Böhlau Verlag, Köln, Wien, pp. 355.

Hahn, J., 1977. Die Spitzbubenhöhle, eine Magdalénien-Fundstelle bei Eselsburg, gemeinde Herbrechtingen kreis Heidenheim. Fundberichte aus Baden-Württemberg, Bd 3, pp. 38-50.

Hahn, J. and Koenigswald, W. Von., 1977. Die Steinzeitlichen Funde und die spätglaziale Nagetierschicht aus der Kleinen Scheuer am Höhlenstein im Lonetal. Fundberichte aus Baden-Württemberg, Bd 3, pp. 51-75.

Hallam, J. S., Edwards, B. J. N., Barnes, B. and Stuart, A. J., 1973. The Remains of a Late Glacial Elk associated with Barbed Points from High Furlong, near Blackpool, Lancashire. Proc. Prehist. Soc., Vol. XXXIX, pp. 100-128.

Halstead, L. B., 1978. New Light on the Piltdown hoax? Nature, Vol. 276, pp. 11-13.

Harrison, R. A., 1976. Two Pieces of Worked Bone from the Pleistocene Deposits at the Chelm's Combe Rock Shelter, Cheddar. Proc. Univ. Bristol Spel Soc., Vol. 14, No. 2, pp. 115-122.

Hawkes, C. J., Tratman, E. K. and Powers, R., 1970. Decorated piece of Rib Bone from the Palaeolithic Levels at Gough's Cave, Cheddar, Somerset. Proc. Univ. Bristol Spel. Soc., Vol. 12, No. 2, pp. 137-142.

Heinzelin de Braucourt, J. de., 1962. Manuel de Typologie des Industries Lithiques. Bruxelles, pp. 74.

Heinzelin de Braucourt, J. de., 1973. L'Industrie du Site Paléolithique de Maisières-Canal. Institut Royal des Sciences Naturelles de Belgique, Mémoire No. 171, pp. 63.

Hibbert, F. A. and Switsur, V. R., 1976. Radiocarbon Dating of Flandrian Pollen Zones in Wales and Northern England. New Phytologist, Vol. 77, pp. 793-807.

Hicks, H., 1886. Results of some Recent Researches in some Bone-caves in North Wales. Q.J.G.S., Vol. 42, pp. 3-11.

Hülle, W. M., 1977. Die Ilsenhöhle unter Burg Ranis/Thüringen: Eine paläolithische Jägerstation. Stuttgart, New York: Gustav Fischer Verlag, pp. 1-XX and 203.

Iversen, J., 1936. Sekundäres Pollen als Fehlerquelle. Ein Korrektionsmethode zur Pollenanalyse Minerogener Sedimente. Danmarks Geologiske Undersøgelse, IV Raekke, Bd 2, Nr 15, pp. 24.

Iversen, J., 1942. En pollenanalytisk Tidfaestelse af Ferskvandslagene ved Nørre Lyngby. Meddelelser fra Dansk Geologisk Forening, Bd 10, pp. 130-151.

Iversen, J., 1946. Geologisk Datering af en senglacial Boplads ved Bromme. Aarbøger, pp. 198-231.

Jackson, J. W., 1945. A Lance-point of Upper Palaeolithic type from Victoria Cave, Settle, Yorkshire. Ant. J., Vol. XXV, pp. 147-148.

Jackson, J. W. and Mattinson, W. K., 1932. A cave on Giggleswick Scars, near Settle, Yorkshire. The Naturalist, pp. 5-9.

Jacobi, R. M., 1975. Aspects of the Post-glacial Archaeology of England and Wales. Unpublished Ph. D. thesis: University of Cambridge.

Jacobi, R. M., 1976. Britain inside and outside Mesolithic Europe. Proc. Prehist. Soc., Vol. 42, pp. 67-84.

Jacobi, R. M., 1978. Northern England in the eighth millennium b.c.: an essay. In P. A. Mellars (ed.) The Early Postglacial Settlement of Northern Europe. Duckworth, pp. 295-332.

Jacobi, R. M., 1979. The Late-glacial settlement of Essex. In D. Buckley (ed.) The Archaeology of Essex to A.D. 1500. C.B.A. Research Report, in press.

Jessen, A. and Nordmann, V., 1915. Ferskvandslagene ved Nørre Lyngby. Danmarks geologiske Undersøgelse, II Raekke, Nr 29, pp. 66.

Johansson, A., 1971. Barmosen-gruppen. Praeboreal boplads fund med skiveøkser in Sydsjaelland. Historisk Samfund for Praestø Amt. Aarborg (1968), pp. 101-170.

John, B. S., 1971. The "Red Lady" of Paviland: a Comment. Antiquity, Vol. XLV, pp. 141-144.

Kerney, M. P., 1977. A Proposed Zonation Scheme for Late-glacial and Post-glacial Deposits using land Mollusca. Journal of Archaeological Science, Vol. 4, No. 4, pp. 387-390.

Klima, B., 1949. Výzkum jeskyně "Nové Drátenické" u Křtin Časopis Moravského Musea V Brně, XXXIV, pp. 1-15.

Kobusiewicz, M., 1973. Problems concerning Hamburgian Culture in Central Europe. Przegląd Archeologiczny, Vol. 21, pp. 65-92.

Kozlowski, J. K., 1971. Les Problèmes du Magdalénien en Europe Centre-est. Actes du VIIIe Congrès International des Sciences Préhistoriques et Protohistoriques (Belgrade), Vol. I, pp. 33-70.

Kozlowski, S. K., 1966. Uwagi o pozyaji stratygraficznej i wieku szczatków ludzkich w Joskini Maszyckiej , pow Olkusz. Archeologia Polski, T XI Z2, pp. 307-312.

Kozlowski, S. K., 1969. Górnopaleolityczne stanowisko w Jaskini Maszyckiej, pow. Olkusz. Sprawozdania Archeologiczne T XX, pp. 25-33.

Kozlowski, S. K., 1975. Quelques Remarques sur le Brommien. Acta Archaeologia, Vol. 46, pp. 134-142.

Lacaille, A. D., 1946a. The Northward March of Palaeolithic Man in Britain. Proc. Geol. Assoc., Vol. LVII, pt 2, pp. 57-81.

Lacaille, A. D., 1946b. Some Flint Implements of special interest from Lincolnshire, Hampshire and Middlesex. Ant. J., Vol. XXVI, pp. 180-186.

Lacaille, A. D. and Grimes, W. F., 1955. The Prehistory of Caldey. Arch. Cambrensis, Vol. CIV, pp. 85-165.

Laet, S. J. De., 1979. Prehistorische Kulturen in het Zuiden der Lage Landen. 2nd Ed. Universa Wetteren, pp. 689.

Lang, G., 1962. Vegetationsgeschichte Untersuchungen der Magdalénienstation an der Schussenquelle. Veröfft Geobot. Inst. E.T.H. Rübel, Nr 37, pp. 129-154.

Lanting, J. N. and Mook, W. G., 1977. The Pre- and Protohistory of the Netherlands in terms of Radiocarbon Dates. Groningen, pp. 247.

Larue, M., Combier, J. and Roche, J., 1956. Le Gisement Périgordien et Magdalénien du Saut-du-Perron (Loire). L'Anthropologie, T. 59, No. 5-6, pp. 401-428.

Leroi-Gourhan, A., Brézillon, M. and Schmider, B., 1976. Les Civilizations du Paléolithique supérieur dans le centre et le sud-est du Bassin Parisien. In H. de Lumley (1976) ed.) La Préhistorie française, Vol. I, pt 2.

Leroi-Gourhan, Arlette and A., 1964. Chronologie des Grottes d'Arcy-sur-Cure (Yonne): Pt I. Climat du Quaternaire récent. Gallia Préhistoire, VII, pp. 1-35.

Leroi-Gourhan Arlette and Renault-Miskovsky, J., 1977. La Palynologie appliquée a L'Archéologie: méthodes, limites et résultats. In Approche écologique de L'Homme fossile. Supplément au Bull. Assoc. Française pour L'Etude du Quaternaire, pp. 35-49.

Leroy-Prost, C., 1974. L'industrie osseuse aurignacienne de la collection Passemard d'Isturitz, Pyrenées Atlantiques. L'Anthropologie, Vol. 78, No. 2, pp. 283-297.

McBurney, C. B. M., 1959. Report on the First Season's Fieldwork on British Upper Palaeolithic Cave Deposits. Proc. Prehist. Soc., Vol. XXV, pp. 260-269.

Mace, A., 1959. An Upper-Palaeolithic Open Site on Hengistbury Head, Christchurch, Hants. Proc. Prehist. Soc., Vol. XXV, pp. 233-259.

Mathiassen, T., 1937. Gudenaa-Kulturen. En mesolitisk inlandsbebyggelse i Jylland. Aarbøger, pp. 1-186.

Mathiassen, T., 1946. En senglacial Boplads ved Bromme. Aarbøger, pp. 121-197.

Matthews, B., 1970. Age and Origin of Aeolian Sand in the Vale of York. Nature, Vol. 227, pp. 1234-1236.

Mauser, P. F., 1970. Die jungpaläolithische Höhlenstation Petersfels im Hegau. Badische Fundberichte Sonderheft. 13, pp. 127.

May, J., 1976. Prehistoric Lincolnshire. History of Lincolnshire I: Lincoln, History of Lincolnshire Committee, pp. 1-XIX and 251.

Mellars, P. A., 1969. Radiocarbon Dates for a New Creswellian Site. Antiquity, Vol. XLIII, pp. 308-310.

Mellars, P. A., 1974. The Palaeolithic and Mesolithic. In C. Renfrew (ed.) British Prehistory—a New Outline. Duckworth, pp. 1-XIV and 348. Vide esp. pp. 41-99.

Menke, B., 1968. Das Spätglazial von Glüsing. Eiszeitalter und Gegenwart, Vol. 19, pp. 73-84.

Mercer, J., 1974. New C14 Dates from the Isle of Jura, Argyll. Antiquity, Vol. XLVIII, pp. 65-66.

Moir, J. Reid., 1930. Ancient Man in the Gipping-Orwell Valley, Suffolk. P.P.S.E.A., Vol. VI, pt III, pp. 182-221.

Molleson, T., 1976. Remains of Pleistocene Man in Paviland and Pontnewydd Caves. Trans. Brit. Cave Res. Assoc., Vol. III, No. 2, pp. 112-116.

Molleson, T., 1977. Skeletal Remains of Man in the British Quaternary. In F. W. Shotton (ed.) British Quaternary Studies: Recent Advances. Oxford. Vide esp. pp. 83-92.

Molleson, T., 1978. A New Date for Goat's Hole Cave. Antiquity, Vol. LII, No. 205, pp. 143-145.

Moore, J. W., 1954. Excavations at Flixton Site 2. In J. G. D. Clark (1954) Star Carr. Vide esp. pp. 192-194.

Mouton, P. and Joffroy, R., 1956. Précisions nouvelles sur les stations de Farincourt (Haute-Marne). Revue Archéologique de l'Est et du Centre-Est, Vol. VII, pp. 193-223.

Movius, H. L., 1960. Radiocarbon Dates and Upper Palaeolithic Archaeology in central and western Europe. Current Anthropology, Vol. I, pp. 355-391.

Movius, H. L. and Judson, G., 1956. The Rock-Shelter of La Colombière. American School of Prehistoric Research, Peabody Museum, Harvard University, Bull. No. 19, pp. 176.

Münnich, K. O., 1957. Heidelberg Natural Radiocarbon Measurements I. Science, Vol. 126, pp. 194-199.

Nédervelde, Br. J. van and Davies, M., 1975. Potter's Cave, Caldey Island. Archaeology in Wales, C.B.A. Group 2 No. 15, pp. 38-39.

Nédervelde, Br. J. van, Davies, M. and John, B. S., 1973. Radiocarbon Dating from Ogof-yr-Ychen, a New Pleistocene Site in West Wales. Nature, Vol. 245, pp. 453-455.

Newell, R. R., 1973. The postglacial adaptations of the indigenous population of the Northwest European plain. In S. K. Kozlowski (ed.) The Mesolithic in Europe. Warsaw, pp. 399-440.

Oakley, K. P., 1968. The Date of the "Red Lady" of Paviland. Antiquity, Vol. 42, pp. 306-307.

Oakley, K. P., 1971. Radiocarbon Dating of Proto-Solutrean in Wales. Nature, Vol. 231, p. 112.

Osborne, P. J., 1972. Insect Faunas of Later Devensian and Flandrain Age from Church Stretton, Shropshire. Royal Soc. Phil. Trans B, Vol. 263, pp. 327-367.

Otte, M., 1976a. Observations sur l'industrie lithique de Maisières et sur ses relations avec les autres ensembles périgordiens de Belgique. B.S.P.F. Tome 73, pp. 335-351.

Otte, M., 1976b. L'Aurignacien en Belgique. U.I.S.P.P. IXe Congrès: Colloque XVI, pp. 144-163.

Otte, M., 1977a. Données Génerales sur le Paléolithique Supérieur Ancien de Belgique. L'Anthropologie, Vol. 81, No. 2, pp. 235-272.

Otte, M., 1977b. Les Sagaies de L'Aurignaco-Périgordien en Belgique. In Méthodologie appliquée a l'industrie de l'os préhistorique. Colloques Internationaux du C.N.R.S., No. 568, pp. 193-203.

Otte, M., 1978. Review of Hülle, W. M. (1977). Die Ilsenhöhle unter Burg Ranis/Thüringen: Eine paläolithische Jägerstation. B.S.P.F., Vol. 75, No. 5, pp. 133-134.

Paddaya, K., 1971. The Late Palaeolithic of the Netherlands—a Review. Helinium, Vol. XI, pp. 257-270.

Paddaya, K., 1973. A Federmesser Site with Tanged Points at Norgervaart prov. of Drenthe (Netherlands). Palaeohistoria, Vol. XV, pp. 167-214.

Palmer, S., 1970. The Stone Age industries of the Isle of Portland, Dorset, and the utilization of Portland Chert as artifact material in southern England. Proc. Prehist. Soc., Vol. XXXVI, pp. 82-115.

Palmer, S., 1971. Second Report on Excavations at Portland Site I, 1967 to 1968. Proc. Dorset Nat. Hist. and Arch. Soc., Vol. 92 (1970), pp. 168-180.

Parry, R. F., 1929. Excavations at the Caves, Cheddar. Proc. Somerset Arch. and Nat. Hist. Soc., Vol. LXXIV (1928), pp. 102-121.

Parry, R. F., 1931. Excavations at Cheddar. Proc. Somerset Arch. and Nat. Hist. Soc., Vol. LXXVI, pt II (1930) pp. 46-62.

Penny, L. F., Coope, G. R. and Catt, J. A., 1969. Age and Insect Fauna of the Dimlington Silts, East Yorkshire. Nature, Vol. 224, pp. 65-67.

Petersen, B. F., 1973. Senpalaeolitiske Flaekkespidser fra Knudshoved Odde, Sydsjaelland, Aarbøger, pp. 211-216.

Polach, H. A. and Golson, J., 1966. Collection of Specimens for Radiocarbon Dating and Interpretation of Results. Canberra: Australian Institute of Aboriginal Studies: Manual No. 2, pp. 42.

Puydt, M. De and Lohest, M., 1887. L'Homme contemporain du Mammouth à Spy (Namur). Annales de la Fédération Archéologique et Historique de Belgique T. II (1886), pp. 207-235.

Riley, D. N., 1978. Risby Warren: An Occupation Site from the Mesolithic to the Early Bronze Age. Lincolnshire History and Archaeology, Vol. XIII, pp. 5-11.

Rowlands, B. M., 1971. Radiocarbon Evidence of the Age of an Irish Sea Glaciation in the Vale of Clwyd. Nature Phys. Sci., Vol. 230, p. 9.

Rust, A., 1943. Die Alt- und Mittelsteinzeitliche Funde von Stellmoor. Karl Wachholtz Verlag, Neumünster, pp. 1-VIII and 240.

Rust, A., 1958a. Die jungpaläolithischen Zeltanlagen von Ahrensburg. Offa Bücher, Band 15: Karl Wachholtz Verlag, Neumünster, pp. 146.

Rust, A., 1958b. Die Funde von Pinnberg. Karl Wachholtz Verlag, Neumünster.

Sacchi, D., 1974. Récherches sur le Paléolithique en Languedoc occidental (1969-70). Cahiers Ligures de Préhistorie et d'Archéologie, No. 19 (1970), pp. 87-96.

Salomonsson, B., 1964. Découverte d'une Habitation du Tardi-Glaciaire à Segebro, Scanie, Suède. Acta Archaeologica, Vol. XXXV, pp. 1-28.

Savory, H. N., 1973. Excavations at the Hoyle, Tenby, in 1968. Arch. Cambrensis, Vol. CXXII, pp. 18-34.

Schild, R., 1973. Esquisse de la Paléogéographie de la Plaine de L'Europe septentrionale dans le Tardiglaciaire: Revue des Problèmes. Przeglad Archeologiczny, Tom XXI, pp. 47-63.

Schild, R., 1976. The Final Palaeolithic Settlements of the European Plain. Scientific American, No. 234, pt 2, pp. 88-99.

Schmider, B., 1971. Les Industries lithiques du Paléolithique Supérieur en Ile de France. VIe Supplément à Gallia Préhistoire: Centre National de la Recherche Scientifique, pp. 218.

Schmider, B., 1977. Un nouveau faciès du Magdalénien final du Bassin Parisien: l'industrie du gisement du Pré des Forges à Marsangy (Yonne). In La Fin des Temps Glaciaires en Europe. Colloques internationaux du Centre National de la Recherche Scientifique, No. 271, pp. 763-771.

Schütrumpf, R., 1958. Die Pollenanalytische Untersuchung an den altsteinzeitlichen Moorfundplätzen Borneck und Poggenwisch. In A. Rust Die Jungpaläolithschen zeltanlagen von Ahrensburg, pp. 11-22.

Schwabedissen, H., 1951a. Zur Besiedlung des Nordseeraumes in der älteren und mittleren Steinzeit. Festschrift für Gustav Schwantes, ed. K. Kersten. Karl Wachholtz Verlag, Neumünster. Vide esp. pp. 59-77

Schwabedissen H., 1951b. Das Vorkommen des Magdalénien im nordwesteuropäischen Flachland. Eiszeitalter und Gegenwart, Vol. I, pp. 152-165.

Schwabedissen, H., 1954. Die Federmesser-Gruppen des nordwesteuropäischen Flachlandes. Offa Bücher, Karl Wachholtz Verlag, Neumünster, pp. 1-VII and 104.

Schwabedissen, H., Schütrumpf, R., and Münnich, K. O., 1957. Das Alter der Federmesser-Zivilisation auf Grund neuer naturwissenschaftlicher Untersuchungen. Eiszeitalter und Gegenwart, Band 8, pp. 200-209.

Schwantes, G., 1923. Das Beil als Schiede zwischen Paläolithikum und Neolithikum. Arkiv. für Anthropologie N.F., Band 20, pp. 13.

Shotton, F. W. and Williams, R. E. G., 1971. Birmingham University Radiocarbon Dates V. Radiocarbon, Vol. 13, No. 2, pp. 141-156.

Shotton, F. W., and Williams, R. E. G., 1973. Birmingham University Radiocarbon Dates VII. Radiocarbon, Vol. 15, No. 3, pp. 451-468.

Simmons, I. G., 1975. Towards an Ecology of Mesolithic Man in the Uplands of Great Britain. J. Arch. Science, Vol. 2, pp. 1-15.

Smith-Woodward, A., 1914. On an apparently Palaeolithic Engraving on a Bone from Sherborne (Dorset). Q.J.G.S., Vol. LXX, pp. 100-103.

Sollas, W. J., 1913. Paviland Cave: an Aurignacian Station in Wales. J.R.A.I., Vol. 43, pp. 325-374.

Sonneville Bordes, D. de., 1969. A propos des pointes pedonculées du Nord de l'Europe: pointe de Lingby et pointe de Teyjat. Quartär, 20, pp. 183-188.

Sonneville Bordes, D. de., 1970. A propos des pointes pedonculées magdaléniennes. Quartär, Band 21, pp. 97-98.

Sonneville Bordes, D. de. and Deffarge, R., 1974. Lames Retouchées Magdaléniennes du Morin (Gironde). Zephyrus T. 25, pp. 95-105.

Stampfuss, R. and Schütrumpf, R., 1970. Harpunen der Allerödzeit aus Dinslacken, Niederrhein. Bonner Jahrbücher, Vol. 170, pp. 19-35.

Stapert, D. and Verlinde, A. D., 1974. Gramsbergen. Nieusbulletin KNOB 73. Archeologisch Nieuws, pp. 246-248.

Stordeur-Yedid, D., 1979. Les Aiguilles à chas au Paléolithique. XIIIe Supplément a Gallia Préhistoire: Centre National de la Recherche Scientifique pp. 215.

Stuart, A. J., 1974. Pleistocene History of the British Vertebrate Fauna. Biol. Review, Vol. 49, pp. 225-266.

Stuart, A. J., 1977a. The Vertebrates of the Last Cold Stage in Britain and Ireland. Phil. Trans R. Soc. London B, 280, pp. 295-312.

Stuart, A. J., 1977b. British Quaternary Vertebrates. In F. W. Shotton (ed.), British Quaternary Studies: Recent Advances. Oxford: Clarendon Press, pp. 69-81.

Sutcliffe, A. J. and Kowalski, K., 1976. Pleistocene Rodents of the British Isles. Bulletin of the British Museum (Natural History) Geology, Vol. 27, No. 2, pp. 33-147.

Swanton, E. W. and Woods, P., 1914. Bygone Haslemere. London: West, Newman and Co., pp. 1-XVI and 394.

Taborin, Y., 1977. Les habitats paléolithiques des bords du Seine: Etiolles (Essone, France). In La Fin des Temps glaciaires en Europe. Colloques internationaux du C.N.R.S., No. 271, Vol. II, pp. 116-132.

Tauber, H., 1970. Danske Kulstof- 14 Dateringer af Arkaeologiske prøver III. Aarbøger, pp. 120-140.

Tauber, H., 1973. Copenhagen Radiocarbon Dates X. Radiocarbon, Vol. 15, pt I, pp. 86-112.

Taute, W., 1963. Funde der Spätpaläolithischen "Federmesser-Gruppe" aus dem Raum zwischen mittlerer Elbe and Weichsel. Berliner Jahrbuch für Vor- und Frühgeschichte, Band 3, pp. 62-111.

Taute, W., 1968. Die Steilspitzen-Gruppen im nördlichen mitteleuropa. Fundamenta, Reihe A, Band 5, Köln/Graz.

Taylor, H., 1928. King Arthur's Cave, near Whitchurch, Ross-on-Wye: second Report: Excavations in 1926 and 1927. Proc. Univ. Bristol Spel. Soc., Vol. 3, No. 2 (1927), pp. 59-83.

Toepfer, V., 1968. Die Weichsel-Eiszeit und ihre paläolithischen Fundplätze II in Gebiet der Deutschen Demokratischen Republik. Ausgrabungen und Funde, Band 13, pp. 9-17.

Tratman, E. K., 1950. Amber from the Palaeolithic Deposits at Gough's Cave, Cheddar. Proc. Univ. Bristol Spel. Soc., Vol. 6, No. 2 (1949-50), pp. 223-227.

Tratman, E. K., 1955. Second Report on the Excavations at Sun Hole, Cheddar. Proc. Univ. Bristol Spel. Soc., Vol. 7, No. 2 (1954-55), pp. 61-75.

Tratman, E. K., 1976. A Late Upper Palaeolithic Calculator (?), Gough's Cave, Cheddar, Somerset. Proc. Univ. Bristol Spel. Soc., Vol. 14, No. 2, pp. 123-129.

Tratman, E. K. and Henderson, G. T. D., 1928. First Report on the Excavations at Sun Hole, Cheddar: levels above the Pleistocene. Proc. Univ. Bristol Spel. Soc., Vol. 3, No. 2 (1927), pp. 84-97.

Tromnau, G., 1975a. Neue Ausgrabungen im Ahrensburger Tunneltal: ein Beitrag zur Erforschung des Jungpaläolithikums im nordwesteuropäischen Flachland. Offa Bücher, Band 33, pp. 105-.

Tromnau, G., 1975b. Die Fundplätze der Hamburger Kultur von Heber und Deimern, Kreis Soltau. Materialhefte zur Ur- und Frühgeschichte neidersachsens, Heft 9, pp. 1-VII and 46.

Tuffreau, A., 1976. Les civilisations du Paléolithique supérieur en Artois. In H. de Lumley, (ed.), La Préhistoire française, Vol. I, pt. 2, pp. 1346-1347.

Vogel, J. C. and Waterbolk, H. T., 1963. Groningen Radiocarbon Dates IV. Radiocarbon, Vol. 5, pp. 163-202.

Vogel, J. C. and Waterbolk, H. T., 1967. Groningen Radiocarbon Dates VII. Radiocarbon, Vol. 9, pp. 107-155.

Vogel, J. C. and Waterbolk, H. T., 1972. Groningen Radiocarbon Dates X. Radiocarbon, Vol. 14, No. 1, pp. 6-110.

Vogel, J. C. and Zagwijn, W. H., 1967. Groningen Radiocarbon Dates VI. Radiocarbon, Vol. 9, No. I, pp. 63-106.

Wasylikowa, K., 1964. Roślinność i klimat późnego glacjalu w środkowej Polsce na podstawie badan w Witowie kolo Lęczycy, Biol. perygl., No. 13, pp. 261-417.

Waterbolk, H. T., 1962. The Lower Rhine Basin. In R. J. Braidwood and G. R. Willey (eds.), Courses toward Urban Life. Chicago: vide esp. pp. 227-253.

Weiss, G., 1974. Magdalenahöhle: Altsteinzeitliche Fundplätze des Rheinlands. Kunst und Altertum am Rheim, Nr 49, pp. 63-65.

Welin, E., Engstrand, L. and Vaczy, S., 1974. Institute of Geological Sciences Radiocarbon Dates V. Radiocarbon, Vol. 16, No. 1, pp. 95-104.

West, S. E., 1973. The Excavation of the Town Defences at Tayfen Road, Bury St. Edmunds, 1968. Proc. Suffolk Inst. of Archaeology, Vol. XXXII, pp. 17-24.

White, G. F., 1970. Excavation of the Dead Man's Cave, North Anston. Worksop Archaeological Research Group. Worksop Corporation: Public Library and Museum, pp. 9.

Williams, R. E. G. and Johnson, A. S., 1976. Birmingham University Radiocarbon DatesX. Radiocarbon, Vol. 18, No.3, pp. 249-267.

Winbolt, S. E., 1929. A Late Pleistocene Flint Point. Ant. J., Vol. IX, pp. 152-153.

Wood, R. H., Ashmead, P. and Mellars, P. A., 1970. First Report on the Archaeological Excavations at Kirkhead Cavern. North West Speleology I, pp. 19-24.

Woodcock, A. G., 1978. The Palaeolithic in Sussex. In P. L. Drewett (ed.), Archaeology in Sussex to A.D. 1500. C.B.A. Research Report, No. 29, vide esp. pp. 8-14.

Woodman, P. C., 1978. The Mesolithic in Ireland. B.A.R. (British Series), No. 58, pp. 360.

Wouters, Aq., 1955. Quelques Artéfactes faits de bois de Renne du Limbourg central. Palaeohistoria, Vol. IV, pp. 1-4.

Wouters, Aq., 1957. Een nieuwe vindplaats van de Ahrensburgcultur onder de gemeente Geldrop. Brabants Heem, Vol. 6, pp. 122-148.

Wymer, J. J., 1971. A possible Late Upper Palaeolithic Site at Cranwich, Norfolk. Norfolk Archaeology, Vol. XXXV, pp. 259-263.

Wymer, J. J., 1977. The Archaeology of man in the British Quaternary. In F. W. Shotton (ed.), British Quaternary Studies. Oxford. Vide esp. pp. 93-106.

Zotz, L. F., 1951. Altsteinzeitkunde Mitteleuropas. Stuttgart, Ferdinand Enke Verlag, pp. 1-VII and 290.

Postscript:

Correspondence with Rogan Jenkins about Anston Stone's Cave suggests that it may not be possible confidently to associate the three radiocarbon determinations (BM 439, 440A and 440B) with the specifically Federmesser artifacts from this site or with the pollen diagram constructed by J. B. Campbell from White's section I at the front of the Cave.

Rogan Jenkins' comments remove the necessity to distort interpretation of the pollen diagram (as in Jacobi (1978) 298-300). There seems no compelling reason therefore not to regard the pollen spectra counted from above and associated with the specifically Federmesser artifacts of section I as of L-De type. There is equally no necessity (as in Mellars (1974) pp. 74-75) to regard the radiocarbon determinations on the reindeer bones and antler as too young. As commented in the main body of the text, they agree well with the date of 7,970 bc ± 130 (on a massacred female reindeer antler) from Gough's (New) Cave. There would thus be no certain evidence for the survival of a Federmesser technology beyond the L-De/Flandrian boundary, since (1) the Federmesser type artifacts from section I at Anston Stone's Cave are regarded as of L-De (?L-De III) age while (2) the age of the other (Federmesser type) artifacts from the Cave remains unresolved but is not demonstrably post Devensian.

Table II has been amended to take account of these observations but it has not been possible to alter comments on Anston Stone's Cave in the body of the text.

CHAPTER THREE

ENVIRONMENTAL CHANGES IN WALES DURING THE HOLOCENE PERIOD

J. A. Taylor

Abstract

This essay attempts to interpret Holocene environmental changes (a) in integrated ecosystematic terms and (b) in relation to the cultural succession. Under (a) the well documented changes in vegetation and inferred variations in climate are interpreted in terms of attendant conditions of soil, hydrology, geomorphology and faunal populations. The inferential nature of the ecosystem models presented must be strongly emphasised. Their validity hinges on the assumption that available pollen-analytical records for Wales permit a reconstruction of interdependent environmental changes. They are clearly liable to future modification and interpretation as research progresses. Under (b) the archaeological periods defined within the context of this volume, and ranging from the Mesolithic to the Iron Age, have been adopted for environmental reference purposes, regardless of their lack of accord with the established sequence of 'Boreal' and 'Atlantic' climatic phases.

It is emerging that the impacts of past climate changes were expressed subcontinentally and regionally and over short-term time-scales rather than hemispherically or globally and over long-term time-scales. Wales, in this context, forms a distinctive region within the British Isles, and within Wales, the effect of altitude and position is to generate three subregions, i.e. the Borders to the east, the mountains and moorlands in the centre and the maritime lowlands to the west. Each sub-region, thus defined, could have responded differently and at different times to the same general trend or change in environment. Parallel deviations in the opportunities for, and constraints against, particular forms of prehistoric settlement and land-use must also have occurred. Thus, a 'human' component has been included in the ecosystem models.

Deviations from the 'Lowland Zone' model

Notwithstanding the now established view that significant and major environmental changes occur over much shorter scales in time and space than has long been assumed, nonetheless the simple fact remains that ten thousand years ago or, more precisely, about 8000 B.C., the last glacial period,

CHRONOLOGY OF ENVIRONMENTAL AND CULTURAL CHANGES IN BRITAIN SINCE CIRCA 12,000 B.P.

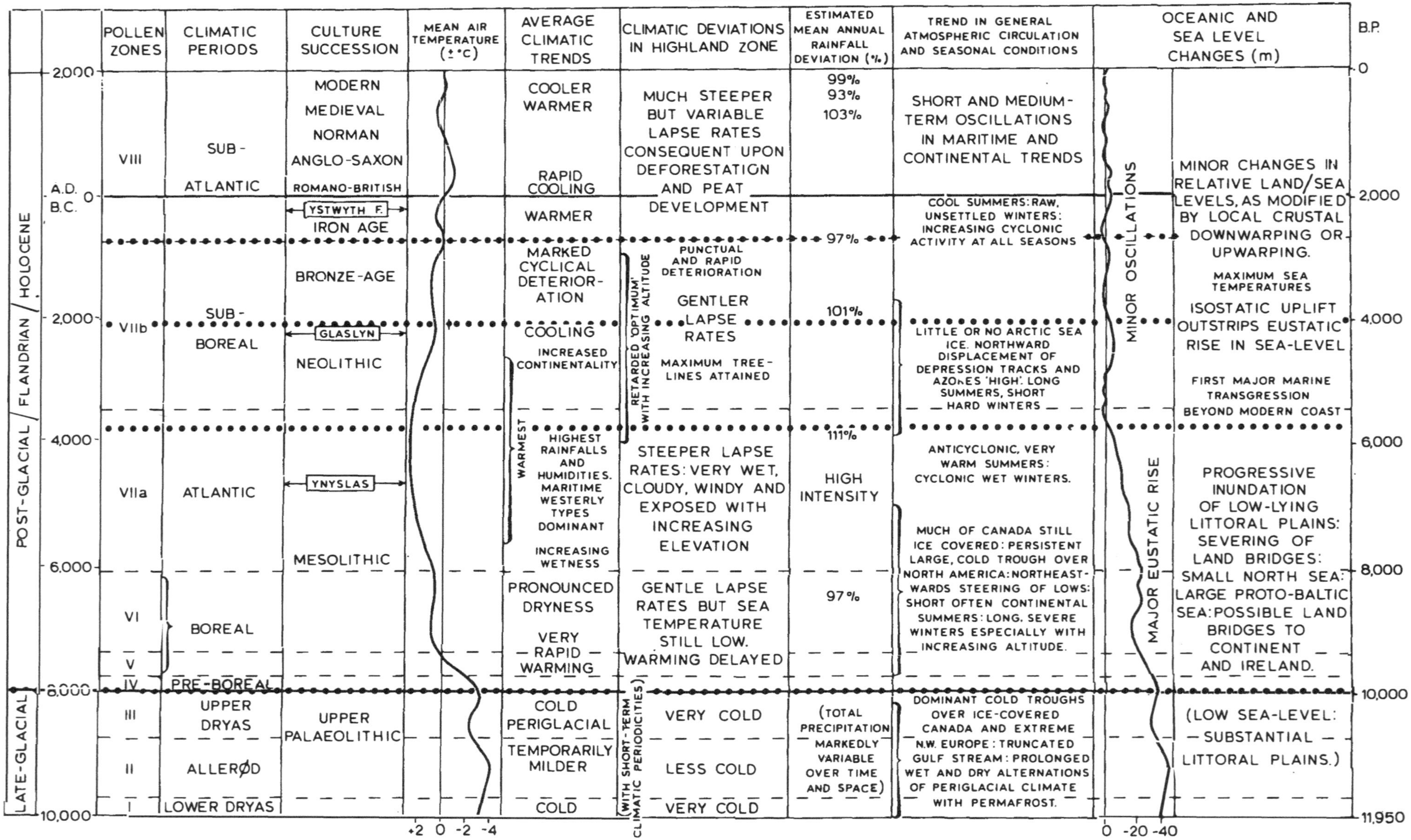

Fig. 3.1 Chart summarising environmental and cultural changes in Britain since circa 12, 000 B.P. (after Taylor, J. A. et al., 1975).

currently termed the Devensian for the British Isles, was finally terminated and, with the permanent establishment of a vegetation cover, eventually dominated by tree species, as the climate improved, marked the beginning of the Holocene period, alternatively termed the Flandrian or Post-glacial period. In archaeological terms, this threshold at 8000 B.C. marks the transition for the Upper Palaeolothic to the Mesolithic. It has survived a number of methodological re-appraisals of preceding and succeeding zonations. Notably, radiocarbon corrections are at a minimum at what is the junction between pollen Zones III and IV, and faunal evidence (Coope 1977; Evans 1972) has certainly corroborated the identity and timing of the cold Zone III, the final pulse of the last glacial period, whilst at the same time imposing new interpretations of Zones II and I as derived from pollen data alone (Bowen, herein) (Ch. 1).

The author elsewhere (Taylor 1975) has demonstrated the deviations in environmental changes which must have occurred in Highland Britain compared with Lowland Britain. Figure 3.1 presents an attempt to incorporate these deviations from the conventional model based on Godwin (1956 and 1975). First of all, research into dendrochronological (Fergusson 1970), Egytological (Michael and Ralph 1970) and radio-carbon dating techniques is indicating a more rapid climatic amelioration through pollen Zones IV, V, VI and early VIIa and an earlier commencement of the so-called 'Post-glacial optimum', by at least 500 years and possibly up to 700 years at about 6,300 B.C. or 8,300 B.P. This implies a longer Zone VIIa and a longer 'optimum' but with the sub-Boreal, Zone VIIb, starting somewhat earlier e.g. just before 3000 B.C. (5,000 B.P.) rather than just after that date. Independently, archaeological research is also proposing earlier initiations of both the Neolithic and Bronze Age periods (see Savory, Lynch and Burgess herein). Again, Smith (1970), Simmons (1975) and Jacobi, herein, have proposed that the Mesolithic man had a greater impact on vegetation and animal resources especially during the later part of the Mesolithic period which may now be identified with the earlier timing of the peak of the climatic optimum at 5000-5500 B.C. or 7000-7500 B.P. (Figure 3.1).

The deviations of Highland Zone environmental changes from those in the Lowland Zone embody five major principles. The first is that of <u>retardation</u> of colonisations north-westwards across Britain from continental sources to the southeast. Plants, animals and man were able to exploit direct land-access routes since the Straits of Dover did not exist in the early Holocene, and the southern part of the North Sea was also dry prior to the major eustatic rise in sea-level (see Figure 3.1). Expressed simply, increasing distance and direction northwestwards across England and into east Wales and southern Scotland meant later and slower colonisations. Secondly, the effect of increasing altitude on the Highland Zone fringes was to <u>enhance</u> this retardation e.g. in the hills of eastern Wales and the southern Pennines. The British climate displays, and always has displayed, an unusually rapid deterioration with altitude, expressed mainly in reduced temperatures, abbreviated growing seasons and increased rain and snowfalls (Taylor 1976).

Thirdly, however, and in marked contrast, the southwestern and western coastal lowlands of the British Isles are remarkable for their climatic earliness. This also applied, relatively, to past climates. Moreover, accessibility from direct southerly routes across western France and originally from the western Mediterranean and the Iberian peninsula, also promoted by the low sea-levels, enabled direct colonisations to western, especially south-western Britain from these sources. Much more accessible, however, were the one-time land bridges to Ireland from different parts of western Britain, not least Wales. Birks (1977) has proposed that very early colonisations of tree species in extreme western Scotland may be explained by the possible survival in refugia in the Western Isles of pre-Devensian flora. If the outer Hebridean islands fulfilled this role, then logically, western, especially south-western and extreme southern Ireland, together with extreme south-west Wales and the South-West Peninsula of England, could also have provided local refugia. This could explain some very early colonisations by hazel (Corylus) for example in coastal Ceredigion (Cardiganshire) in west central Wales, as will be discussed later.

Fourthly, the impact of both zonal and regional climate changes is subject to modification in timing and scale by strictly local conditions of topography, soil, flora and fauna, —and indeed human activities, which themselves may be instrumental in changing the climatic subsceptibilities of the natural environment e.g. by selective or large-scale deforestation. For example, slope and aspect of the ground, the sandiness or clayeyness of the soil, the presence or absence of peat-cover, the presence or absence of trees or woodland cover—all these factors may cause major local deviations in the impact of climatic variations. Moreover, pedogenic or biological factors may themselves be capable of initiating environmental changes within the ecosystem and independently of the climate (Smith and Taylor 1969; Taylor and Smith 1972).

Wales, as a major, compact peninsula on the west side of Great Britain is unique in combining a central geographical position with a simple topographic pattern of upland core and lowland peripheries on all sides, plus two major westwards peninsulas, those of the Llŷn and Pembrokeshire, which allow between them the generous but penetrating sweep of Cardigan Bay. Thus, Wales witnesses the maximum climatic confrontation between the intensely maritime, early, west-slope environments along that Bay and the more extreme, cold, late habitats of the Cambrian Mountains which are succeeding eastwards by distinctly less maritime, harder climates of the Welsh Borderland where the climatic deterioration with altitude is markedly less than west of the main watershed (Taylor 1975). Again, north Wales is exposed to the impact of cold northerly-derived Polar air, rendering Snowdonia very wet and, at altitude, sub-Arctic in regime, and north-east Wales, as a consequence, the snowiest part of the Principality with the hardest winters. In contrast, South Wales is accessible to warm tropical air flows from the south and especially south-west, which renders the tips of the Pembrokeshire peninsulas almost virtually frost-free and marginally sub-tropical in regime; at least it must have been so at the climatic optimum, as will be explained later. Thus, over short distances and over small increases in elevation or small changes in slope or soil, Wales includes, and always has included, sharp changes in environment—even by British standards.

Against this blackcloth of the varying zonal, regional and local scales of environmental changes, the general sequence of Welsh Holocene will now be summarised and interpreted, prior to a discussion of individual periods and phases.

The Chronology of the Holocene in Wales

The 10,000 years of the Holocene period in Wales may be subdivided into three broad stages; first, a warming phase of about 3,500 years till about 6500 B.C.; then an 'optimum' phase of similar duration of about 3,500 years till 3000 B.C.; and, finally, a cooling phase which has been maintained for almost 5,000 years up to the present-day. This major climatic cycle (which, alternatively, Smith (1970) has suggested may merely be nothing more than a natural vegetational succession, ipso facto), already asymmetrical over time, included many medium- and short-term oscillations over the centuries and decades, respectively, and occasional reversals, some major, e.g. the Medieval warming of the eleventh and twelfth centuries, some minor, e.g. the mild phase occupying the first forty years of this century. Dendrochronological research will eventually enable the identification of similar reversals for the prehistoric climatic record (Pearson, et al 1977).

The warming phase and the earlier half of the 'optimum' phase are synchronous with pollen Zones IV, V, VII and the first half of VIIa: this also represents the broad time span of the Mesolithic settlement of Wales. The second half of the 'optimum' phase coincides with the second half of pollen Zone VIIa and with the transition to, and establishment of, the Neolithic culture in Wales. The third cooling phase embraces pollen Zones VIIb and VIII and extends from the Neolithic/Early Bronze Age transition and through the remainder of the Bronze Age and the Iron Age into the historic period. Thus, in broad terms, the period of climatic deterioration over the past 5,000 years or so has been accompanied by an increasing facility for prehistoric man to interfere with, and generally deplete, the environment as well. Conversely, during the 5,000 years from 8000 B.C. to 3000 B.C. the climatic warming and 'optimum' were broadly coincident with the far less abrasive Mesolithic culture. Therefore, flora, fauna and soils were allowed to maintain themselves within a relatively stable ecosystem which was in reasonable balance with man's contemporary needs. From Neolithic times onwards this balance was to be upset but it is of especial significance to add that Pearsall's (1952 and 1965) original hypothesis, subsequently corroborated by Smith and Taylor (1969), provides evidence for a process of natural soil degradation, especially under persistent birchwood cover, to have taken place during the later Mesolithic and Neolithic periods such that Bronze Age man had to cope with a secondary type of woodland. Early Boreal soils may be envisaged as having high nutrient status and high pH, e.g. of the order of 6.0, compared with the present-day. Soil profiles were very probably of the brown earth variety, perhaps of the brown podzolic in the wetter uplands. The variably brown (often yellow-orange-brown) 'B' horizons of Welsh upland podzols today may be interpreted as residual horizons of these essentially pre-Neolithic soils. These horizons yield pH's today of about 5.0 or rather less (in contrast to the values of 4.0 or less for the leached and acidic organic horizons above) which indicates the extent to which soil degradation has proceeded.

Climatically, the major amelioration included a series of phases making up the Boreal period. The successive warmings of climate, at first very rapid then much slower, were dominated by a continental type of regime with relatively low precipitation, low humidity, harder winters, sunnier summers and reduced windiness. This contrasts with the succeeding Atlantic period,

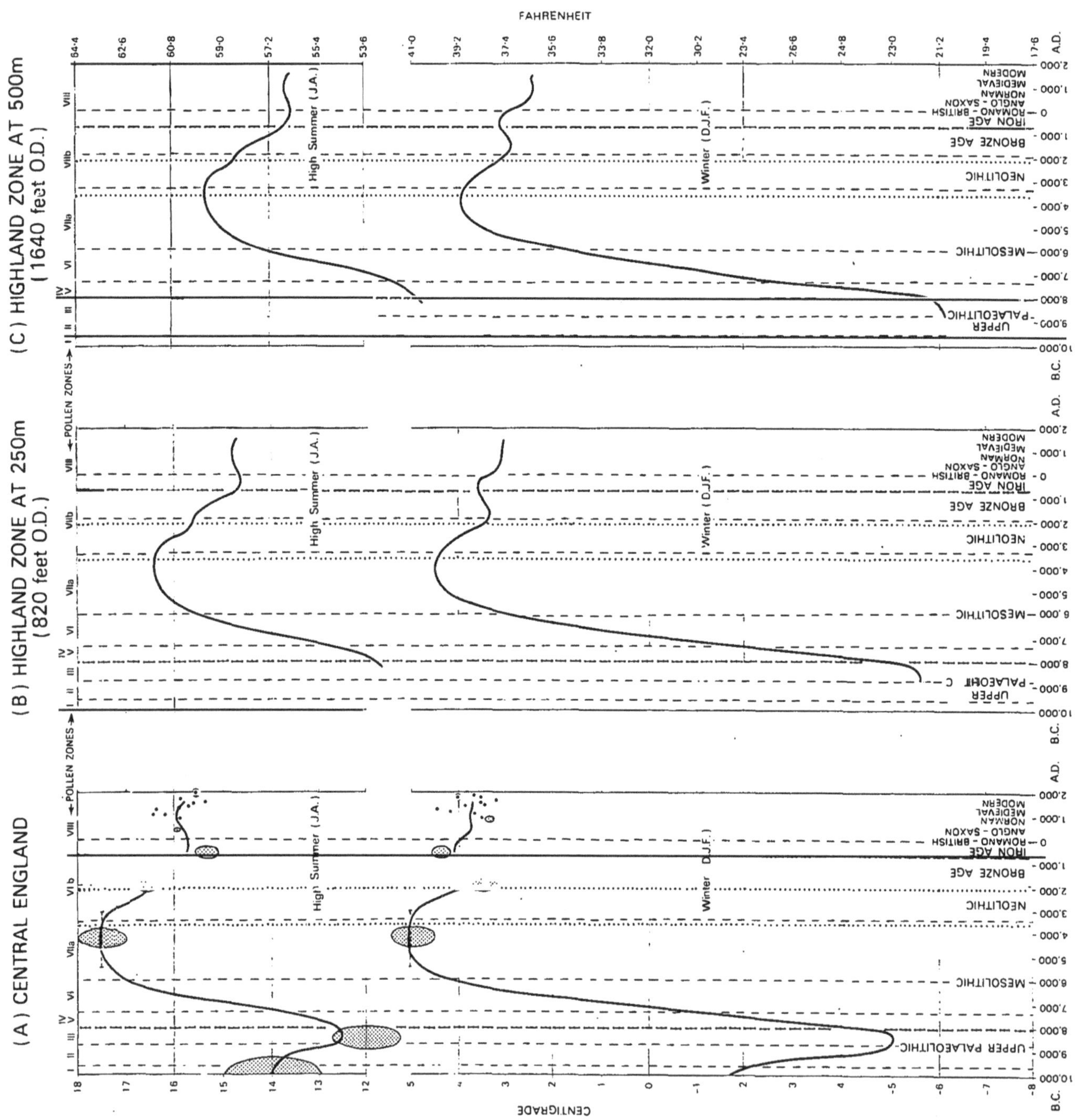
(A) CENTRAL ENGLAND
(B) HIGHLAND ZONE AT 250m
(820 feet O.D.)
(C) HIGHLAND ZONE AT 500m
(1640 feet O.D.)
POLLEN ZONES
FAHRENHEIT
CENTIGRADE
High Summer (J.A.)
Winter (D.J.F.)
B.C.
A.D.
UPPER PALAEOLITHIC
MESOLITHIC
NEOLITHIC
BRONZE AGE
IRON AGE
ROMANO - BRITISH
ANGLO - SAXON
NORMAN
MEDIEVAL
MODERN

Fig. 3.2 (opposite) Average summer and winter mean air temperature curves since _circa_ 12,000 B.P. (a) in the Welsh Borderland, (b) in Wales at _circa_ 250 m (820'O.D.) and (c) in Wales at _circa_ 500 m (1640'O.D.). (After Taylor, J. A. _et al._, 1975).

Note (1) The temperature curves in (A) show the course of the _supposed_ (or last millenium calculated) 1000 year averages. The dots in (A) refer to individual century averages. The dots in brackets refer to the period 1900-1965 (66 years). The shaded ovals indicate the approximate ranges within which estimates of temperature lie at δ 2 error margins of dating (carbon-14 method) on either side of the most probable date. The horizontal lines through the ovals indicate the supposed duration of the epoch to which the 'observed' mean value applies (after Lamb _et al._, 1966).

Note (2) The corrections in (B) and (C) are based on the estimated variations in the range of temperature lapse-rates with altitude as synthesised by Harrison (1974) for available data for Britain, Europe and comparable environments. These range between 5.6^{o} C and 7.6^{o} C per 1000 m and have been adjusted appropriately for 'Boreal' and 'Atlantic' phases, respectively (after Taylor, 1975).

Date (A.D./B.C.)	CLIMATE	POLLEN ZONES	CULTURE PERIODS	(a) CENTRAL ENGLAND S	(a) CENTRAL ENGLAND W	(b) WELSH BORDERLAND S	(b) WELSH BORDERLAND W	(c) CARDS. BAY COASTLANDS S	(c) CARDS. BAY COASTLANDS W	(d) AT SEA-LEVEL S	(d) AT SEA-LEVEL W	(e) AT 250m (820' O.D.) S	(e) AT 250m (820' O.D.) W	(f) AT 500m (1640' O.D.) S	(f) AT 500m (1640' O.D.) W
										IN WALES, *ceteris paribus*					
1000 A.D.	SUB-ATLANTIC	VIII	DARK AGES												
0			ROMANO-BRITISH	15.7 (60.3)	4.0 (39.2)	14.7 (58.5)	4.4 (40.0)	14.3 (57.7)	5.2 (41.3)	15.6 (60.1)	4.0 (39.2)	14.1 (57.4)	3.5 (38.3)	13.5 (56.3)	2.9 (37.2)
			IRON AGE	15.5 (59.9)	4.1 (39.4)	15.1 (59.1)	4.6 (40.2)	14.6 (58.3)	5.3 (41.5)	16.0 (60.8)	4.1 (39.4)	13.7 (56.7)	3.6 (38.5)	13.5 (56.3)	3.0 (37.4)
				15.6 (60.1)	4.2 (39.6)	15.2 (59.3)	4.6 (40.3)	14.8 (58.6)	5.3 (41.6)	16.5 (61.7)	4.2 (39.6)	14.7 (58.5)	3.7 (38.7)	13.6 (56.5)	3.1 (37.6)
1000 B.C.	SUB-BOREAL	VIIb	L.B.A. / M.B.A.	L.B.A. / IRON AGE TRANSITION **1200 B.C. Bronze Age 'Crisis Point'**											
			E.B.A.	16.4 (61.5)	3.7 (38.7)	16.0 (60.8)	4.3 (39.7)	16.1 (61.0)	5.0 (41.0)	17.3 (63.1)	3.9 (39.0)	15.9 (60.6)	3.3 (37.9)	14.6 (58.3)	2.8 (37.0)
2000				NEOLITHIC / BRONZE AGE TRANSITION **2500 B.C. Neolithic 'Crisis Point'**											
3000	ATLANTIC	VIIa	NEO-LITHIC	17.4 (63.3)	5.0 (41.0)	17.0 (62.6)	5.4 (41.8)	16.6 (61.8)	6.2 (43.1)	17.4 (63.3)	5.3 (41.5)	16.3 (61.3)	4.5 (40.1)	15.3 (59.5)	3.8 (38.8)
4000				MESOLITHIC / NEOLITHIC TRANSITION **4350 B.C. Mesolithic 'Crisis Point'**											
5000			OPTIMUM / MESOLITHIC	17.6 (63.7)	5.2 (41.4)	17.2 (63.0)	5.7 (42.2)	16.8 (62.2)	6.4 (43.5)	17.5 (63.5)	5.4 (41.7)	16.4 (61.5)	4.6 (40.3)	15.3 (59.5)	3.8 (38.8)
6000	BOREAL	VIa		17.0 (62.6)	4.3 (39.7)	16.6 (61.9)	4.7 (40.4)	16.2 (61.2)	5.4 (41.7)	17.0 (62.6)	4.6 (40.3)	15.6 (60.1)	3.2 (37.8)	14.2 (57.6)	1.7 (35.1)
7000		VIb / VIc / V		13.7 (56.7)	−1.0 (30.2)	13.3 (56.0)	−0.6 (31.0)	12.9 (55.2)	0.2 (32.3)	13.5 (56.3)	−1.5 (29.3)	12.5 (54.5)	−2.1 (28.2)	11.5 (52.7)	−2.7 (27.1)
8000		IV		12.6 (54.7)	−4.5 (23.9)	13.1 (55.5)	−4.9 (23.2)	13.4 (56.2)	−4.2 (24.5)	12.4 (54.3)	−5.1 (23.0)	11.6 (52.9)	5.4 (22.3)	10.8 (51.4)	−5.8 (21.6)
AVERAGE MEAN 1901 – 1930				16.4 (61.5)	4.6 (40.3)	16.0 (60.8)	5.0 (41.0)	15.4 (59.7)	5.9 (42.6)	15.4 (59.8)	5.7 (42.2)	14.4 (58.0)	4.7 (40.4)	13.4 (56.2)	3.7 (38.6)

Fig. 3.3 (opposite) Chart showing estimated mean average summer and winter temperatures since circa 12,000 B.P. for (a) Central England, (b) the Welsh Borderland, (c) the Cardigan Bay Lowlands and (d) for sea-level, 250 m (820' O.D.) and 500 m (1640' O.D.) ceteris paribus in Wales. (N.B. Summer data refer to means for July/August, and winter data refer to means for December/January/February: see Figure 3.2).

identified with 'optimum' phase, when a more maritime and warmer climate persisted with an abundance of wet, windy weather at all seasons. The third phase of cooling climate inaugurated a second continental type of climate in the Sub-Boreal period but temperatures were lower generally than in the early-Holocene Boreal period.

In general terms, it may be assumed that during the two continental periods of the Holocene climatic succession, the lapse rates of temperature with altitude were consistently smaller than those operating during the two maritime periods i.e. the Atlantic just referred to above and the more recent Sub-Atlantic phase which began about 600 B. C. and is still in operation, so far as can be judged. The implication of this contrast for growing seasons, tree-limits, settlement limits and agricultural limits is simply that altitudinal constraints were likely to have been more severe at the warmest 'optimum' phase than for either the preceding or succeeding cooler and continental periods. It is possible then that the highest altitudinal ceilings for tree growth, settlement and agriculture were attained either after, or even before the 'optimum' in environments such as those of upland Wales.

In fact, three broad but discrete regional environments in Wales may be differentiated in terms of their bioclimates and their individual responses to the same climate change. They are, firstly, the Welsh Borderland which stretches in modern reference terms, from the Dee estuary to the coast of Gwent and includes eastern Clwyd, eastern Powys, Gwent and the adjacent portions of the English counties of Cheshire, Salop and Herefordshire. This zone grades from clay-covered, rain-shadowed lowlands in the east, which were directly accessible to colonisation from the south-east, and developed dense oak-alder woodlands at the climatic optimum which had to await the Romans before through-access routes were established. Wide valleys penetrate deeply westwards e.g. the Severn, Wye and Dee, thus extending this environment into central Wales. This borderland possesses the least maritime, i.e. the most continental climate, of the Principality—and it always has had presumably at all previous stages of the Holocene. Its slopes, therefore in addition to the effect of their broadly easterly aspect, present relatively small lapse rates and surprisingly good ecological, agricultural and settlement potentialities even above 1000' O.D. (305 m) (see Taylor 1976).

The second zone is the mass of the Welsh uplands, lying roughly north-south across central Wales from Snowdonia and the Hiraethog Moors to the Brecon Beacons and the Preseli Hills. This series of high plateaux is so extensive and so compact as to have imposed its own timing and expression of individual climatic trends. In the British context, uplands respond late to a warming climate but early to a cooling climate. Thus, the Welsh moorlands were climatically retarded in the early Holocene but climatically advanced in the late Holocene and during the historic period. The argument applies equally to 'late' sites or soils, e.g. peaty, wet soils, as compared with 'early' sites or soils, e.g. sandy, well-drained ones. The effect of the development of the peat cover in upland Britain was clearly to reinforce this mesoclimatic differentiation between upland and lowland, as will be expanded upon later.

The third zone, which may be termed the Welsh 'Maritime' or 'Atlantic' Wales (Fleure 1951; Bowen 1959; Evans 1975), includes Anglesey, coastal Gwynydd, including the Llŷn Peninsula, coastal, and especially extreme southwest, Dyfed, i.e. the Pembrokeshire peninsula, the lower Vale of Towy, the Gower and the extreme southern fringe of the Vale of Glamorgan. This is the mesoclimatically 'early' region of the Principality which would respond early to a climatic amelioration but which is backed inland by one of the most rapid temperature lapse rates, per unit altitude rise, in the world, —as it merges quite rapidly into the upland zone two described above. This zone is clearly accessible to, and orientated to, the southwest and west whence came many of its early Holocene colonisations.

Within this three-fold subdivision, local topography and soil etc., impose further variation, sometimes enhancing, sometimes reducing, the broad contrasts and potentialities. The details of the environmental succession in Wales throughout the Holocene will now be presented with reference to each of the established archaeological stages.

Early Mesolithic environments (Fig. 3.5)

Prior to Early Mesolithic colonisations in Wales about 6700 B.C. in pollen Zone VIb (see Jacobi, Ch. 2, herein), the climate and soil had been fully emancipated from the cold grip of Zone III to allow a full 'Boreal' forest cover to develop, However, although the temperatures rose quite rapidly through Zones IV and V, soil development was slower, and, as previously explained, the response in the uplands was retarded. For example, the periglacial conditions of Zone III probably persisted into Zone IV, especially in the form of persistently severe winters and repetitively short summers (see Figs. 3.2 and 3.3). In contrast, extreme southwest Dyfed would have been the earliest part of Wales, with its then extended coastal plains, including the one-time plain of Cardigan (Taylor 1973), to respond to the rising air temperatures and, precisely off these coastlands, slowly rising sea temperatures. Thus, the scene was set at the dawning of the Holocene for the sharp dichotomy between responsive coastal lowlands and cold, sluggish uplands.

Moore (1970) and Handa and Moore (1976) adopt the Juniper curve as an indicator of the climatic transition from Late-glacial to early Holocene. The beginning of the Holocene is revealed in the expansion of Juniperus, notably at the Cledlyn pingo sites (Handa and Moore, 1976) and at Elan Valley (Moore, 1970) and for two sites, Glenallt and Llewsig, near Capel Curig (Crabtree, 1972) where the Juniper peak is allocated to a discrete Zone III/IV transition. At other sites Zone IV is marked by the survival and then the demise of Juniper as extensive birch woods, with hazel and local pine, willow and alder stands, expanded (e.g. Tregaron, (Hibbert and Switzur, 1976); Clarach (Taylor, 1973); and Gwarllyn (Moore and Chater, 1969 and Moore, 1972). However, at Aberaeron (Taylor, 1973) Juniperus pollen is conspicuous by its absence from a continuous Late-glacial/Holocene stratigraphy, and at Clarach is peaked in Zone III and was well on the decline in Zone IV. These two sites on the modern coastline reflect very early Holocene conditions at the change of slope of the plain of Cardigan which must have been very wide at this early stage. Excessive ground wetness and continually high atmospheric moisture levels at the change of slope may explain the relative rarity of Juniperus. Walker (1966) also reports lower values for Juniperus for lowland sites in Cumberland compared with adjacent upland sites.

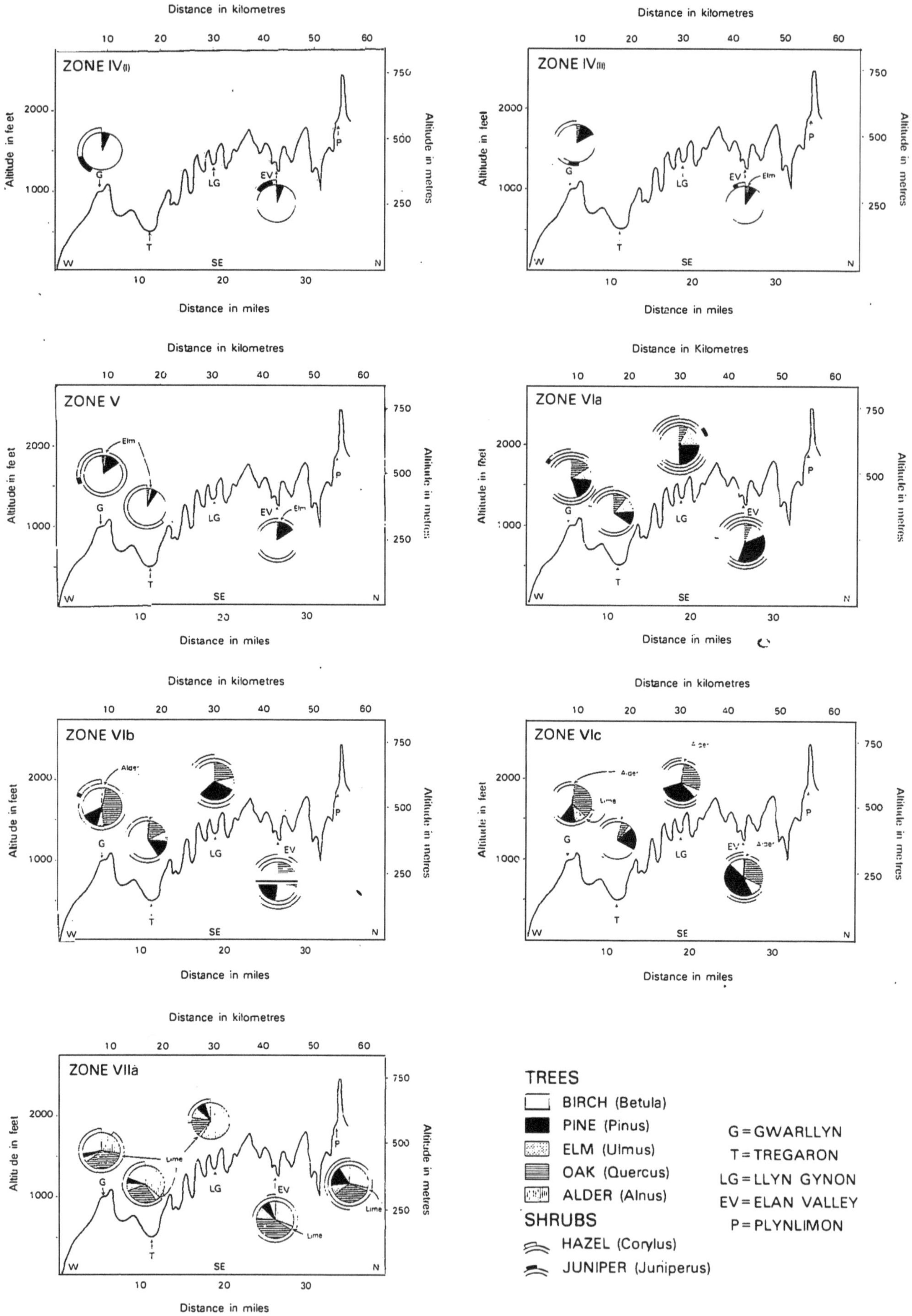

Fig. 3.4 Forest composition along an altitudinal transect in West Central Wales during Pollen Zones IV-VIIa, inclusive: (adapted from Moore, P. D. (1972)).

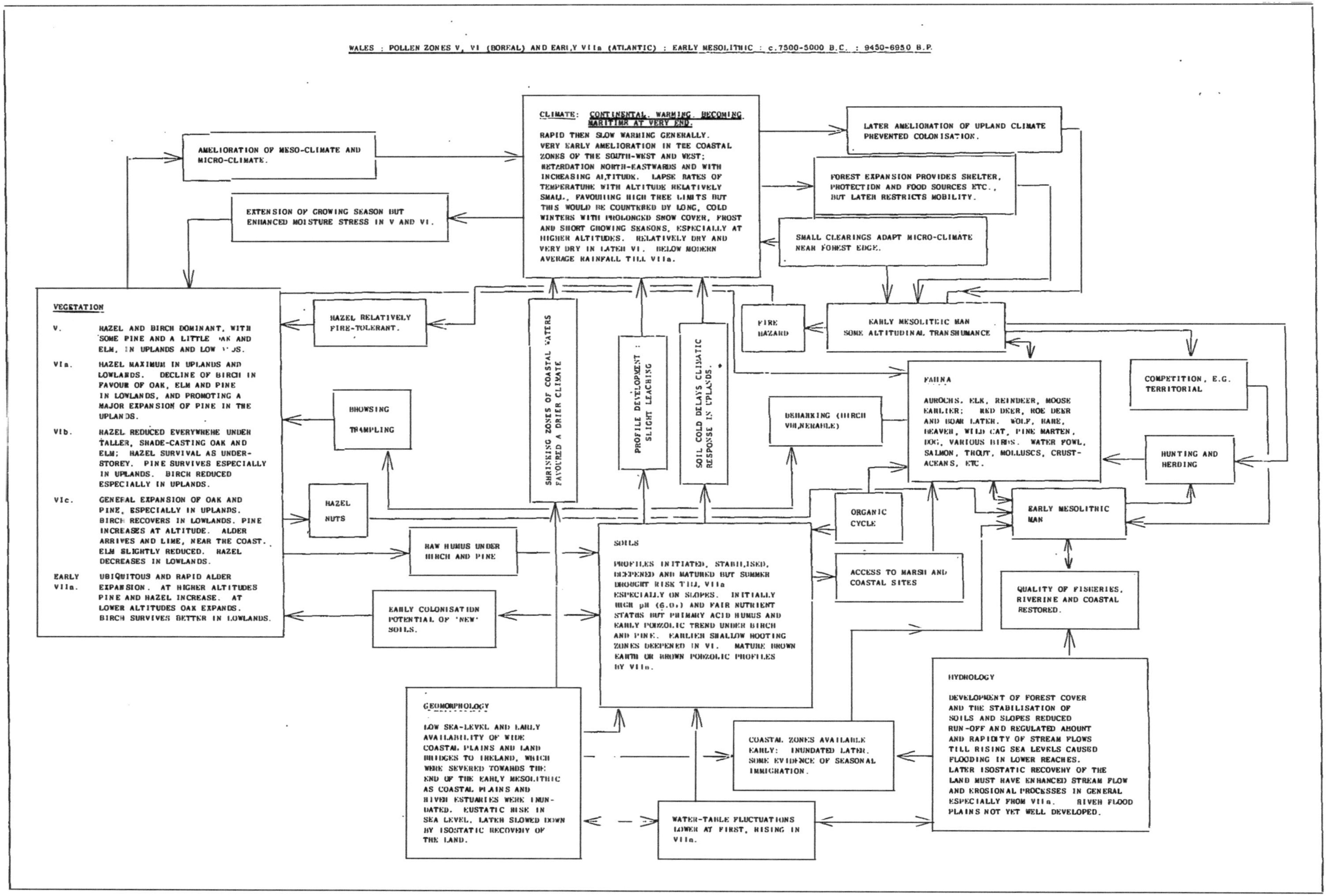

Fig. 3.5 Ecosystem model for the Early Mesolithic

The establishment of Zone IV, however, is witnessed in a rapid and early expansion of threes species, notably birch, hazel, pine and willow. The radiocarbon date of 9,747 BP for the arrival of Corylus at Tregaron (Hibbert and Switzur, 1976) is one of the earliest for Britain, and is identified with Zone IV and timed only about three centuries after the start of the Holocene. At Gwarllyn, some 312m (1030 'O.D.) and located in the hilly area of Mynydd Bach west of Tregaron Bog and generally south of Aberystwyth, the Corylus curve starts to rise before the Juniperus peak (Moore, 1972 and see Fig. 3.4). This indicates that hazel had ascended the coastal plateaux just east of the modern coastline of Cardigan Bay almost 10,000 years ago at the Zone III/IV transition. The explanation lies, firstly, in the availability westward of the modern coast of the one-time plain of Cardigan (Taylor, 1973), with probable refugia enabling pre-Devensian flora to survive in, and colonise eastwards from, the west, and, sec ondly, in the exceptional early climatic response of that plain and what are now the adjacent coastal zones of west and south-west Wales. Now at the Cledlyn pingos (Handa and Moore, 1976), Nant Ffrancon and Llyn Dwythwch, Caernarvonshire (Seddon, 1962; Hibbert and Switzur, 1976), the Corylus rise occurs immediately after the fall in Juniperus. However, at Elan Valley on the major north-south watershed of the Central Wales uplands at 384 m (1260 'O.D.) (Moore and Chater, 1969; Moore, 1970), and in a lake-basin at 257 m (843' O.D.) north of Capel Curig, Snowdonia, (Crabtree, 1972) the rise in Corylus does not occur till Juniperus had finally declined and disappeared in late Zone IV and early Zone V.

These data, taken together, enable the migration pattern of Corylus to be constructed. From the central and exposed plain of Cardigan, the pollen rain, heavily charged, with Corylus, must have been carried eastwards, north-eastwards and south-eastwards by the dominant winds but was delayed somewhat at higher altitudes. Moore (1972) discusses the further migrations northwards to the Lake District and Scotland and north-eastwards into the Cheshire plain (see also Deacon, 1974). Clearly, the one-time plain of Cardigan, being intensely maritime, was less liable to diurnal and seasonal temperature extremes and had earlier, longer summers and a reduced frost risk. This environment favoured Corylus in its competition with Betula and Juniperus.

By late Zone IV, however, some oak had arrived on Mynydd Bach and at Elan Valley (see Figure 3.4) where elm too was recorded for the first time. Pine had increased slightly since early Zone IV (Figure 3.4) but was probably adapted to localised sites e.g. littoral sandy stretches, frost hollows or exposed sites at altitude. Again, by Zone V (Figure 3.4) pine was still a minor component but with Corylus and Betula expanding and dominating. Oak and elm were present in small quantities with the latter expanding, indicating the continuing improvement in climate and the increasing availability of new, nutrient-rich soils.

Zone VIa witnessed the climax of the Corylus expansion in the uplands as well as the lowlands, as the upland growing season finally lengthened to about its present-day dimensions and winter severity eased again to about present-day standards. At all sites birch had declined in favour of oak, elm and pine in the lowlands, and in favour of a major expansion of pine in the uplands. It was this combination of mixed hazel-birch woodlands in the lowlands and

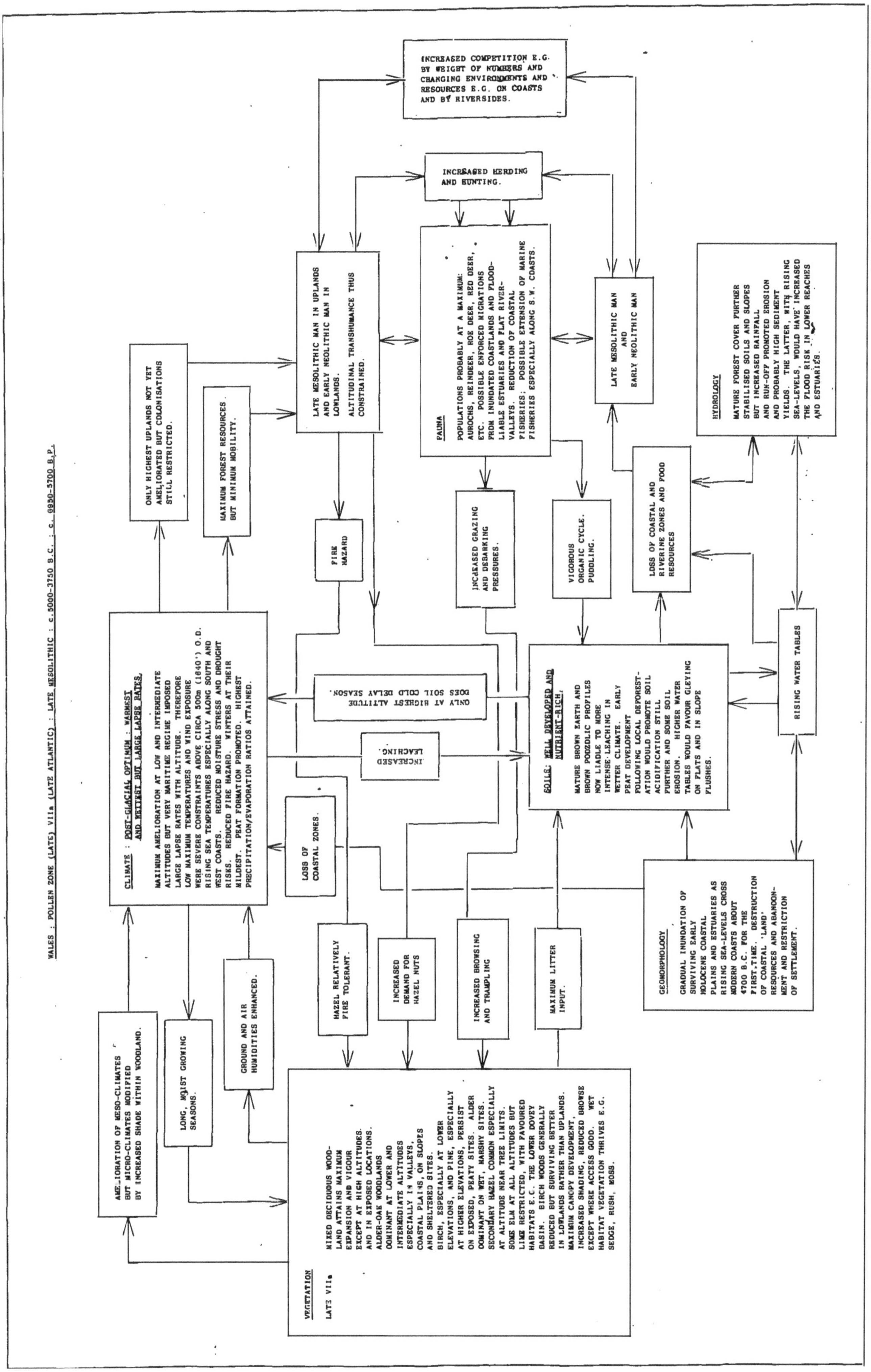

Fig. 3.6 Ecosystem model for the Late Mesolithic

mixed birch-pine woodlands in the uplands which greeted Early Mesolithic man in Wales between about 6700 and 5500 B.C. Both formations would have presented canopies of moderate density, allowing adequate light penetration for ground vegetation to develop, providing browse and also access for animals and man. In Figure 3.5 an attempt is made to reconstruct the entire natural ecosystem of the Early Mesolithic period in Wales and to depict the role of man within it. The individual physical and biological components are conveniently boxed but it is the interlinkages between and among them, basic energy conversions or flows, which, expressed in the form of arrows, are the functional network of the ecosystem. For example, it is the degree of penetrability of the forest canopy which allows the understorey and ground flora development which provides that combination of browse and access for the fauna to be hunted or herded by Early Mesolithic man. Again, the quality of river and coastal fisheries is dependent on a combination of hydrological, climatic, vegetational and pedological factors and the relative changes in land/sea levels. However, it was the early extensive, fire-resistant hazel woods which provided the major habitat for the Welsh Early Mesolithic settlements, concentrated, in the light of current knowledge, in coastal sites in south-west Wales (Jacobi Ch. III herein). Abundant and varied fauna provided food for early Mesolithic man, on woodland edge, river-bank and foreshore alike. It is probable that heavy browsing and debarking by aurochs and deer in particular exerted local control on forest expansion and may well on occasions have forced the forest to retreat, especially when systems of herding and semi-domestication were introduced (Simmons, 1975). The burning of the forest edges to improve browsing resources and also to promote hazel growth for nuts for human consumption was probably quite widespread. However, the general impact on the woodland cover, with maturing soils and a warming climate, was minor compared with later stages. As the Holocene climatic optimum approached about 6000 B.C. the warm but wet Atlantic climatic phase became established, and the scene was set for the Late Mesolithic culture to develop.

The Late Mesolithic (Fig. 3.6)

The environmental and cultural relationships are expressed ecosystematically in Figure 3.6. Although Smith (1972) has pointed out the forest succession through Zones IV to VIIa of juniper-hazel-birch-pine-oak-elm-alder is as much a natural sequence as necessarily a consequence of climatic amelioration, it is clear that any valid interpretation of events must involve both hypotheses. However, by far the major climatic change in VIIa was the onset of much wetter but windier conditions which, in combination with warmer summers and milder winters, induced exceptional growth of mixed alder-oak woods with elm, occasional lime and ash. These taller trees with a thicker canopy outshaded hazel and birch, which, however, survived substantially, the former significantly in the uplands (where canopies were that bit less dense), and the latter, maintaining its continuity in the lowlands (see Figure 3.4) but merely as a secondary species. Pine, too, survived better, as a minor species in the uplands, especially in the more sterile and more exposed locations where competition with other tree species was least.

It is a moot point whether the Welsh Holocene woodlands attained their greatest altitudinal extent during the climatic optimum. This was a time of intensive maritime climate and rapid lapse rates of temperature, and therefore of growing season, with increasing elevation. Even allowing for the generally increased temperature levels at all seasons during the Atlantic period it is possible that the slow lapse rates of temperature operating during the preceding Boreal period and/or the succeeding sub-Boreal period enabled trees to reach higher altitudes than in the Atlantic. Moreover, trees like birch, pine and hazel during the Boreal were colonising, initially with little competition, soils at altitude which had, up to that stage of the Holocene, supported only an Arctic-alpine of 'maritime-tundra' type of vegetation. The temperature regime, then as now, must have been akin to sub-Arctic near the summits but the rainfall regime was quite wet, although not as wet as now when rainfalls above 610 m (2000' O.D.) are greater than 2,540 mm (100"). The wind factor, a major constraint against tree growth, would have been reduced during the continental Boreal and Sub-Boreal than during the intervening Atlantic. During the Sub-Boreal phase, however, temperatures were falling and Bronze Age man was increasingly operative along woodland margins and in clearings and in direct deforestation. In contrast, the Zone VI woodlands had indeed developed progressively with minimal interference from Early Mesolithic man for as long as 2000 years at least from very early colonisation, especially in South and West Wales. This implies a natural capacity to expand to the absolute altitudinal limits which must have approximated to the 610 m ± 61 m contour ± (2000' ± 200') depending on aspect, slope, soil etc. The Forestry Commission today have discovered that planted Sitka spruce (*Picea sitchensis*) will survive and grow very slowly at 2200' (671 m) on the lee side of Plynlimon summit but *not* higher. In contrast, the celing for commercial timber growth approximates to the 518 m (1700' O.D.) which is closely identified with the modern lower limit of Arctic-alpine flora and of zones where bilberry (*Vaccinium* spp) has become dominant. Environmentally and espcially climatically, these areas are the most extreme of the Welsh uplands, and they always were.

With the maximum development and density of the mid-Holocene woodlands having been attained in the late Mesolithic period, their penetrability was reduced especially at lower elevations and on lower slopes. Moreover, the eustatic rise in sea-level crossed modern coastlines about 4700 B.C. and flooded coastal and estuarine sites. Late Mesolithic man, therefore, must have been constrained to concentrate on upland locations and tolerate a restricted system of transhumance. Eventually, the lowlands witnessed the establishment of early Neolithic settlements, which confined the Mesolithic peoples to the hills where the highest ground suffered the increased exposure of the maritime climate (Taylor, 1975). All this implies increased competition for the best habitats and increased 'internal' pressure, both environmental and cultural, on land, floral and faunal resources as human populations, Mesolithic and Neolithic, increased quite sharply. The trend in soil development and the hydrological stage reached were both critical in that any extensive local deforestation would increase the risk of erosion, enhanced run-off, flash floods and reduced water quality due to high sediment yield. This would have put riverine and coastal fisheries at some increased risk, however temporary.

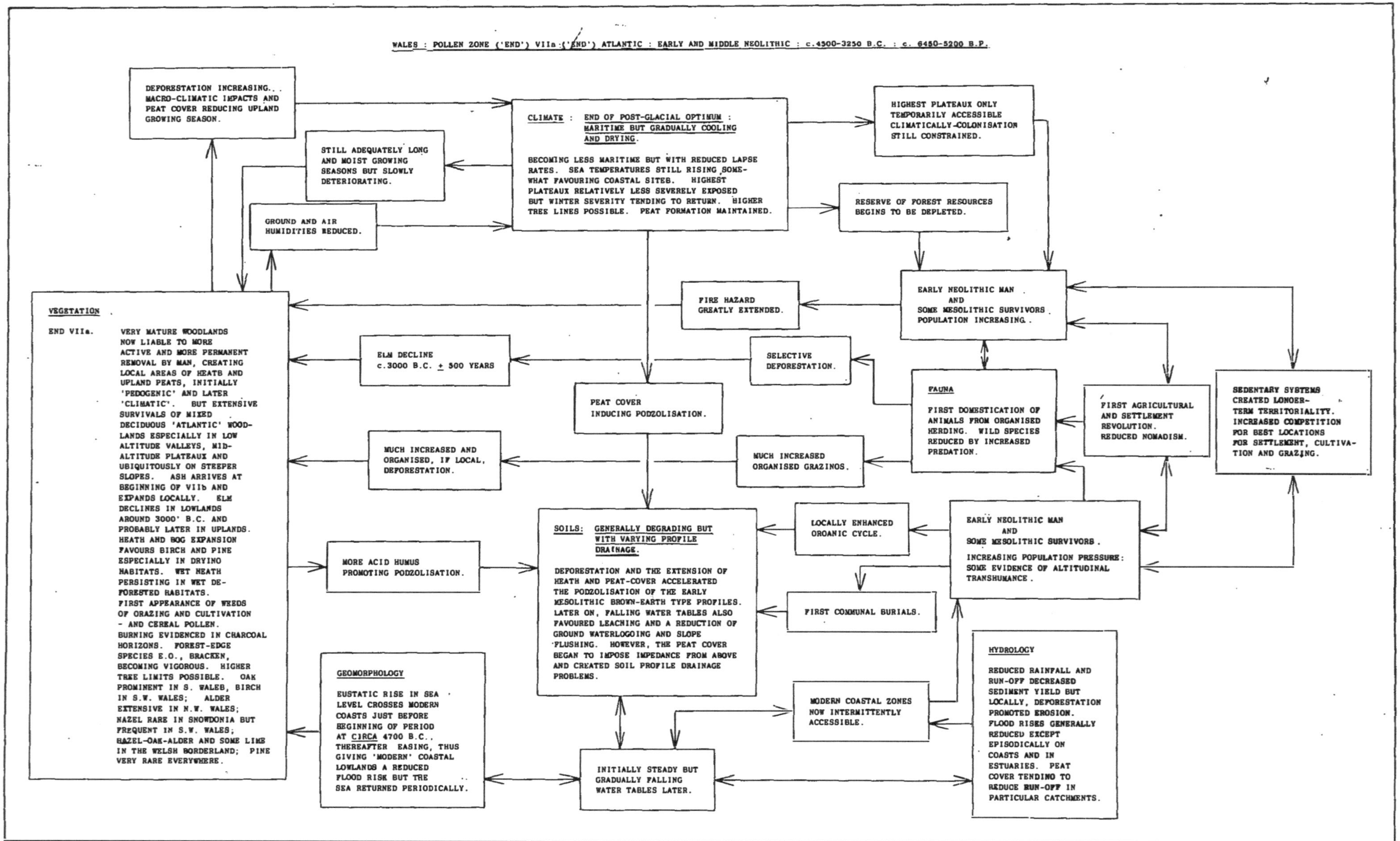

Fig. 3.7 Ecosystem model for the Early and Middle Neolithic

Early peat development in the uplands also dates from these early clearances as brown-podzolic profiles became more exposed to, and leached by, the direct impact of a wet, and at altitude a rather cool, 'Atlantic' climate. Ground wetness must have persisted and spread, making seasons later and shorter at ground, and narrowing the options available for settlement and herding. But both environment and culture at about 3750 B.C. were poised for changes of direction and some of these changes were fundamental and, later, revolutionary in nature. The sedentary Neolithic culture was to replace the Mesolithic nomadism and the climate was to take a downward turn, first cooler, then drier, as we enter the next stage.

The Early and Middle Neolithic (Fig. 3.7)

The arrival of Neolithic man marked the first sedentary forms of settlement and the first major domestications of animals and plants which prepared the way for the first permanent agriculture. This implied longer-term territorial occupation and specialisation and more drastic manipulation of the environment in the more intensely used areas. In the uplands, deforested zones began to receive the impact of a cooling climate which, in the uplands, remained wet enough to promote peat formation and soil leaching, although it became generally drier in the Late Neolithic, the stage to be considered next. The range and numbers of wild animals were reduced by increased and often selective predation.

Isostatic recovery of the land was now in the ascendant over the eustatic sea-level rise, but a number of short-term oscillations appear to have occurred (Tooley, 1974) but it is likely that these were local rather than national in occurrence and mainly affected those coasts with flattish hinterlands and gently sloping foreshores. Here some flood risk would have been operational. Otherwise, the hydrological balance between a reducing rainfall, shrinking forest cover and an expanding peat cover was probably reasonably well maintained with only occasional hazards (flash floods, slope instability) affecting lower reaches and upstream sections, respectively.

On the whole, then, Neolithic man inherited an environment which was only just beginning to be depleted, and its resources initially met most of his requirements. But the stage was set for an accelerated depletion of the environment, especially where local pressure was severe and repetitive, and as the climate continued to cool. The state of the vegetation cover about 5000 years ago has been recreated in a series of maps compiled by Birks et al., 1975 (Fig. 3.8). Using the available, and admittedly unevenly distributed, selections of radio-carbon dated pollen sites, isopollen lines have been extrapolated, somewhat bravely, for percentage values of total tree pollen for seven major species, i.e. pine, birch, oak, elm, alder, hazel and lime. The absence of pine is remarkable and characteristic for Welsh sites, and regionally, the extensive alder woods with subsidiary birch and oak in northern Wales are in contrast to the birch - hazel - oak dominance in southern Wales. North-west Wales is under-represented in terms of hazel and extreme South-west Wales in terms of alder. The Welsh Borderland is represented by hazel - oak - alder and some lime in the south grading into more dominantly alder - oak - hazel and some birch in the north. Only in West Central Wales did elm occur and the lime is concentrated in East Central Wales. It must be emphasised

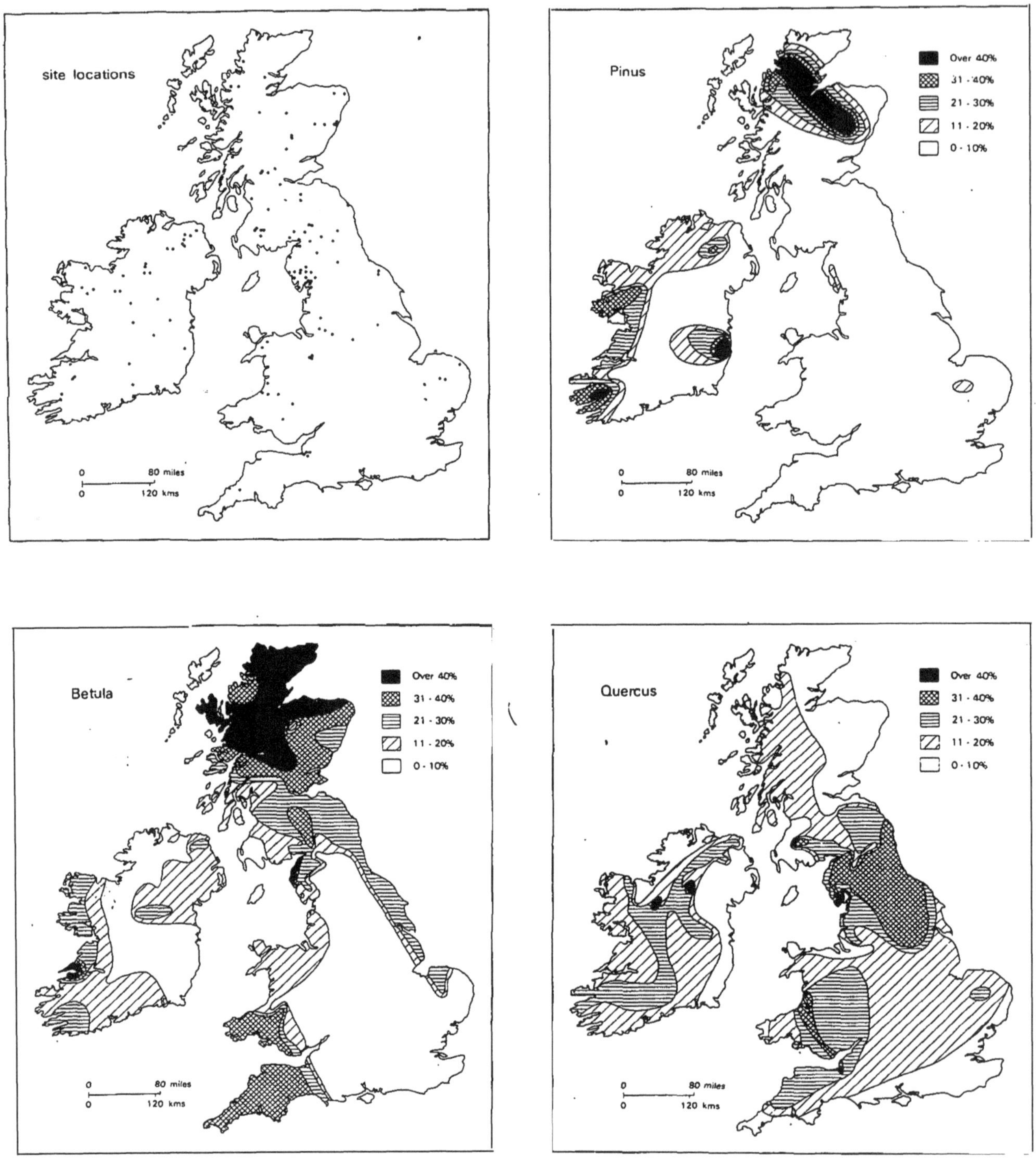

Fig. 3.8 (a) Isopollen maps for the British Isles for circa 5,000 B.P. (A) Site locations, Pinus, Betula and Quercus. (B) Ulmus, Almus, Corylus and Tilia.

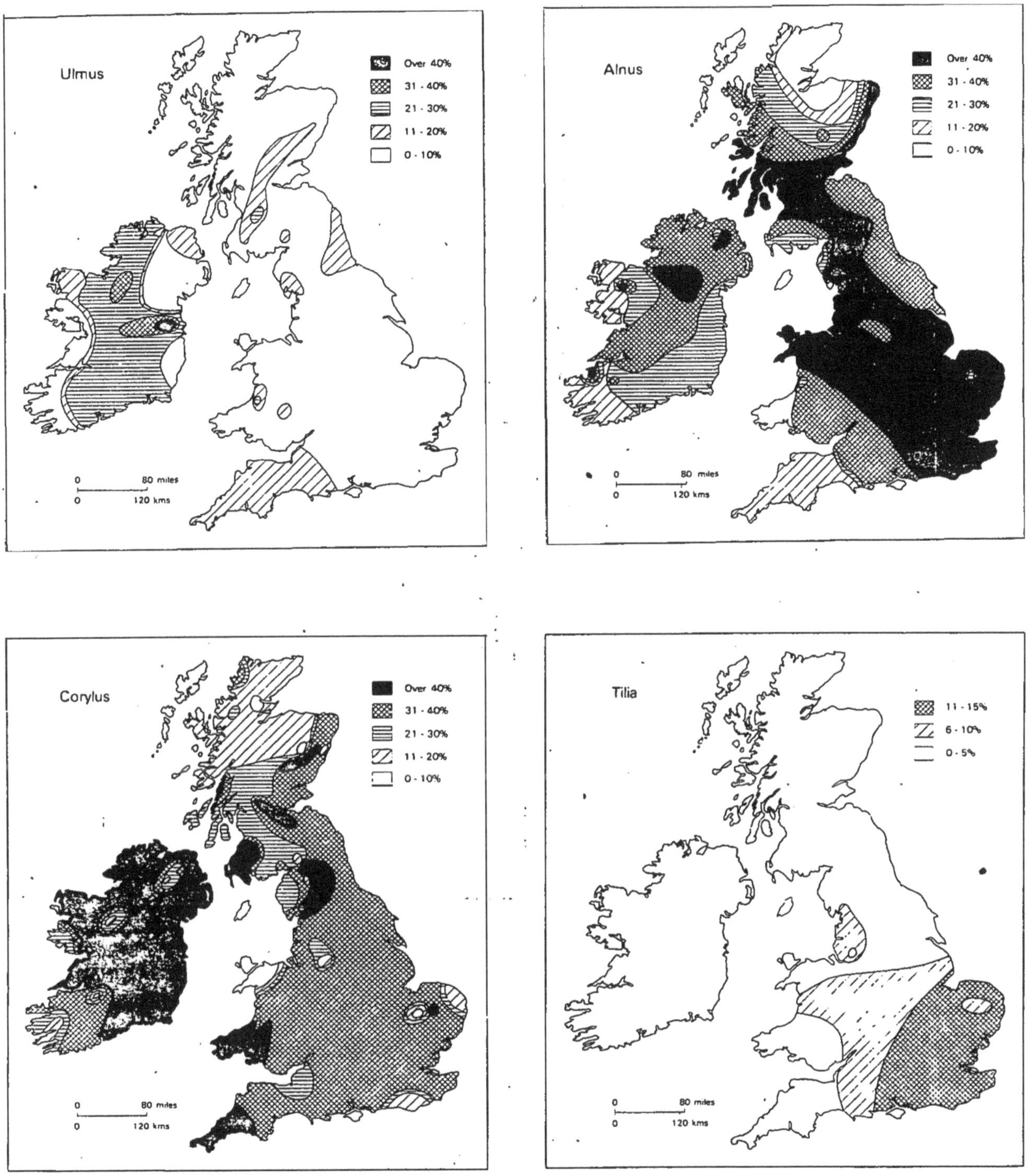

Fig. 3.8 (b) Isopollen maps for the British Isles for circa 5,000 B.P. (A) Site locations, Pinus, Betula and Quercus. (B) Ulmus, Alnus, Corylus and Tilia.

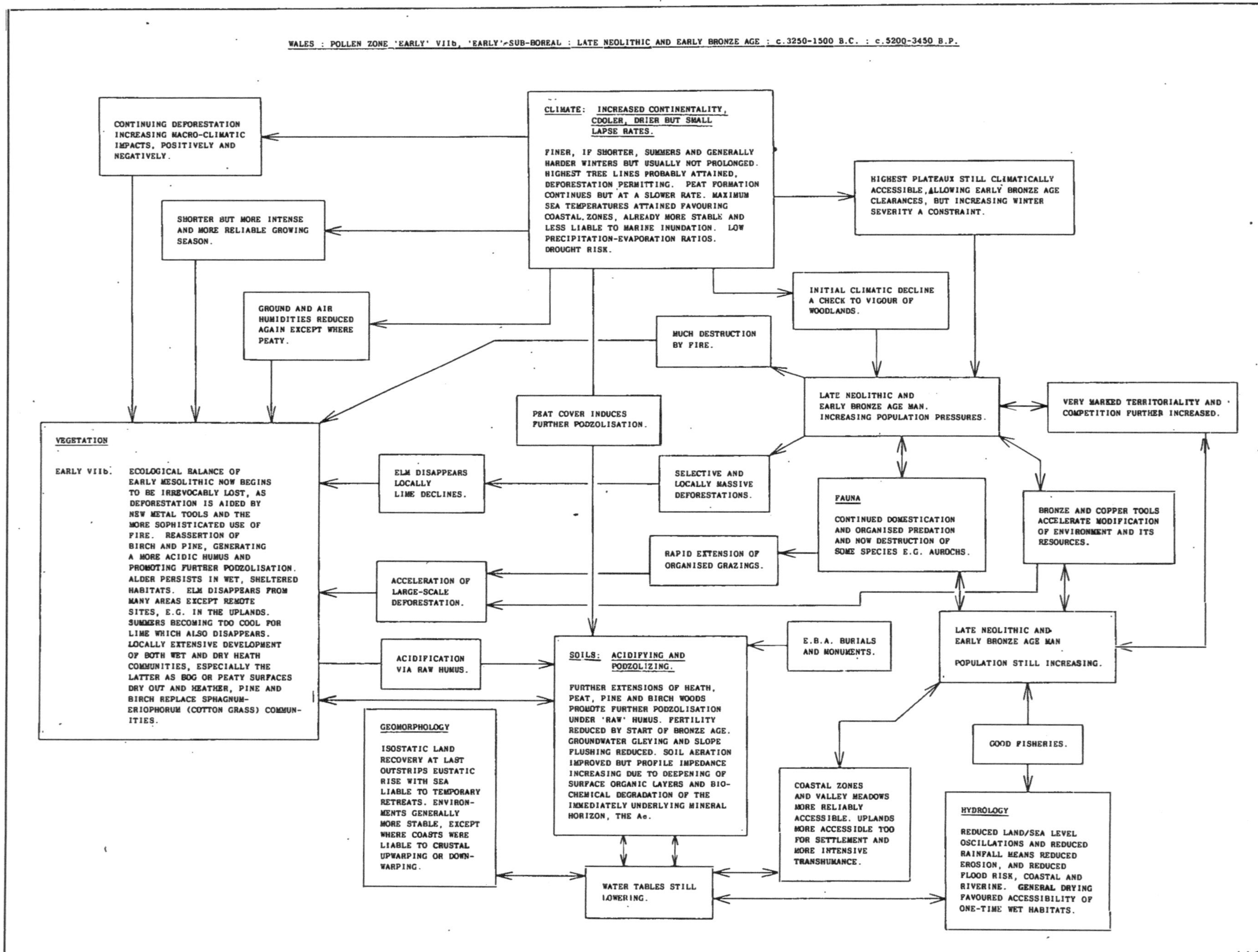

Fig. 3.9 Ecosystem model for the Late Neolithic and Early Bronze Age

that this vegetation pattern is extrapolated and based on fragmentary evidence. Clearly, it must be tested by new data and future revisions. Nonetheless, some preliminary indications of the regional variations in forest composition are shown.

The date 5000 BP has been adopted by many palaeobotanists as that of the 'elm decline'. Many pollen diagrams for the British Isles and adjacent N.W. Europe suggest a correlated and broadly synchonrous disappearance of the elm about this date (e.g. Smith and Pilcher, 1973). However, Seddon (1967) and Taylor (1973) have challenged the view that the elm decline was synchronous over an area as vast as a subcontinent and over environments as contrasted as Wales and East Anglia. A variability of ± 500 years rather than the proposed ± 250 years provides a greater range and allows for the incroporation of the view that in situ prescribed removal of the elm could have been due to, and was therefore timed by, the arrival of Late Neolithic or Early Bronze man on the spot who, either directly by selection or indirectly by its selection by his grazing animals, depleted the elm (see Smith, A.G., 1975). A site at Glaslyn (SH. 827933) some 12½ km north of the summit of Plynlimon, reve aled a basal peat with an elm decline sampled and radio carbon-dated at 3870 ± 100 years B.P. (I.4928) (Taylor, 1973). Although elm decline appears to be secondary rather than primary (the buried-soil pollen record below the peat is not unequivocal), the fact remains it is very late. The settlement here was early Bronze Age. Only future research can confirm whether this later date (by 1000 years or so) for the elm decline in upland Wal es is characteristic or exceptional.

The late Neolithic and Early Bronze Age (Fig. 3.9)

This extensive cultural transition occupied some 1,750 years which reveals the simplistic nature of the conventional cultural boundaries over time which, like discontinuities in pollen diagrams, may often claim discrete segments of prehistoric time in their own right. The Late Neolithic and Early Bronze Age are identified with the first half of the Sub-Boreal climatic phase, with its drier but again increasingly cooler continental type of seasonal combination of shorter but quite fine and vigorous summers and generally harder, longer winters. Lapse rates with altitude were small, relatively, and it is very possible that the highest tree limits in Wales at circa 2000 ± 200' O.D. (609 m ± 61 m) were attained at this stage. Summer occupation of the high ground would clearly have been favoured but winters would have imposed further constraints on access.

Selective and locally massive deforestations must have occurred once during the first thousand years of this transitional period, where we can place successive elm declines. Peat growth was maintained on the higher wetter plateaux and soil podzolisation, often under a resurgence of birch and pine woods, was decelerated. Smith and Taylor (1969) have argued that Bronze Age man in Wales was greeted by a naturally evolving secondary woodland on degraded soils where decaying mineral 'A' horizons and iron-pan development aggravated the drainage of soil profiles which were already inhibited by the accumulations of surface 'mor' humus and, later on, peats. Hydrological balances, with reduced precipitation-evaporation ratios and reduced oscillations in land-sea level, were generally enhanced and, consequently, coastal and riverine - and

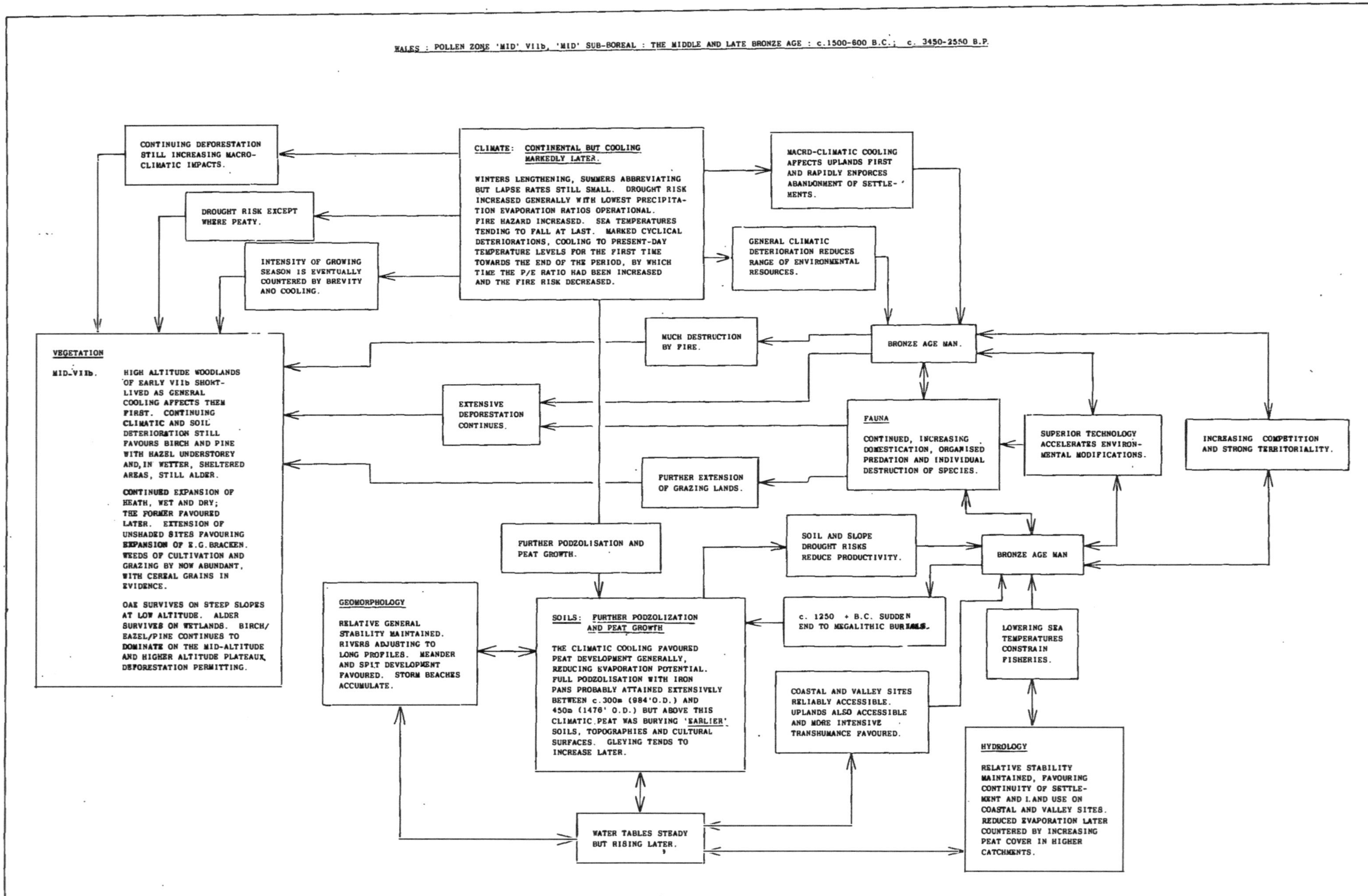

Fig. 3.10 Ecosystem model for the Middle and Late Bronze Age

also one-time marshy-habitats became available for settlement, agriculture and pastoralism. Inevitably, many species of animals became the victims of organised predation and several approached extinction e.g. the aurochs.

The Middle and Late Bronze Age (Fig. 3.10)

The cooling climatic trend continued markedly so at the end of this period as the Sub-Atlantic phase was inaugurated during the last millennium B.C. It is highly likely that the phase of rapid cooling occupied a run of several centuries and was episodic and variable. Nonetheless, a sequence of bad years or simply one bad decade would have had severe impact especially in the uplands where the highest plateaux, above say 1600' O.D. (457 m) initially and even above 1000' O.D. (305 m) later, were becoming severely constrained, climatically, due to the rapidity of the temperature lapse rates in an increasingly maritime regime, a shortening of the growing season and a prolongution and severity of winters with long snow-lie and persistent frost. But, on the credit side, summers were more reliable if short, giving a combination of seasons not unlike, but not necessarily similar to, the decade of the 1940s in Britain during the present century. But all this applies to the end of the period under discussion. The Middle Bronze Age still embraced the later stages of the continental Sub-Boreal which was a time of relative environmental stability. Rivers were adjusting, uninterrupted, to their long profiles and it is possible that many large meanders systems were created in the major 'mature' valleys of the Principality. Coastal and riverine sites maintained reasonable access and the uplands, during their shorter summers, were available for transhumant flocks and temporary settlements. However, the general depletion of forest and soil resources continued, and it is probable that the peaty-gley-podzols of the uplands became mature with full iron-pan development at this stage. The lowering of temperatures favoured continuing, if rather slow, peat accumulation. Heath vegetation, moorland grasses and clearance species, such as bracken, were all bound to expand as the forests were removed. Oak eventually was to survive only on the steeper, remote slopes whilst the plateau woodlands were dominantly birch/hazel/pine with alder prominent in wet habitats at low and intermediate elevations.

In the Late Bronze Age, the severe climatic deterioration further depleted the environmental resources, especially at altitude. Little wonder that Iron Age man, to follow, became adapted to intermediate and low altitude habitats where his superior technology enabled to cope with, and as required, remove the denser woodlands.

The Iron Age (Fig. 3.11)

Environmental stress and environmental hazard during the last 600 years B.C. must have rendered much of the Welsh uplands untenable. Cyclical cooling was, however, temporarily interrupted in the mid-Iron Age between about 300 and 100 B.C. Thereafter, the climate at the end of the Iron Age became substantially colder than it is today. The uplands bore the brunt of this and tree limits, agricultural limits and settlement limits were all reduced to their lowest elevations since ZoneV, the Pre-Boreal. Peat formation now becomes rampart in the uplands with the impact of the worst of the Holocene climates on deforested areas and wet soils—which were becoming even wetter.

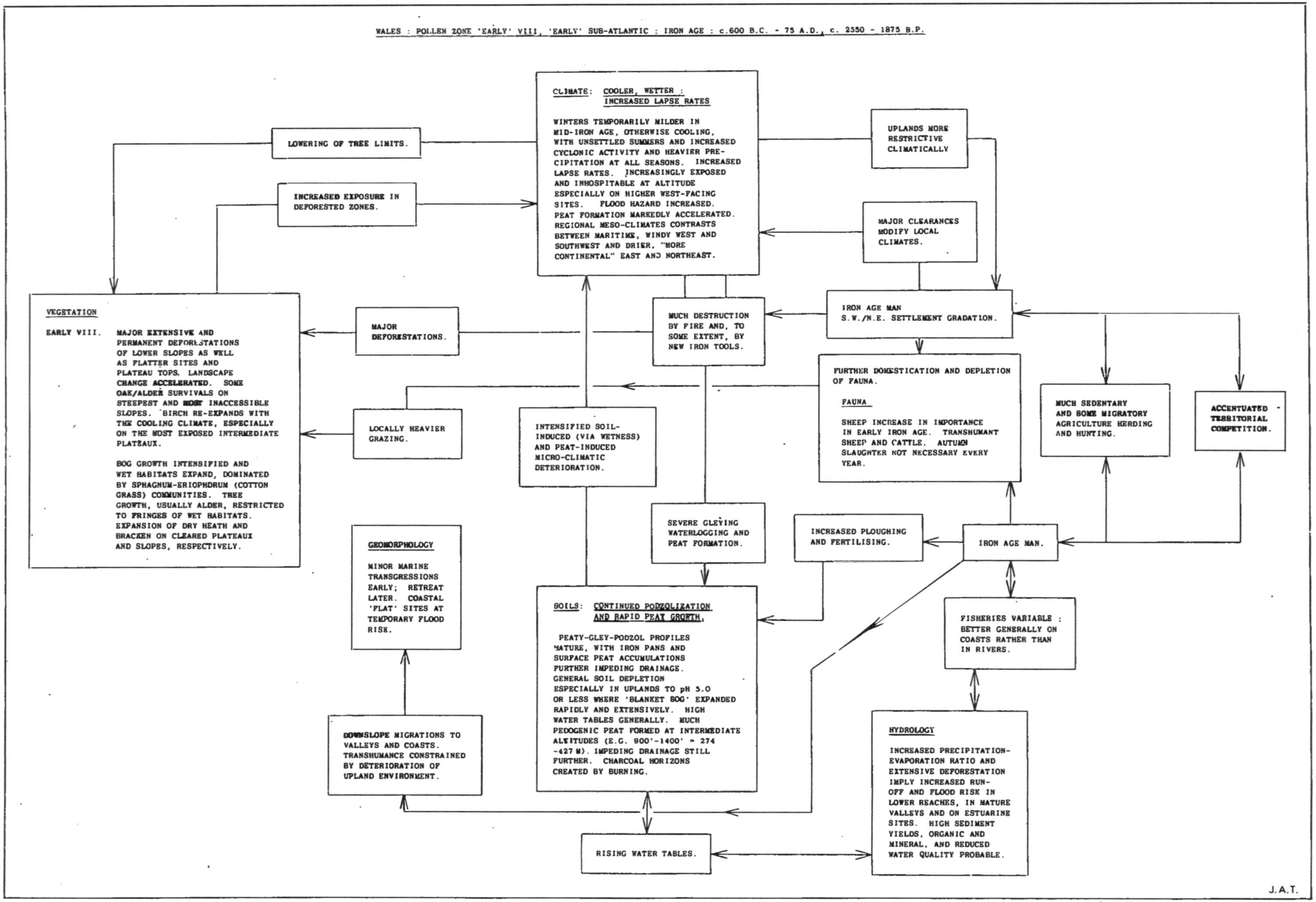

Fig. 3.11 Ecosystem model for the Iron Age

Increased run-off and erosion risks, increased flood risks, increased marshiness— all those factors point to an avoidance of flat, concave or valley bottom sites and a preference for convex "shedding" hill-top sites—aside from their greatest natural defensibility, i.e. the "hill-forts". Sea coasts and their fisheries were probably superior to riverine or exclusively estuarine sites. It should be recorded that sea-temperatures probably attained a retarded maximum during the later, rather than the earlier, Sub-Boreal, thus giving, ceteris paribus, a relatively superior water environment for coastal fisheries rather than inland fisheries, where the chilling effects of cooling climate—especially a cooling upland climate where most of the precipitation was falling —would be more directly felt.

Therefore, at the terminus of the Holocene both man and climate were contributing jointly to create one of the most critical phases of environmental depletion in the Welsh prehistoric record. Iron Age men, with the most superior technology to date, attacked the lowland forests. The Sub-Atlantic climate deteriorated everywhere but particularly in the uplands where peatiness and wetness prevailed.

Conclusion

It should be made clear that the ecosystem models presented above (Figs. 3.5, 3.6, 3.7, 3.9, 3.10 and 3.11) are all basically inferential and assume that climatic swings did take place over the last 10,000 years in Wales, as elsewhere in the British Isles and beyond. On the weight of available evidence of contemporary and past climatic oscillations, our climates always have been subjected to short-term variabilities over decades which may change in scale and intensity over the centuries. When biological or human tolerance thresholds are crossed, leading to migrations, evacuations, extinctions (—or, on the positive side, new colonisations) then the climate may be held to have changed. On three occasions such changes may be charted through the Holocene. The first was the very rapid recovery from periglacial conditions in Pollen Zones V and VI during the seventh millennium B.C. in the Early Mesolithic. The second was the ever so gradual build-up and later decline from the 'optimum' of Pollen Zone VIIa, the Atlantic, at the Mesolithic-Neolithic transition. The third was the cyclical deterioration which occurred in late Pollen Zone VIIb, the Sub-Boreal, affecting the Middle and Late Bronze Age and, finally, the Iron Age. When such climatic changes are negative and coincide with severe environmental depletion by man, the consequences are clearly at their most severe—as at this third stage.

REFERENCES

Birks, H. J. B., 1977. The Flandrian forest history of Scotland: a preliminary synthesis in Shotton, F. W., ed., British Quaternary studies: recent advances. Clarendon Press, Oxford, 119-35.

Birks, H. J. B., Deacon, J. and Pegler, S., 1975. Pollen maps for the British Isles 5000 years ago. Proc. R. Soc. Lond. B. 189, 87-105.

Bowen, E. G., 1959. Le pays de Galles. Trans. Inst. Brit. Geogrs., 26, 1-23.

Coope, G. R., 1977. Quaternary Coleoptera as aids in the interpretation of environmental history in Shotton, F. W., ed., British Quaternary studies: recent advances. Clarendon Press, Oxford, 55-68.

Crabtree, K., 1972. Late-glacial deposits near Capel Curig, Caernarvonshire. New Phyt. 71, 1233-1243.

Deacon, J., 1974. The location of refugia of Corylus avellana L. during the Weichselian glaciation. New Phyt. 73, 1055-1063.

Evans, E. E., 1975. Highland landscapes: habitat and heritage in Evans, J. G. et al (eds.) The effect of man on the landscape: the Highland Zone. Res. Report No. 11 C.B.A. London 1-5.

Evans, J. G., 1972. Land snails and archaeology. London, Seminar Press.

Fergusson, C. W., 1970. Dendrochronology of bristle cone pine, Pinus aristata. Establishment of a 7484 year chronology in the White Mountains of eastern-central California, U.S.A. in Olsson, I. U., ed. Radiocarbon variations and absolute chronology. Proc. 12th Nobel Symp. Uppsala, Sweden, 237-59.

Fleure, H. J., 1951. The natural history of man in Britain. London, Collins. (Fontana edition with M. Davies in 1971).

Godwin, H., 1956 and 1975. History of British Flora C.U.P.

Handa, S. and Moore, P. D., 1976. Studies in the vegetational history of mid-Wales, IV, Pollen analysis of some pingo basins. New Phyt. 77, 203-23.

Harrison, S. J., 1974. Problems in the measurement and evaluation of the climatic resources of upland Britain in Taylor, J. A. (ed.), Climatic Resources and Economic Activity, Newton Abbott, David & Charles, 47-63.

Hibbert, F. A. and Switzur, V. R., 1976. Radiocarbon dating of Flandrian pollen zones in Wales and northern England. New Phyt. 77, 793-807.

Lamb, H. H., 1966. The changing climate: selected papers. London, Methuen.

Lamb, H. H., Lewis, R. P. W. and Woodruffe, A., 1966. Atmospheric circulation and the main climatic variables between 8000 and O.B.C.: meteorological evidence in Sawyer, J. S. (ed.), World Climate from 8000 to O.B.C. Roy. Met. Soc., 174-217.

Manley, G., 1964. The evolution of our climatic environment in Watson, J. W. and Sissons, J. B., eds., The British Isles: a systematic geography, London, Nelson, 152-76.

Meteorological Office., 1952. Climatological Atlas of the British Isles. H.M.S.O.

Michael, H. N. and Ralph, E. K., 1970. Correction factors applied to Egyptian radiocarbon dates from the era before Christ, in Olsson, I. U. (ed.), Radiocarbon variations and absolute chronology, Nobel Symp., No. 12, Uppsala, Sweden, 109-20.

Moore, P. D., 1970. Studies in the vegetational history of mid-Wales II: the Late-glacial period in Cardiganshire. New Phyt. 69, 363-75.

Moore, P. D., 1972. The influence of Post-glacial climatic fluctuations upon forest structure and development in mid-Wales in Taylor, J. A., ed., Research papers in forest meteorology, Cambrian News, Aberystwyth, 20-30.

Moore, P. D. and Chater, E. H., 1969. Studies in the vegetational history of mid-Wales I The Post-glacial period in Cardiganshire. New Phyt. 68, 183-96.

Olsson, I. U. (ed) 1970. Radiocarbon variations and absolute chronology. Proc. 12th Nobel Symp., Uppsala, Sweden.

Pearson, G. W., Pilcher, J. R., Hillam, J. and Baillie, M. G. L., 1977. Absolute radiocarbon dating using a low altitude European tree-ring calibration, Nature, 270, 25-29.

Pearsall, W. H., 1952. The pH of natural soils and its ecological significance. J. Soil Sc., 3, 41-51.

Pearsall, W. H., 1965. A discussion of the development of habitats in the Post-glacial. Proc. Roy. Soc. 'B', 161, 293-4.

Seddon, B., 1962. Late-glacial deposits at Llyn Dwythwch and Nant Ffrancon, Caernarvonshire. Phil. Trans. Roy. Soc. B, 244, 459-81.

Seddon, B., 1967. Prehistoric climate and agriculture: a review of recent palaeoecological investigations in Taylor, J. A., ed., Weather and Agriculture, Pergamon Press, Oxford, 173-85.

Simmons, I. G., 1975. The ecological setting of Mesolithic man in the Highland zone in Evans, J. G., et al, eds., The effect of man on the landscape: the Highland Zone. Res. Report No. 11, C.B.A., 57-63.

Smith, A. G., 1970. The influence of Mesolithic and Neolithic man on British vegetation: a discussion, in Walker, D. and West, R. G., eds., Studies in the vegetational history of the British Isles. C.U.P., 81-96.

Smith, A. G., 1975. Neolithic and Bronze Age landscape changes in northern Ireland in Evans, J. G. et al, eds., The effect of man on the landscape: the Highland Zone. C.B.A. Res. Report No. 11, 64-74.

Smith, A. G. and Pilcher, J. R., 1973. Radiocarbon dates and the vegetational history of the British Isles. New Phyt., 72, 903-14.

Smith, R. T., 1972. A reconsideration of the role of climate in the development of post-Weichselian forest types in Taylor, J. A. (ed.), Research Papers in Forest Meteorology, Cambrian News, Aberystwyth, 1-19.

Smith, R. T. and Taylor, J. A., 1969. The Post-glacial development of soils and vegetation in northern Cardiganshire. Trans. Inst. Brit. Geogrs., 48, 75-96.

Taylor, J. A., 1973. Chronometers and chronides: a study of palaeoenvironments in west central Wales in Progress in Geography, Vol. 5, Arnold, London, 247-334.

Taylor, J. A., 1975. The role of climatic factors in environmental and cultural changes in prehistoric times in Evans, J. G. et al, eds., The effect of man on the landscape: the Highland Zone. Res. Report. No. 11 C.B.A., 6-19.

Taylor, J. A., 1976. Upland Climates in Chandler, T. J. and Gregory, S. eds. The Climate of the British Isles, Longmans, London, 264-87.

Taylor, J. A. and Smith, R. T., 1972. Climatic peat—a misnomer? Trans. Fourth Int. Peat Congress, Helsinki, Finland, Vol. 1, 471-84.

Tooley, M. J., 1974. Sea-level changes during the last 9000 years in north-west England. Geog. J., 140, 18-42.

Walker, D., 1966. The late Quaternary history of the Cumberland lowland. Phil. Trans. Roy. Soc. London, B., 251, 1-210.

CHAPTER FOUR

THE EARLY HOLOCENE SETTLEMENTS OF WALES

Roger M. Jacobi

ABSTRACT

In this chapter the Mesolithic of Wales is examined in two parts—the Early and the Later Mesolithic, with a suggested boundary between the two falling somewhere within the 7th millennium b.c. While many of the Early Mesolithic find spots in Wales are in what are at present cliff-top situations, consideration of past sea-level patterns indicates that these would have been closer to the centres of substantial dry-land territories at the time of their occupation. Later Mesolithic assemblages can be recognized from coastal, lowland and upland situations and after a discussion of the role of, and evidence for, coastal exploitation along the west coast of Britain, sites in upland and lowland situations are linked into a single economic model.

When we come forward in time to the Post-glacial there is a sharp increase in the quantity of archaeological data at our disposal. Thus while there is only a total of 91 Late-glacial findspots from England and Wales (Fig. 4.1), the number of findspots known for the Mesolithic, that period of the Post-glacial falling between about 8,000 and 3,400 radiocarbon years b.c., can be estimated as something in the order of ten thousand (Jacobi 1975, Wymer 1977). Again, while 41 (or 45%) of the Late-glacial findspots are in caves or rock-shelters, only 58 Post-glacial sites are similarly located. Of these some 35% are in south-eastern England (Fig. 4.2). The scale of the evidence is thus vastly different.

Fig. 4.1, a map of Late-glacial sites, differs from that presented by Campbell (1977 Map 34) in the very much larger number of open sites identified—particularly in south-east England—even though omitted, from our map, are surface finds of "Zinken"-like artifacts from northern England, the real age of which, it has been suggested, must remain problematic (see chapter two). In both maps, however, central England is without findspots, but rather than interpreting this absence as indicative of an unoccupied "socio-economic buffer zone" (Ibid map and vol. I, p. 160), it more probably reflects the poor quality of archaeological survey in this area.

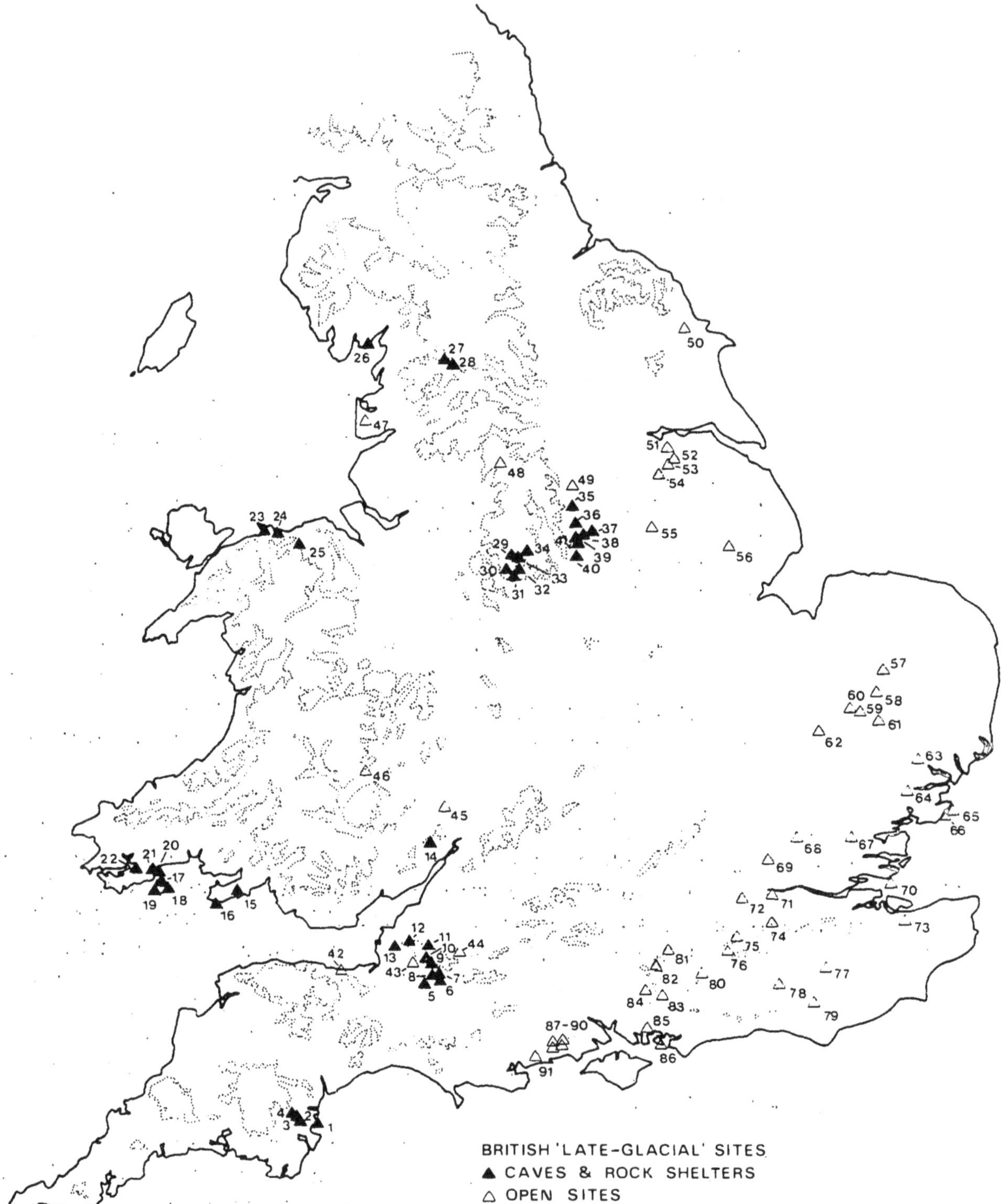
BRITISH 'LATE-GLACIAL' SITES
▲ CAVES & ROCK SHELTERS
△ OPEN SITES

4.1 (opposite) "Distribution of British Late-glacial Sites"

Key:

Cave and Rockshelter Sites.

1. Kent's Cavern
2. Tornewton.
3. Torbryan Caves.
4. Three Holes Cave.
5. Flint Jack's Cave.
6. Great Oone's Hole.
7. Soldier's Hole.
8. Gough's (New) Cave.
9. Sun Hole Cave.
10. Chelm's Coombe.
11. Aveline's Hole.
12. Banwell Cave.
13. Hutton Cave.
14. King Arthur's Cave.
15. Cathole.
16. Paviland.
17. Nanna's Cave.
18. Ogof yr Ychen.
19. Potter's Cave.
20. Little Hoyle.
21. Hoyle's Mouth.
22. Priory Farm Cave.
23. Upper Kendrick's Cave.
24. Ogof Tan-y-Bryn.
25. Plas-yn-Cefn.
26. Kirkhead Cave.
27. Kinsey Cave.
28. Victoria Cave.
29. Dowel Hall Cave.
30. Ossom's Cave.
31. Elder Bush Cave.
32. Thor's Fissure.
33. Fox Hole Cave.
34. Stoney Low (Sheldon).
35. Dead Man's Cave (North Anston).
36. Lob Wells Shelter.
37. Mother Grundy's Parlour.
38. Robin Hood's Cave.
39. Pin Hole Cave.
40. Langwith.
41. Church Hole Cave.

Open Sites.

42. Daniford Cliff.
43. Callow Hill.
44. Herriots Bridge.
45. Nr. Ledbury.
46. Arrow Court.
47. Poulton le Fylde.
48. Cupwith Hill.
49. Edlington Wood.
50. Flixton.
51. Winterton.
52. Risby Warren.
53. Brumby Wood.
54. Messingham.
55. Morton (Gainsborough).
56. Salmonby.
57. Cranwich.
58. Great Fakenham.
59. Mildenhall.
60. Undley.
61. London Botton.
62. Swaffham Prior.
63. Bolton & Laughlins Pit, Ipswich.
64. Nr. Manningtree.
65. Stone Point.
66. Walton on Naze.
67. Widford Brickpit.
68. Broxbourne.
69. Tolpitts Lane, Rickmansworth.
70. Shoeburyness.
71. Wandsworth.
72. Syon Reach.
73. Oare.
74. Wallington.
75. Horsell.
76. Pyrford.
77. Stonewall Park.
78. Old Faygate.
79. Newhouse Farm.
80. Bunch Lane (Haslemere).
81. Headley.
82. Fir Hill, Kingsley.
83. Sheet.
84. Butser Hill.
85. Wiccot Shore.
86. Long Island.
87. Hengistbury Head A.
88. Hengistbury Head B.
89. Hengistbury Head C1.
90. Hengistbury Head C2.
91. Rush Corner (Bournemouth).

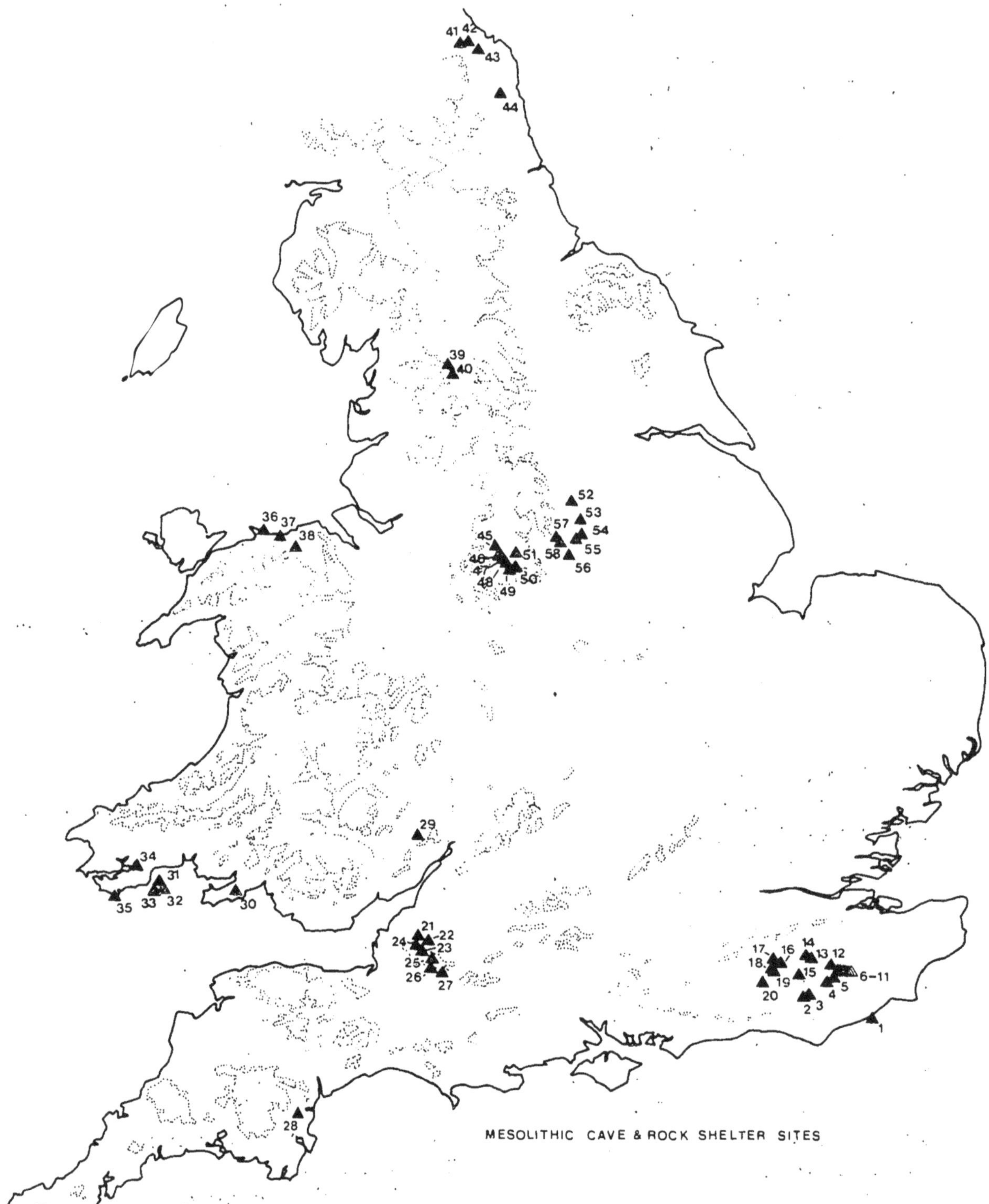
41 42
43
44
39
40
52
53
54
55
56
57
58
45
46
47
48
49
50
51
36 37
38
29
34
31
33
32
35
30
21
22
23
24
25
26
27
17
16
14
13
12
18
15
6–11
19
20
5
4
2
3
1
28
MESOLITHIC CAVE & ROCK SHELTER SITES

4.2 (opposite) "Distribution of Mesolithic Cave and Rock-shelter sites in England and Wales".

Key:

1. Cliff End, Pett.
2. Hermitage Rocks A.
3. Hermitage Rocks B.
4. Bowles Rocks.
5. Harrison's Rocks.
6. High Rocks C.
7. High Rocks D.
8. High Rocks E.
9. High Rocks F.
10. High Rocks G.
11. High Rocks J.
12. Happy Valley Rocks.
13. Stonewall Park A.
14. Stonewall Park B.
15. Blund's Hole.
16. Stone Rocks.
17. Ridge Hill.
18. Chittingly Woods.
19. West Hoathley.
20. Balcombe Rock Shelter.
21. Wotle Cave.
22. Aveline's Hole.
23. Rowberrow Cavern.
24. Haywood Rock Shelter.
25. Sun Hole Cave.
26. Gough's Cave.
27. Totly Pot.
28. Three Holes Cave.
29. King Arthur's Cave.
30. Cathole Gower.
31. Nanna's Cave.
32. Ogof yr Ychen.
33. Potter's Cave.
34. Priory Farm Cave.
35. Ogof Garrog Hir.
36. Lloches yr Aft.
37. Ogof Pant y Wennol.
38. Gop Cave.
39. Victoria Cave.
40. Calf Hole Cave.
41. Goatscrag A.
42. Goatscrag B.
43. Bowdon Doors.
44. Corby's Crags.
45. Dowel Hall Cave.
46. Seven Ways Cave.
47. Wetton Mill Mirror.
48. Dafur Ridge Cave.
49. Elder Bush Cave.
50. Fox Hole Cave.
51. Sheldon.
52. Thorpe Common Shelter.
53. Ash Tree Cave.
54. Mother Grundy's Parlour.
55. Pin Hole Cave.
56. Langwith.
57. Whaley I.
58. Whaley II.

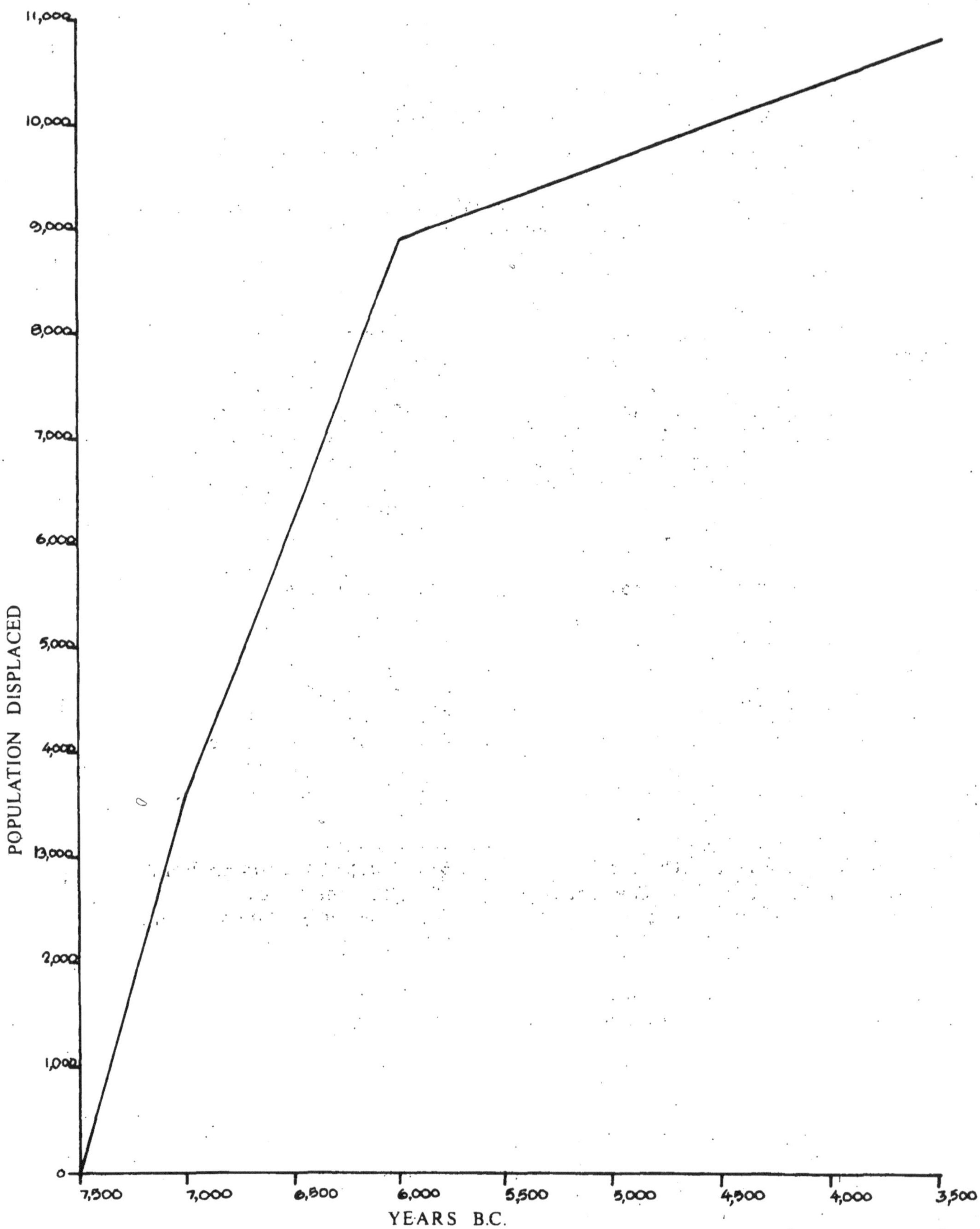

NORTH SEA BASIN.

CURVE OF POPULATION DISPLACEMENT.

Fig. 4.3 North Sea Basin. Curve of Population Displacement

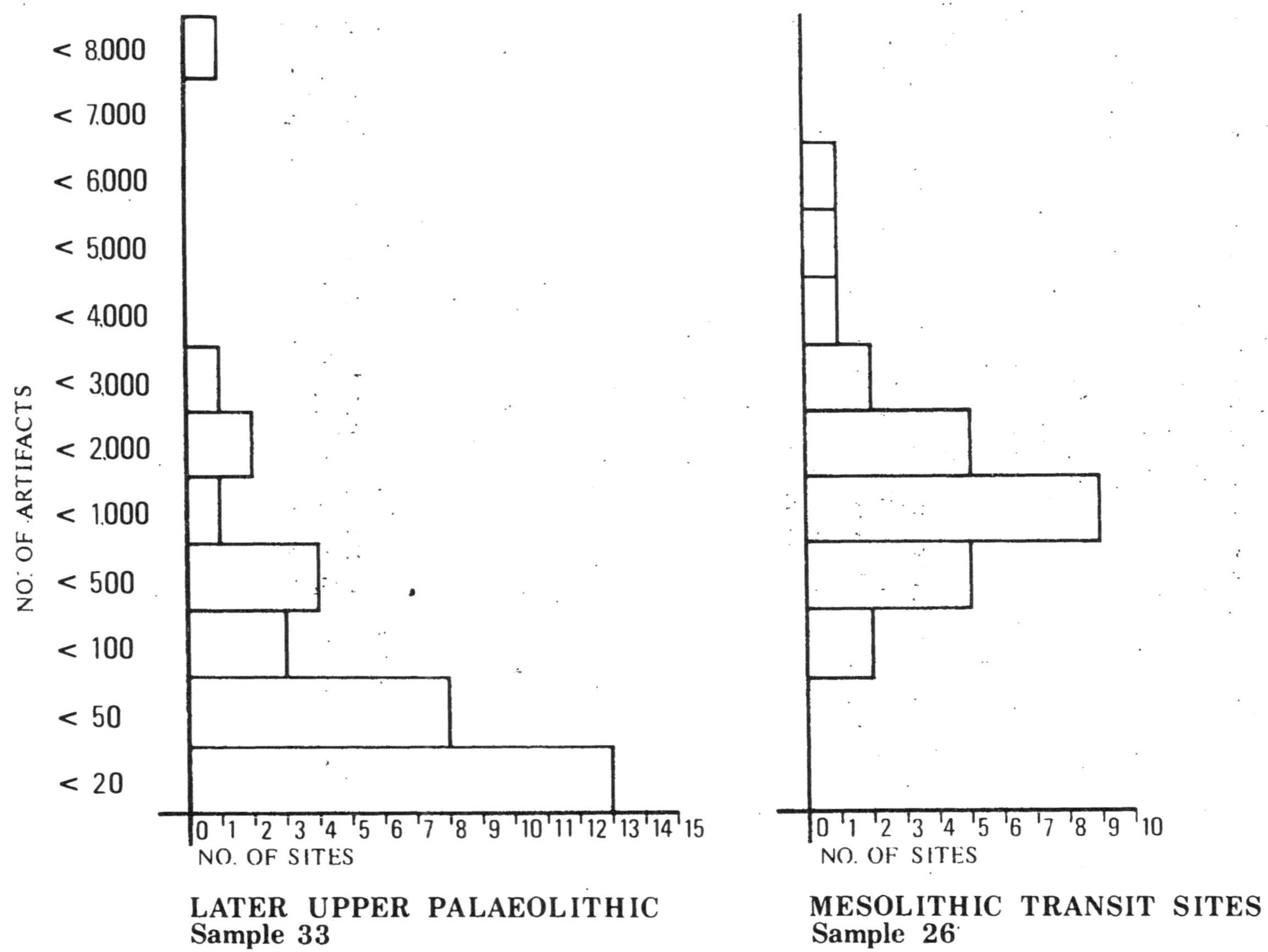

Fig. 4.4 Numbers of struck flints on Late-glacial and Early Post-glacial sites

Similar plotting of findspots attributable to the first one and a half millennia of the Post-glacial (in preparation) reveals very great similarities in distribution to that of finds covering the Late-glacial. Extensions of settlement within the British Isles at this time seem particularly slight with just the appearance of the first sites in the central Midlands (Blacklow Hill (Warwicks) and Ascott under Wychwood (Oxon), in Cornwall, perhaps as far west as New Shop in the parish of St. Buryan near Land's End, and on Anglesey. There is also along the eastern side of Britain a spread of settlement onto the Clevelands and perhaps into the area of present-day County Durham (Jacobi, 1977). This distribution, however, taken overall, is so close to that of Late-glacial findspots that it is no longer possible, as mistakenly in the past (Jacobi, 1973, p. 245), to use their respective distributions as evidence of a social break at the moment of the Late- to Post-glacial transition. Indeed, almost the opposite may seem to be implied although, until we know more about the archaeology of Zone III, useful conclusions are difficult to draw, and as we have already suggested technological changes at this moment appear particularly abrupt and show no obvious patterning confirmative of a technological continuity.

It would be tempting to explain the spread of human settlement into Ireland and northwards along both the east and west coasts of Scotland during the 7th millennium b.c. as a consequence of eustatic rise of sea-level around Britain. This land loss reached its greatest acceleration, not at the end of the Mesolithic, but during the 8th and 7th millennia b.c. (Fig. 4.3 expressed in terms of population displaced).

While the radiocarbon record for the Upper Palaeolithic is markedly discontinuous, this reflecting the episodic nature of most of this Later Pleistocene settlement of Britain as much as the very small number of immediately relevant determinations, plotting onto a single diagram—at a single standard deviation—of all the radiocarbon determinations available for Post-glacial archaeological occurrences reveals no moment of time not covered by a date or its standard deviation. There is thus, it could be suggested, no point within the Mesolithic when there was not an exploiting population in some part or another of Britain.

There is again a difference in the range of site types representing, respectively, Late- and Post-glacial activity. Thus, some Early Mesolithic sites such as Star Carr, Oakhanger, Downton and Thatcham, exhibit ground plans with areas that suggest use of the site by a number of family groups while the numbers of finished tools recovered run into several hundreds. Plotting, however, of the total number of struck flints recovered from the thirty-three most thoroughly investigated Late-glacial cave and rock shelter sites, together with the totals for Hengistbury Head sites C1 and C2 reveals (Fig. 4.4) that the number of flints is markedly fewer in most instances than the totals recovered from completely investigated upland Mesolithic sites interpreted only as transit camps. The contrast between the two halves of the diagram is immediately striking especially in that both groups are producing an almost identical repertoire of end scrapers, burins, truncated pieces and backed tools, Thus, of the thirty-three Late-glacial sites, twenty-one have less than fifty worked pieces while nineteen of the twenty-six Mesolithic sites

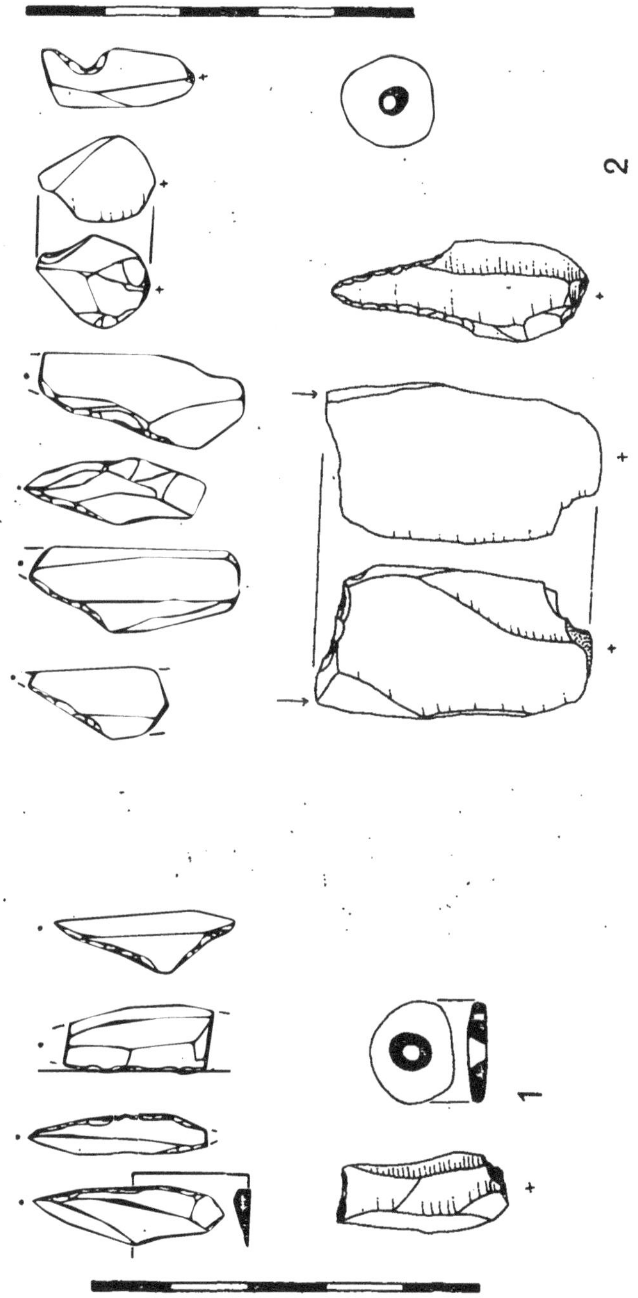

Fig. 4.5 Finds of beads: (1) Linney Burrows. (2) Freshwater East: Site A6

Fig. 4.6 Map of Early Mesolithic Sites in Wales

have more than five hundred. The obvious exception, of course, is Gough's (New) Cave with over seven thousand struck flints recorded by Parry and an unknown quantity removed by Gough during his digging and blasting begun in 1892. It should be remembered, however, that Parry's excavation covered 19 six inch spits representing an unknown length of time.

One interpretation of this diagram is to suggest that while archaeological exploration has picked up Late-glacial "transit sites", most probably casual kill and butchery locations, we are missing anything interpretable as a base location. For the Early Mesolithic, by contrast, it could be suggested, excavation has sampled both base and transit camps, and both summer and winter sites are known for the earliest Post-glacial

The Earliest Mesolithic of Wales

As in England, a study of the microlithic material from Welsh sites suggests that this can be divided into two chronological groupings, the first and older of which takes in such assemblages as Rhuddlan sites E and M (Miles 1972A and B) and Trwn Ddu (Aberffraw) on Anglesey. These assemblages compare closely to the Maglemosian or Early Mesolithic industries of south-eastern and eastern England which now have numerous age-estimates of between 8,000 and 6,800 radiocarbon years b.c. (Jacobi, 1976, pp. 68-70). Joining with these recently excavated sites in North Wales will, it is suggested, be Daylight Rock on Caldey.

Fragments of what were clearly similar assemblages have been recovered from Pencilan Head on the Lleyn Peninsula, the Nab Head in northern Pembrokeshire (see below), Freshwater East, in Southern Pembs. (site A5: Leach (1913) pp. 396-400) and from the present island of Burry Holms (Glamorgan: Rutter (1948) p. 33: Grimes (1951) pp. 10-11 and Wainwright (1963) p. 119, Fig. 12). In addition to worked flints—obliquely blunted points, microburins, an awl, a burin and a pair of end scrapers (Wainwright (1963) Fig. 5, nos 11 and 26)—the site at Freshwater East has produced a single perforated shale bead (Fig. 4.5. right-handside: see also Leach (1918) p. 59 and (1933) pp. 233-234). Other findspots included on Fig. 4.6, a sketch map of Early Mesolithic findspots in Wales, are a pair of small surface sites: Valley Field on Caldey and Palmerston Farm near Haverfordwest (J. Rees: pers. comm.). Aberystwyth also is incorporated (Thomas and Dudlyke 1925), although the surviving collection appears to include both such potentially early material as large, obliquely blunted points and triangles, as well as such later microlith forms as "rods" (microlithic backed bladelets) and scalene micro-triangles. This apparent mixture need occasion little surprise when it is recalled that the excavations producing the collections with which we now have to work were conducted on two totally distinct sites—the Hospital and Harbour-Bluffs (Ibid, pp. 74-79). The finds are in no way separated in the original report and no way now exists of attributing what survives to one site or the other. Preservation of this material has been only partial, and it is apparent that many of the smaller, obviously later microliths, illustrated in this report (Ibid Pl. XVI, Fig. I, line 2) have disappeared. While the bulk of the finds are clearly Mesolithic, a leaf arrowhead (Ibid Pl. XVII, Fig. 2, line 2) and a flake from a polished axe indicate later activity on the site.

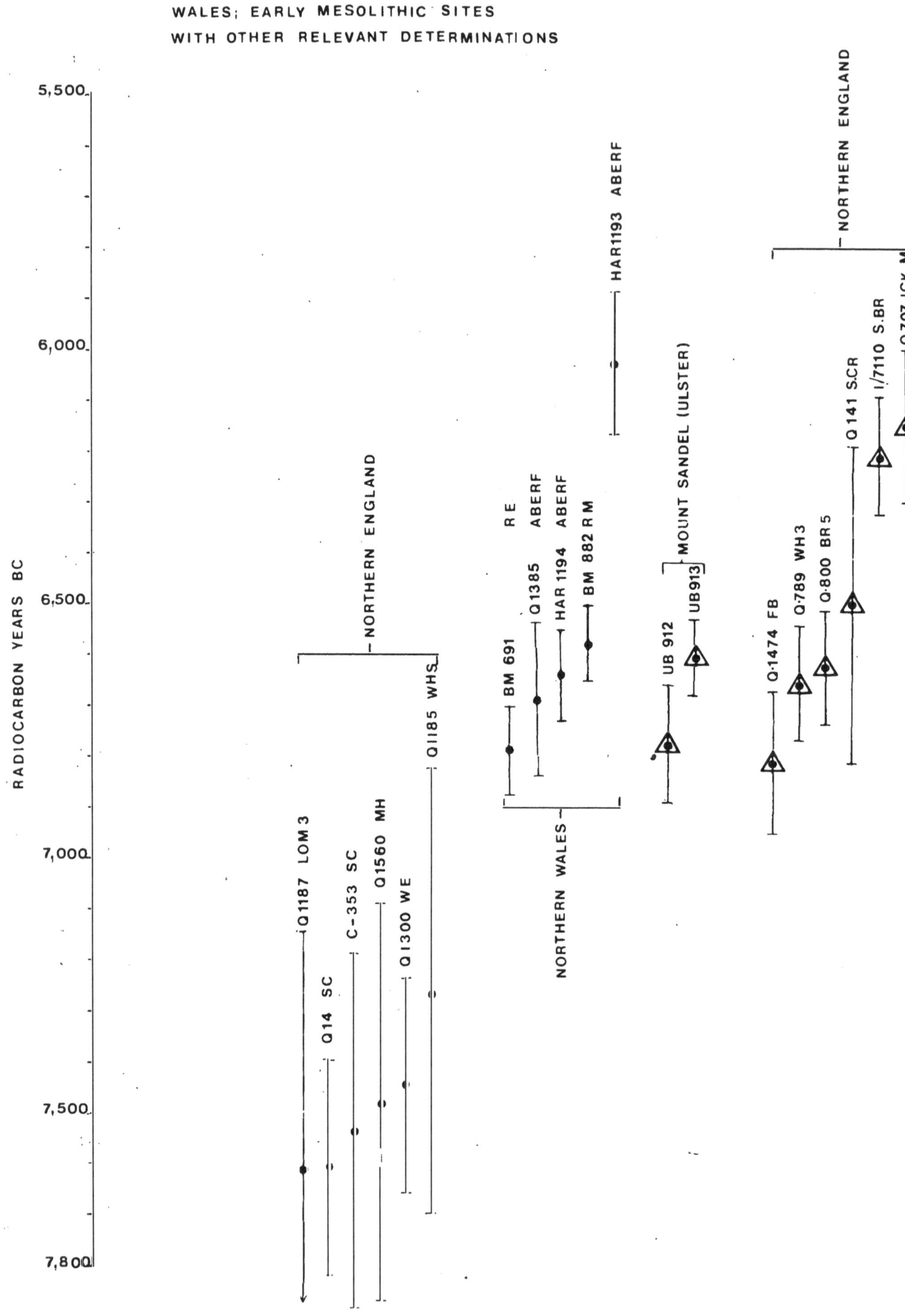

Fig. 4.7 Radiocarbon dates for Early Mesolithic sites in Wales compared to dates for early Post-glacial assemblages from Northern England and Northern Ireland

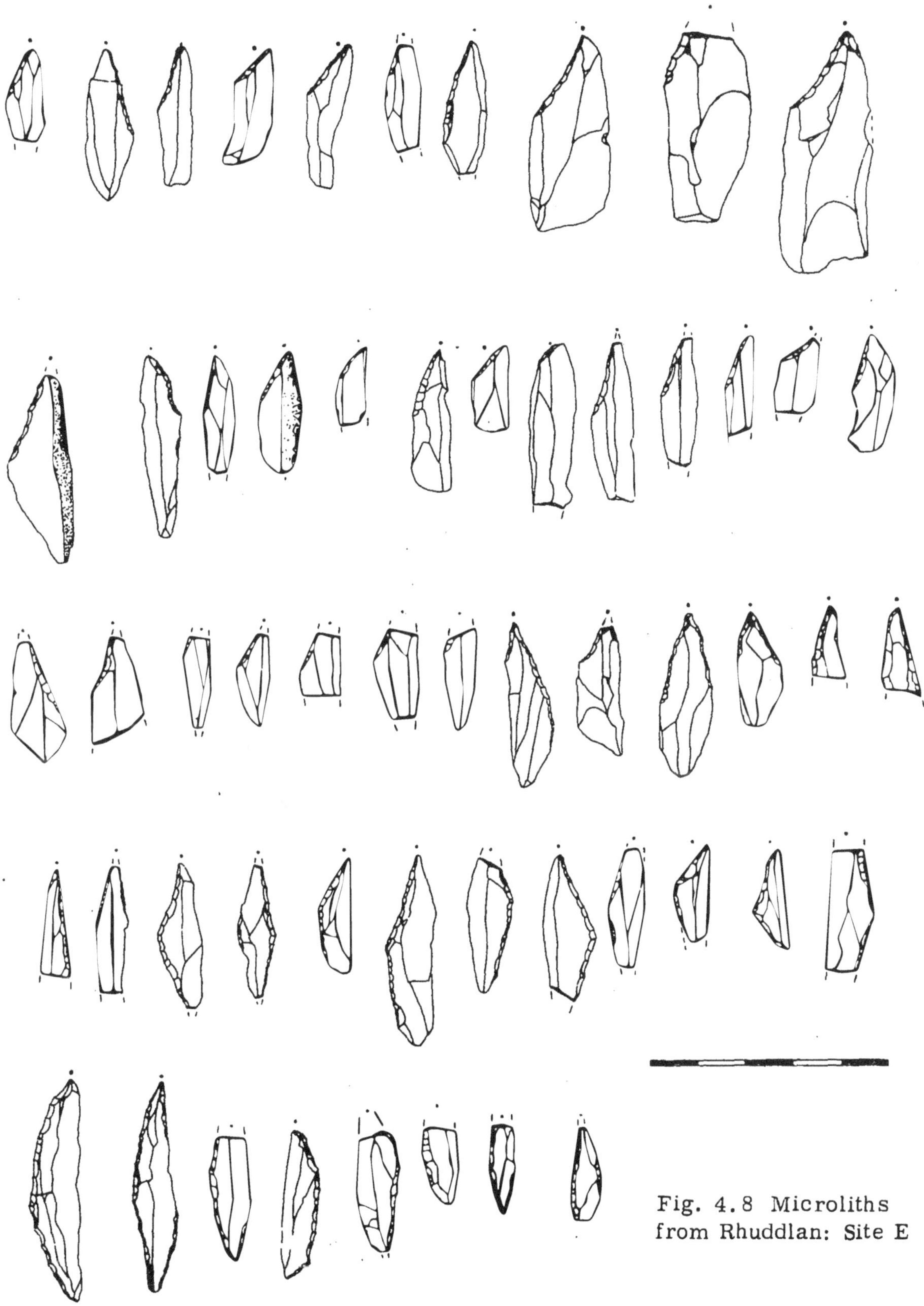

Fig. 4.8 Microliths from Rhuddlan: Site E

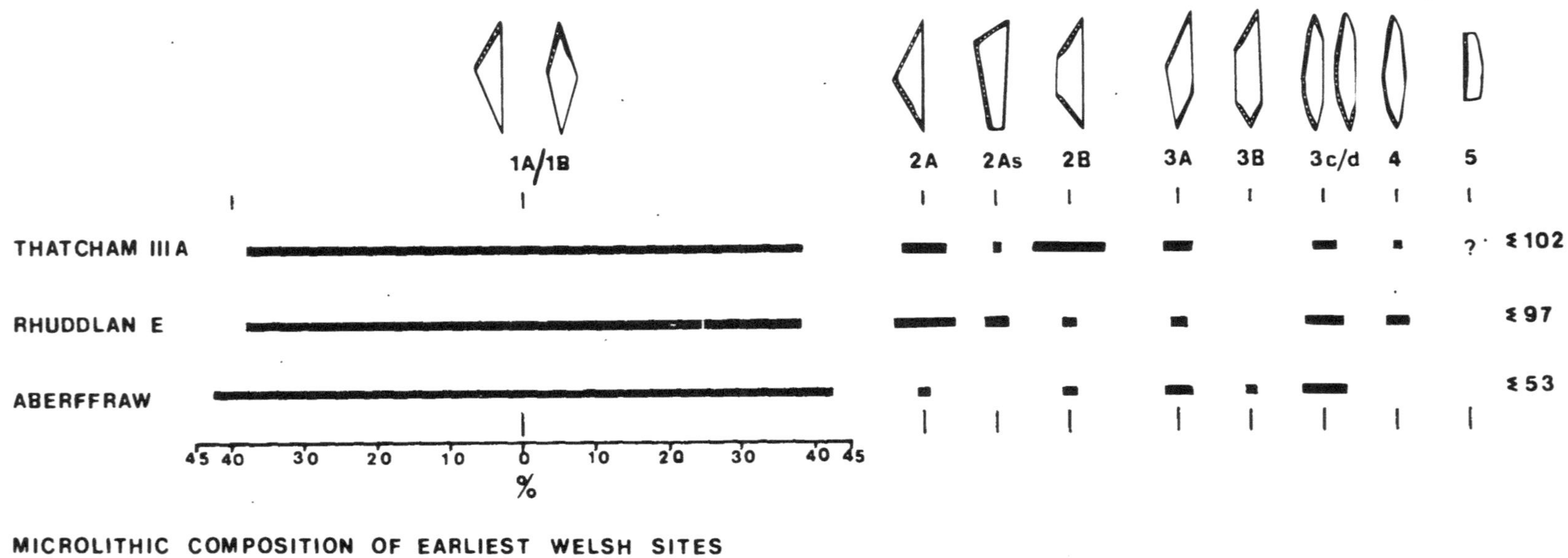

Fig. 4.9 Comparison of the microlithic industry from Thatcham IIIA with those from Rhuddlan E and Trwn Ddu (Aberffraw)

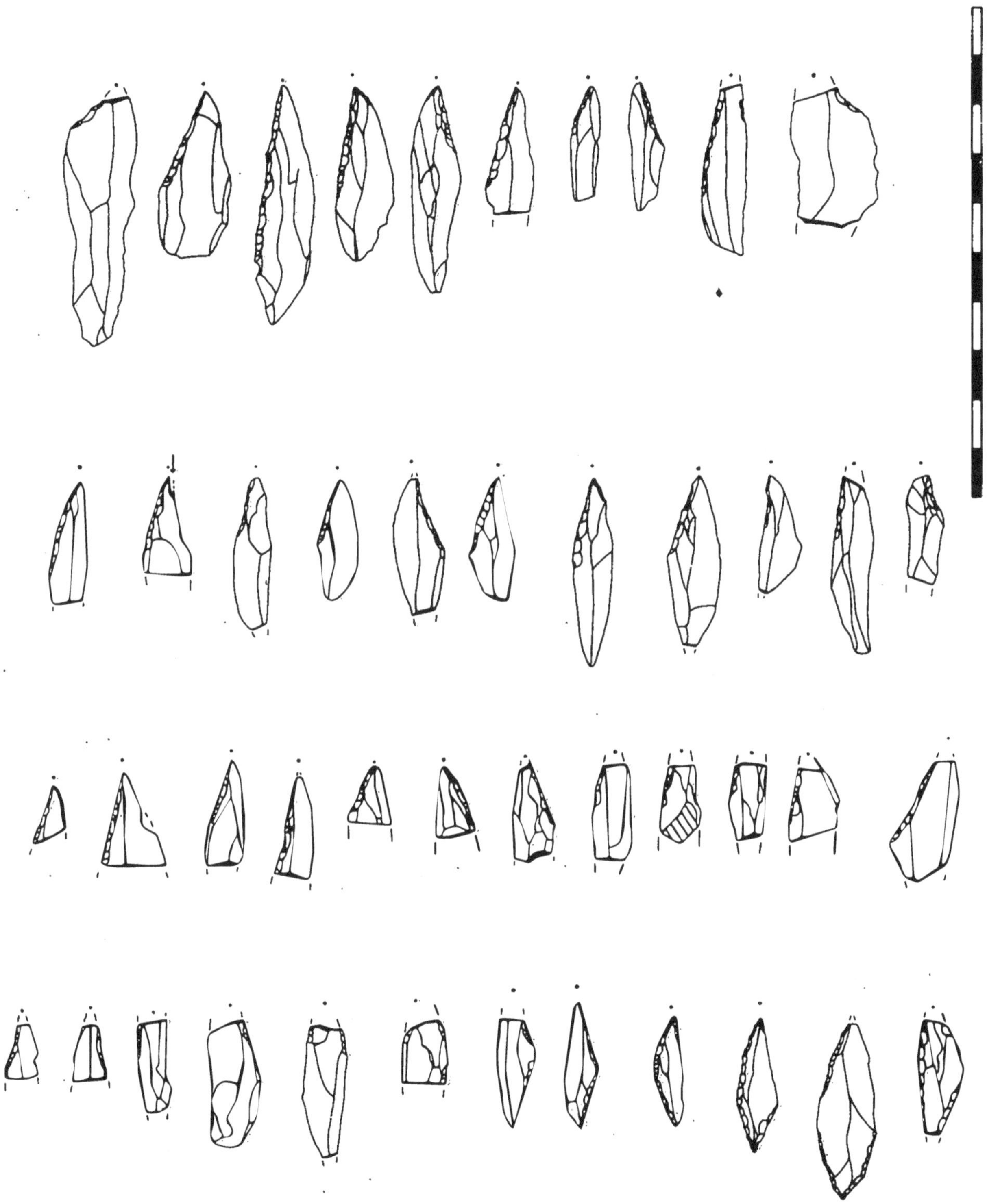

Fig. 4.10 Microlithic from Trwn Ddu, (Aberffraw) Anglesey

MICROLITHS →

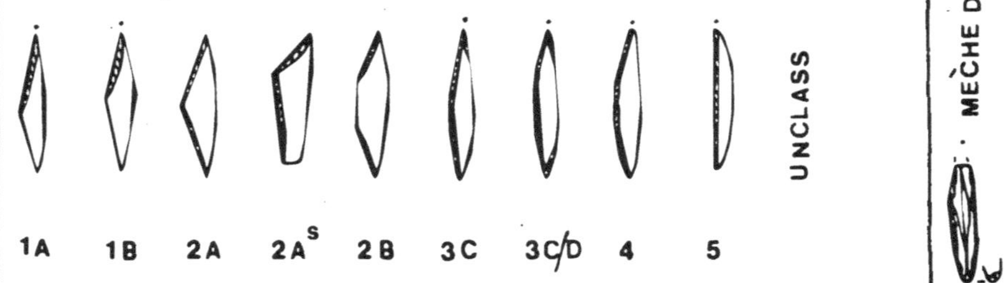

SITES ↓	1A	1B	2A	2A^S	2B	3C	3C/D	4	5	UNCLASS		MÈCHE DE FORET
RHUDDLAN M (UNDISTURBED)	12	7	1	1		1	1	1		8	≤32	
RHUDDLAN M (DISTURBED)	7	4				1		1	1	5	≤19	1

Fig. 4.11 Table of Microliths from Rhuddlan site M

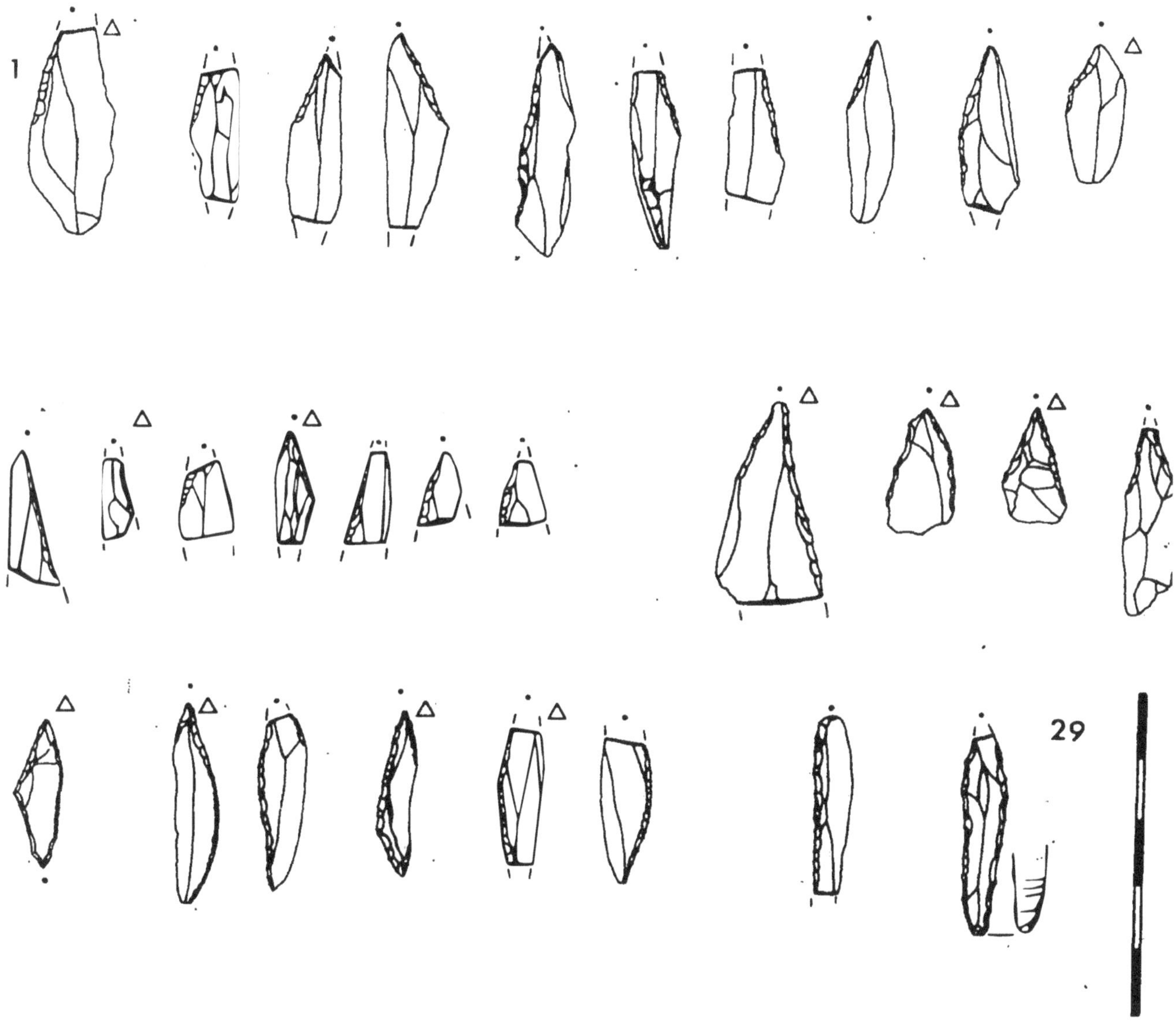

Fig. 4.12 Microliths from Rhuddlan site M: those marked with a △ found in association with radiocarbon date

Also incorporated into Fig. 4.6 are stray finds of transversely sharpened core-adzes since evidence suggests that these are to be associated in Wales with an Early Mesolithic technology. The distribution of these isolated adzes is effectively the same as that presented by Wainwright (1963: p. 112), except that it is possible to add an adze found near St. David's in 1911 (National Museum Acc. No. 70.24H).

Given the paucity of radiocarbon determinations it is difficult to document the chronology of the appearance in Wales of this technology in which the commonest form of microlith is the obliquely blunted point and where what variation there is between assemblages is in the relative proportions of the rarer and more elaborate microlith shapes (our classes 2 to 4). The dates we do have are all from north Wales the earliest (Fig. 4.7) falling close to 6,700 b.c. with determination of BM 691 = 6,789 b.c. ± 86 for Rhuddlan site E (Microliths: Fig. 4.8) and of Q 1385 = 6,690 ± 150 for Trwn Ddu (White (1975) p. 25). Broadly concordant would be a second determination from Trwn Ddu (Har 1194) and a single date for Rhuddlan site M of BM 882 = 6, 598 b.c. ± 73. The third date from Trwn Ddu (Har 1193) is believed to be too young and appears out of sequence within the local site stratigraphy (White, R. B. (1975) pers. comm.).

While apparently a millennium later than Thatcham site IIIA, our earliest British Maglemosian assemblage, Fig. 4.9 reveals the quite clear similarities between the microlithic technology of this site and that of our two earliest Welsh sites (see also Fig. 4.10 for illustration of microliths collected at Trwn Ddu prior to the recent excavations). In each case some 80% of the microliths are obliquely blunted points and, as we suggested earlier, what variability there is occurs among the other shapes, most noticeably the rather higher representation of triangles at Rhuddlan E (Fig. 4.8, line 4) and the absence of convex backed (class 4) microliths in the sample from Trwn Ddu.

The group from undisturbed contexts at Rhuddlan site M consists of only some 32 microliths, and the 20 classifiable examples include (Fig. 4.11) no form not present at Rhuddlan E or Trwn Ddu. The more complete microliths actually associated with the determination of BM-882 = 6,578 b.c. ± 73 are indicated on Fig. 4.12 by open triangles. From disturbed contexts comes a similar range of microlith forms together with a single rod-like piece, perhaps later Mesolithic in date (Fig. 4.12, no 28) and an awl retouched steeply down both sides (no. 29).

Two points immediately emerge from any consideration of the dates presented on Fig. 4.7. The first is that dates for these North Welshdates appear markedly more recent than estimates for comparable early ("Broad Blade") assemblages in northern England whose ages—plotted on the same diagram—even given the extreme standard deviations imposed by the smallness of the charcoal samples available, would seem to cluster close to 7,500 b.c. (Switsur and Jacobi 1975). Secondly, the Welsh dates overlap in age the earliest determinations for later industries characterized by micro-triangles from both Ulster (Woodman, 1974A) and for northern England (Jacobi, 1976, pp. 71-72)) the two land units geographically closest to North Wales. At the same time, taken in an overall European context, they appear to be among the latest dates for a "Maglemosian" technology anywhere obtained.

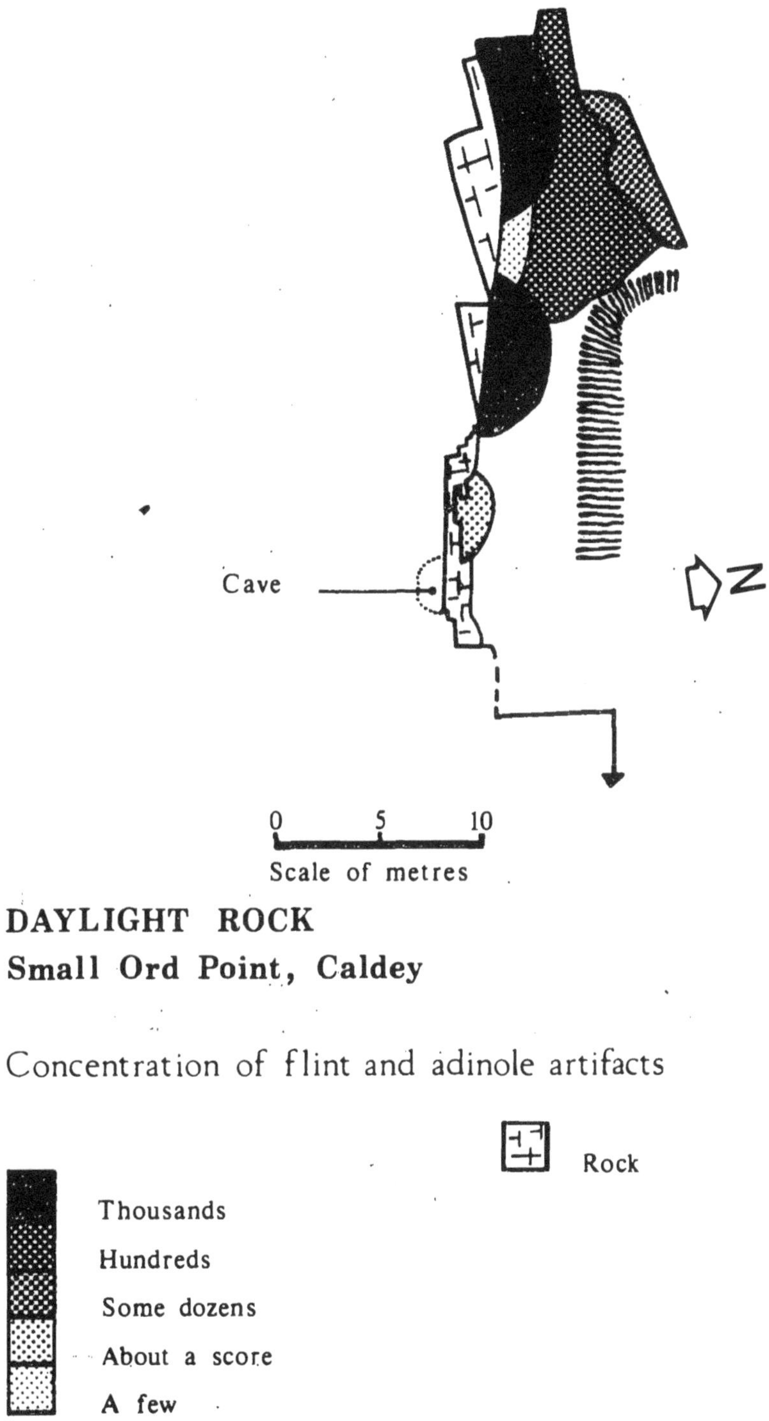

Fig. 4.13 Plan of Mesolithic Site at Daylight Rock, Caldey

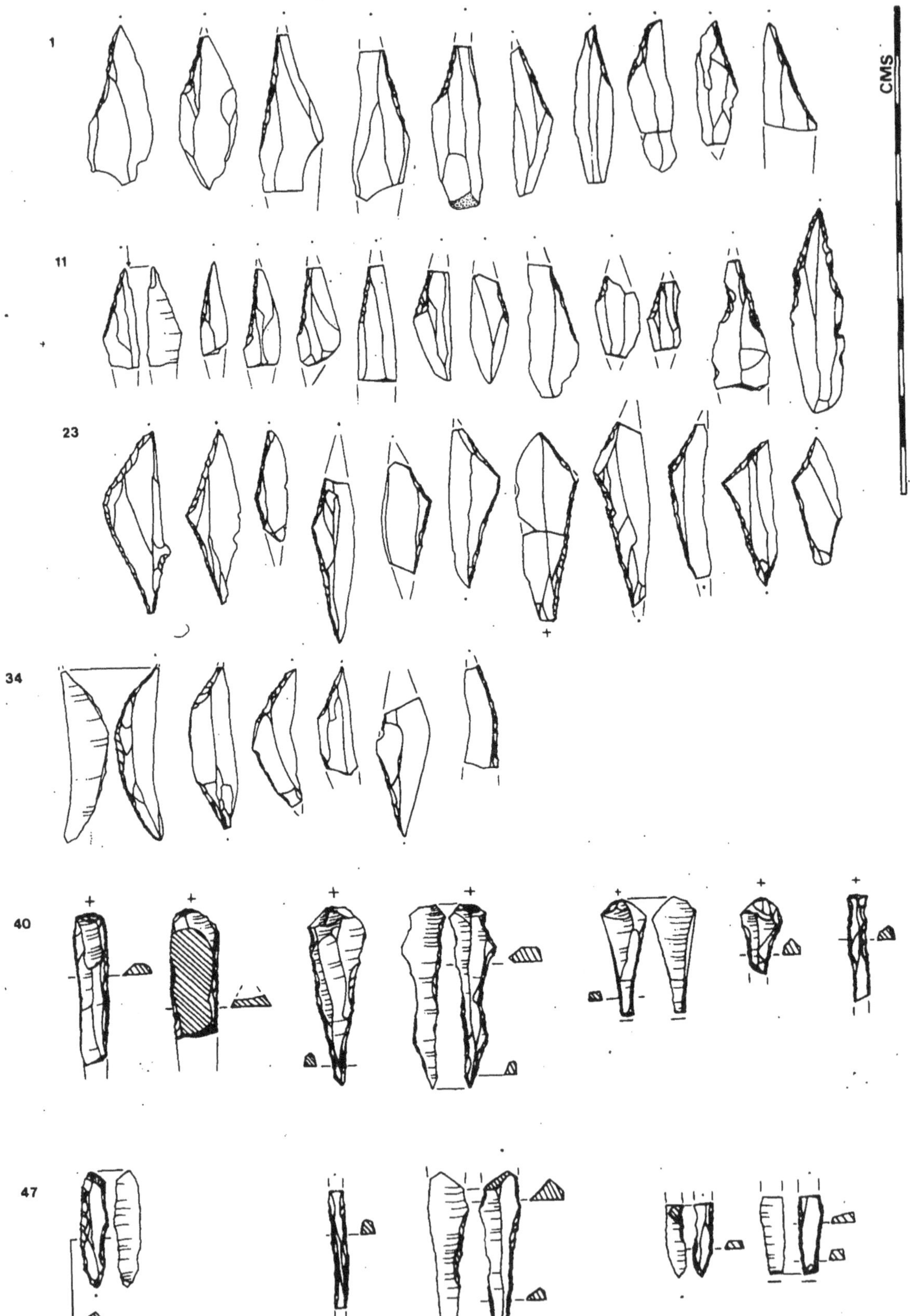

Fig. 4.14 Microliths and Mèches de Foret from Daylight Rock, Caldey

If, on typological and geographical grounds, we wish to relate the earliest microlithic industries of North Wales to the Early Mesolithic settlement of northern England, then it is important to recognise a gap in the radiocarbon record either side of 7,000 b.c. for one area or the other.

It would seem unreasonable to attempt to extrapolate from these radiocarbon estimates the date at which a microlithic technology appeared as far away as south Wales and around the Bristol Channel on either its present northern or southern coasts (Harding (1950), Smith (1963), Norman 1975)), and it appears important to consider the evidence from this area on its own merits. On the Welsh side of the Bristol Channel relevant to this problem are not only the very small assemblages of artifacts from Burry Holms and Freshwater East (about which little can be said except that they could fit anywhere within the early Mesolithic) but also the assemblages from Daylight Rock, and less directly than from the Nab Head. In view of the considerable discussion which has surrounded these two latter assemblages (Lacaille and Grimes (1961) pp. 48-50, Wainwright (1963) pp. 125-126, Houlder (1977)) it is perhaps appropriate to look in quite some detail at the archaeological content of these two, now "cliff-top" sites. Of these two assemblages , both undated by physical means, that from Daylight Rock must be of the greater value in that it was excavated at one time from a single site with maximum dimensions of 19 m by 9 m and a total area calculated at just over 110 m^2 (Fig. 4.13: Lacaille and Grimes (1955) pp. 131-136). The elongated oval distribution of the artifacts reflects local topography with optimum use being made of the shelter provided by the cliffs, while within the oval there are two semi-circular concentrations of artifacts against the rock, 8 m x 3 m and 8.5 m x 3 m, separated by a 3 m belt with a sharply lowered density of worked flint.

In considering this material, it is of course a necessary assumption that only one population of artifacts is represented. A total of 68 small, steeply retouched artifacts are present among the surviving samples of excavated material, of which 54 are certainly microliths. Of these 27 are obliquely blunted points (Fig. 4.14, nos. 1-22). Three of these (Nos. 21 and 22) have additional retouch on the leading edge opposite the oblique truncation while one (No. 11) is damaged at the tip, a spall being broken away on impact, presumably upon the bone of a prey animal. Thirteen microliths are triangles, five isosceles (Nos. 23-26) and eight large scalene forms (Nos. 27-34). Similar large scalene triangles are known from such eighth millennium contexts as Star Carr (Clark (1954), Fig. 35, no. 34), Flixton Carr (Moore (1950), Fig. 4, nos. 54. and 55), Thatcham site II (Wymer 1958) and Broxbourne Site 104 (in preparation). At the two latter sites, as at Daylight Rock, these large scalene triangles which obviously bear a close resemblance to the Creswell points of the Late-glacial outnumber triangles isosceles in outline. Four blades have a double oblique truncation to give a trapezoidal outline (nos. 35-38) and one has a double truncation to give a rhomboidal outline (Lacaille and Grimes (1955) Fig. 23, no. 15), while the last (no. 39) appears to be a fragment from a point with convex blunting down its right hand margin. Eight microliths are too fragmentary for classification.

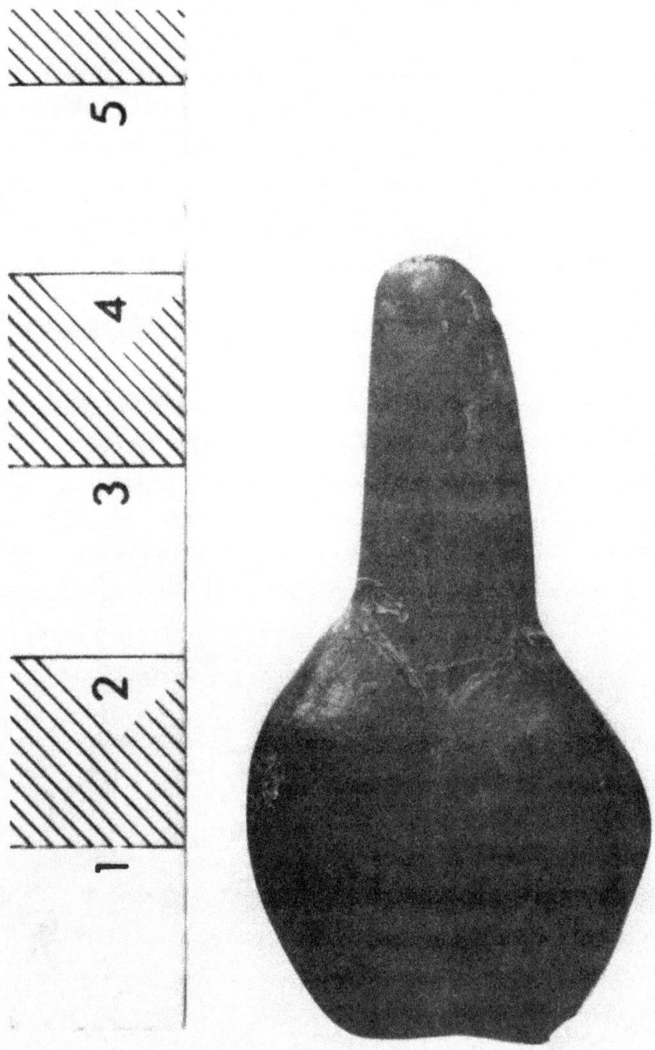

Plate 4.I "Venus" figurine from the Nab Head: front view: (inches scale)

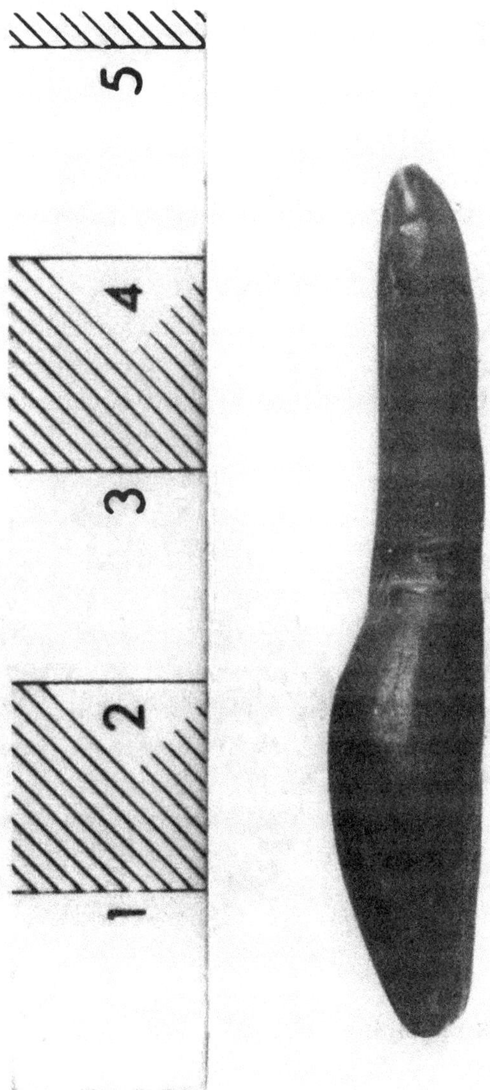

Plate 4.11 "Venus" figurine from the Nab Head: side view: (inches scale)

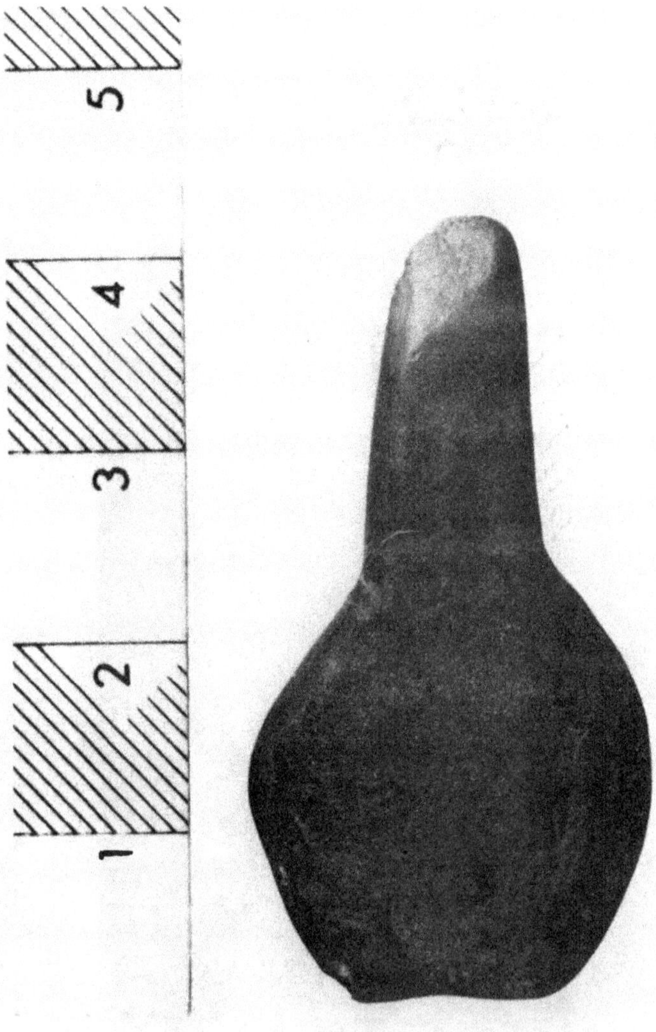

Plate 4.III "Venus" figurine from the Nab Head: rear view: (inches scale)

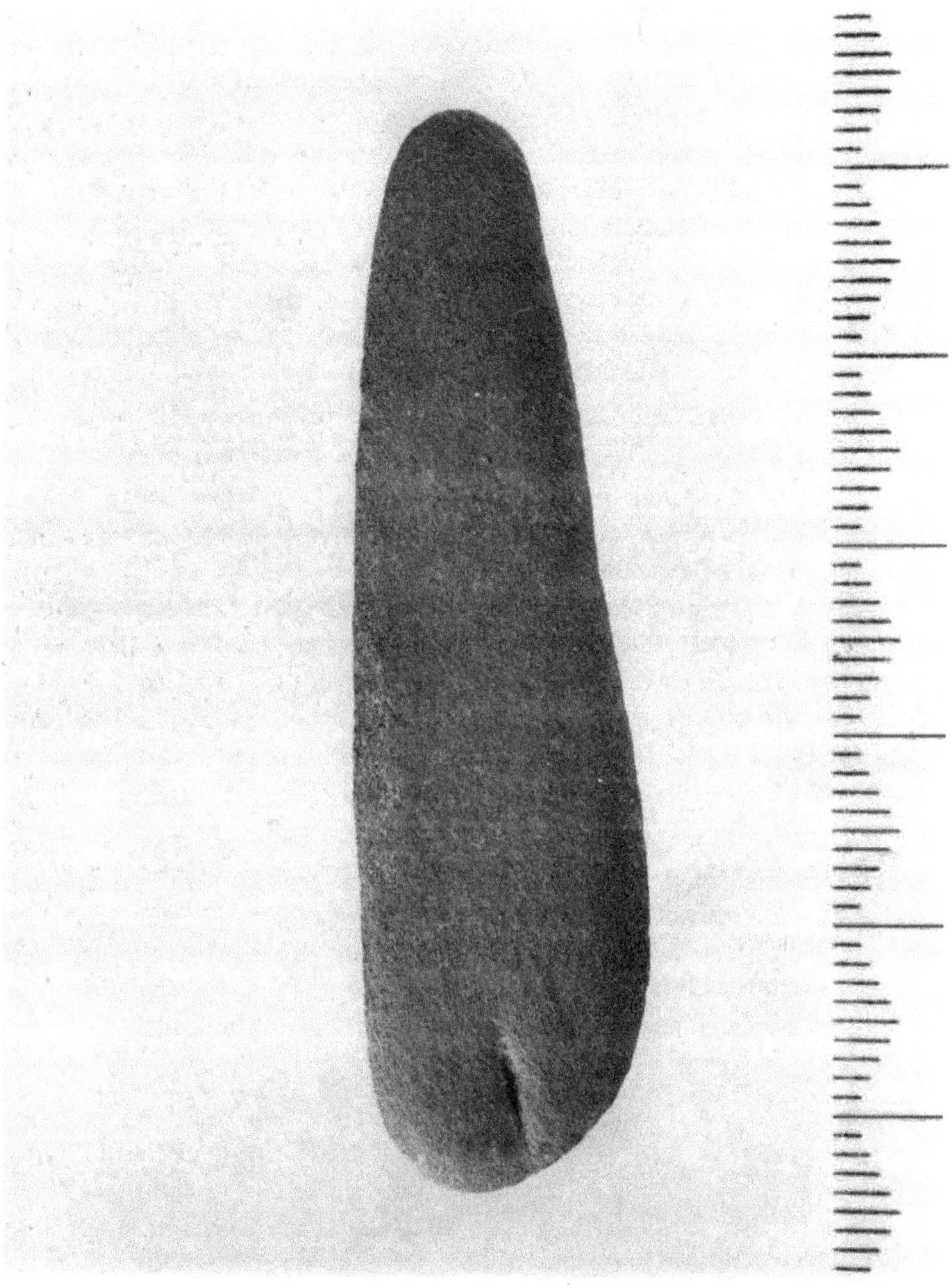

Plate 4.IV "? Stylised Venus" from the Nab Head: rear view: (inches scale)

While none of these shapes would appear out of place in an Early Mesolithic context the heavy representation of triangles, particularly scalene forms, sets this material apart from the bulk of English and Welsh Early Mesolithic (Maglemosian) assemblages. This impression is confirmed by computer cluster analysis (using both Average Linkage and Ward's method clustering on Euclidean distances calculated from an arcsine transformed percentages: forthcoming) which suggests that the number of microlithic assemblages to which that from Daylight Rock bears any close quantitative similarities are indeed relatively few but that this group includes both Star Carr and Flixton Carr as well as a pair of upland Cleveland's sites (Pointed Stone sites II and III: Jacobi 1977) and Broxbourne site 104 (Jacobi et al. in preparation). Consideration of the available radiocarbon and pollen evidence for Star Carr, Flixton Carr and Broxbourne suggest that all three sites belong within the 8th millennium b.c.—in the case of Star Carr to the middle, and of Flixton to the end, of the millennium, and it would seem possible to argue therefore, on the basis of the restricted typological comparison offered, that Daylight Rock also belongs within this time bracket, and could therefore be rather older than either Rhuddlan or Aberffraw.

The remaining 14 "steeply retouched artifacts" from Daylight Rock call for more detailed comment. All could be described as microblades with vertical (or near vertical) blunting down both sides. Of complete or near complete examples, seven retain distinct butts and bulbs (Fig. 4.14, nos. 40-46) with points developed on their distal ends. Two (no. 47) have the points worked at the bulbar extremity while there are a pair of medial fragments (nos. 48 and 49) and two broken tips, one still pointed (no. 50) the other blunted by use (no. 51).

It would be simple, as did Wainwright (1963, pp. 108, 114 and 120), to regard these as microliths ("Needle-shaped points") and compare them to specimens found in Later Mesolithic assemblages. In such assemblages, however, they are never more common than scalene micro-triangles (our class 7) and straight rod-like microliths blunted down one side (class 5), neither of which category is represented at Daylight Rock.

An alternative point of comparison, however, for these fourteen steeply retouched pieces might be among the 'category' of 'mèches de Foret' (Brinch) Petersen (1966) p. 84: Clark (1975) p. 108) or steeply retouched awls which appear over a wide area of Northern Europe in the mid 8th millennium b.c. Tools equivalent in shape and size to those from Daylight Rock are recorded almost always as isolated examples, from a large number of British Early Mesolithic assemblages—Thatcham (Wymer (1962) Fig. 7, nos. 61 and 76). Downton, Hillwood (Epping), Dozmary Pool (Wainwright (1960) Fig. 4, nos. 49 and 50), Trwn Ddu Anglesey: Recent excavations) and Rhuddlan site M (Fig. 4.12, no. 29).

At Daylight Rock, however, these "Mèches de Foret" make up some 10% of the tool kit and it may not be completely coincidental, given the comparisons already made, that the only English site where they achieve an equivalent representation is Star Carr with a value of 12%. The tips of many of these awls from Daylight Rock show scaling and crushing comparable to that observed on the isolated specimens from Iping, Rhuddlan and Dozmary Pool, and on

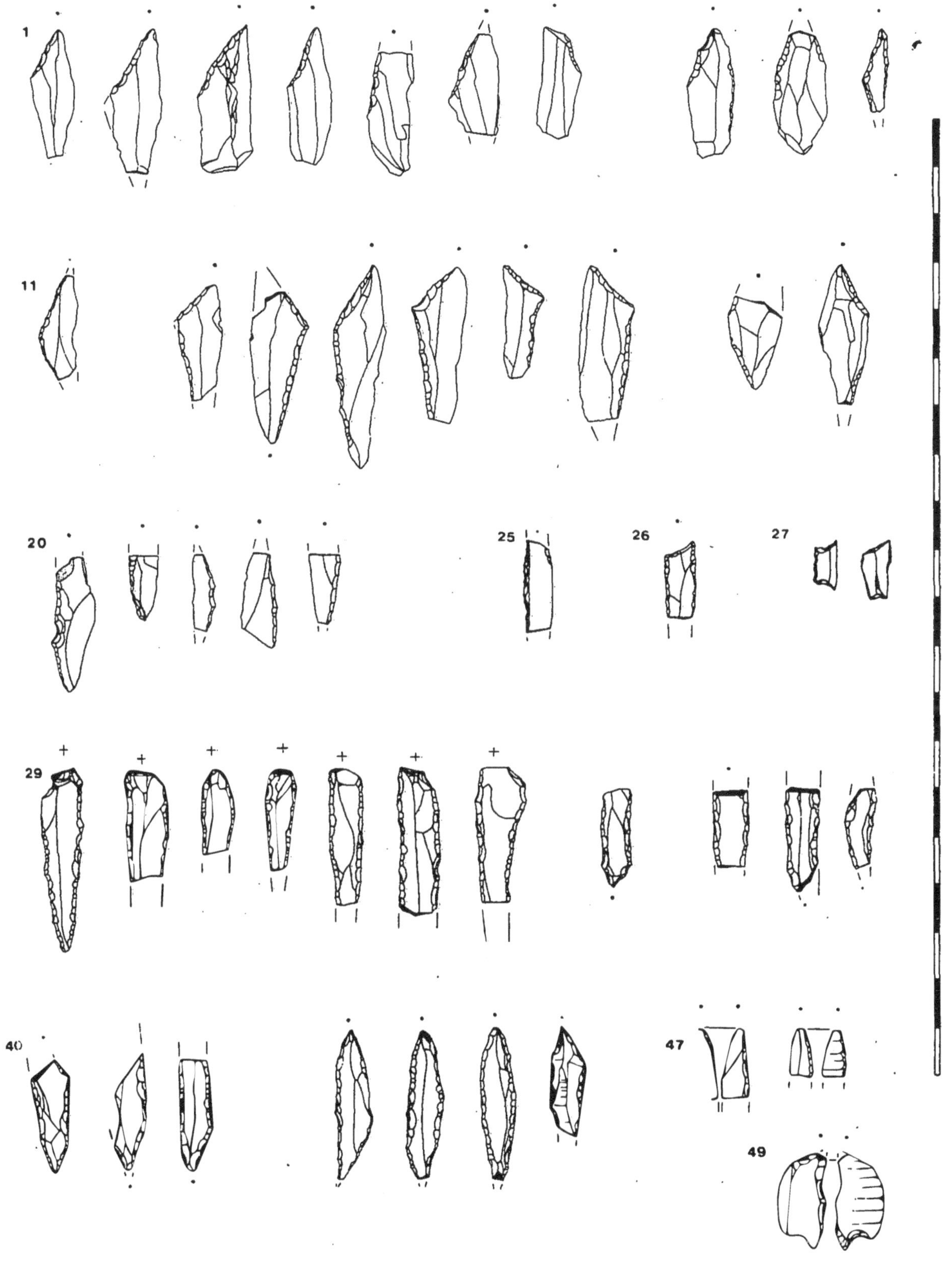

Fig. 4.15 Microliths and Mèches de Foret from the Nab Head

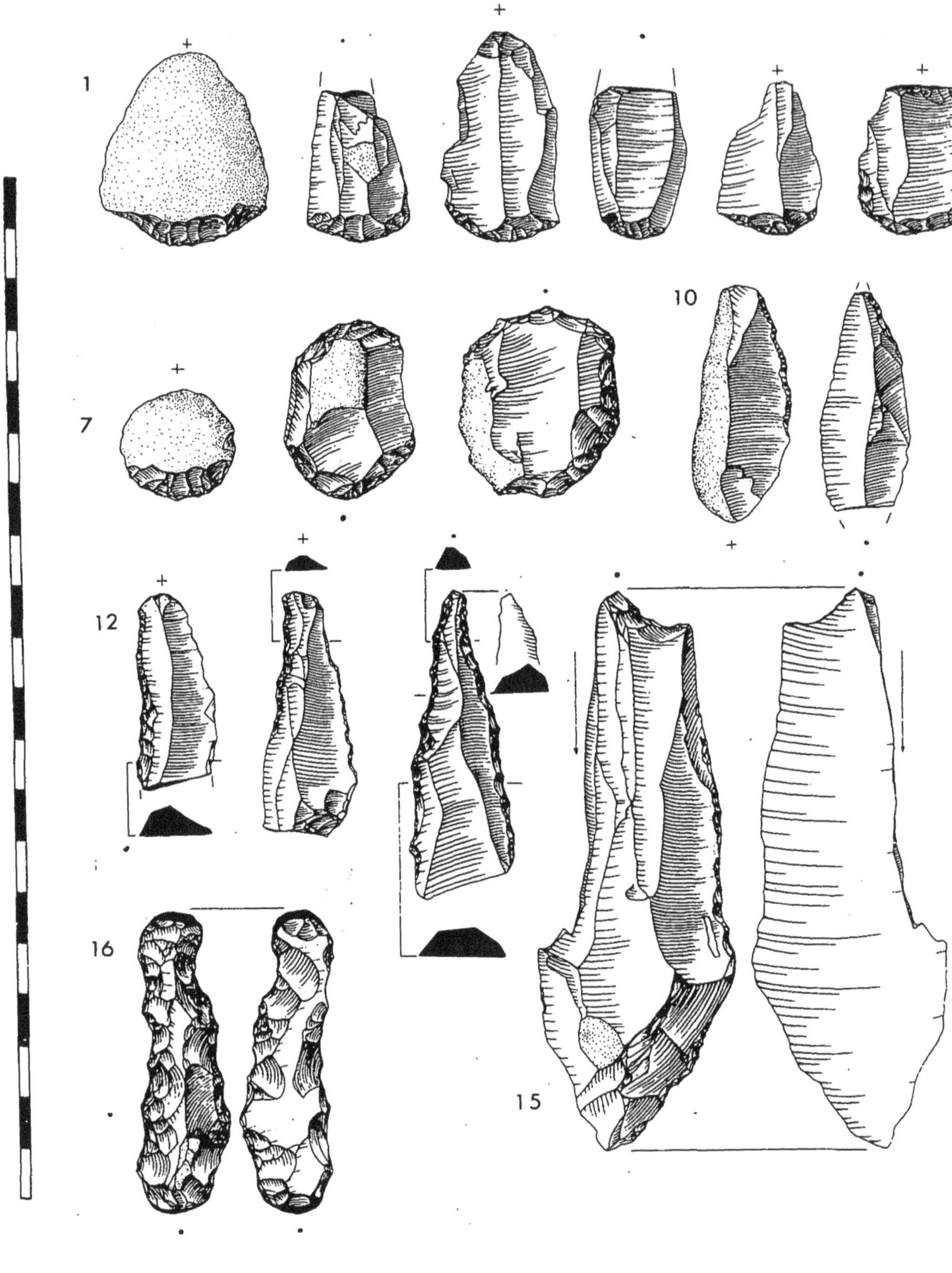

Fig. 4.16 Flint tools from the Nab Head

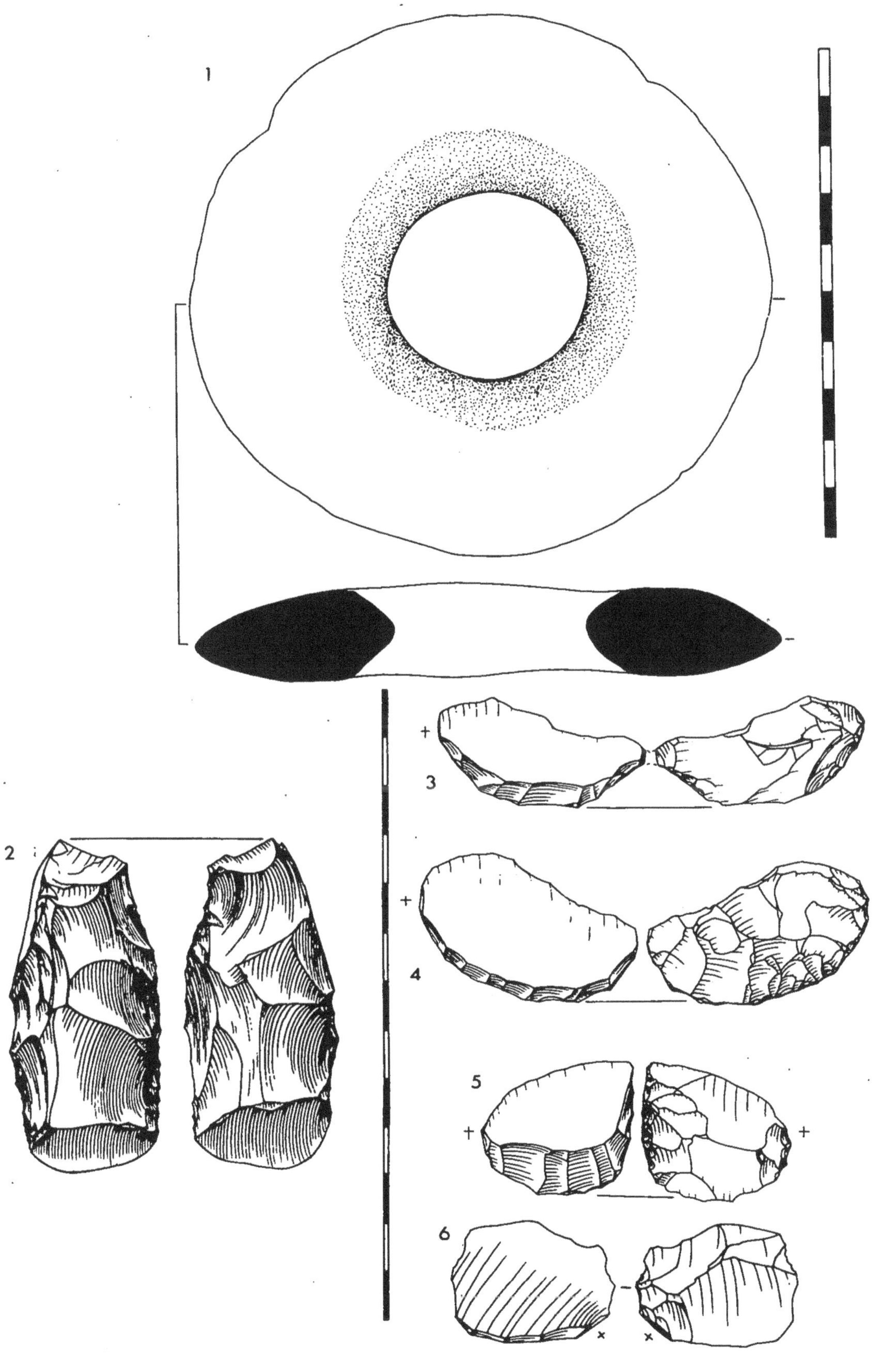

Fig. 4.17 Stone ring, flint adze and adze-sharpening flakes from the Nab Head

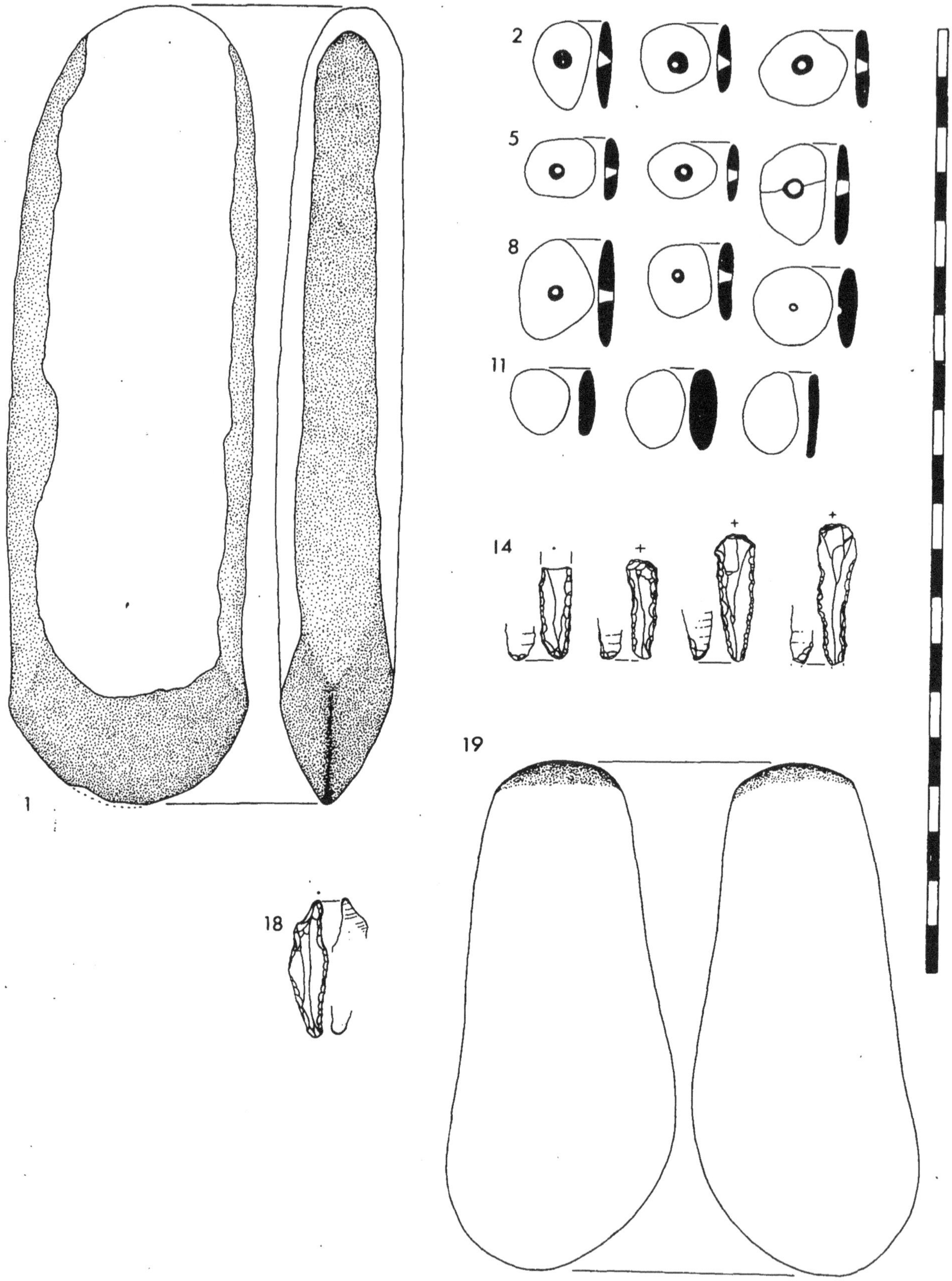

Fig. 4.18 Pecked stone adze, beads, Mèches de Foret and bevelled pebble from the Nab Head

very many of the examples from the Nab Head where these specialized "awls" make up some 17% of the total tool kit. This damage, which certainly lies outside the range of impact damage so commonly encountered on microlithic projectile equipment, resembles far more the utilization traces observed on hand-turned drill bits, only one example from the Nab Head actually being worn smooth (Fig. 4.18, no. 18) as if by rotation in a high speed precision "pump" or "bow" drill, for which the perforated stone weights ("Mace-heads") appear in Northern Europe, including Britain, rather before 7,000 b.c.

At the Nab Head at least, it is tempting to associate these tools with the perforation for beads of natural shale discs collected from beaches between one and two miles from the site, and certainly unperforated discs, as well as one half drilled have been recovered from the site (see below). No such beads were, however, recovered from Daylight Rock, and this cannot be the explanation of their presence here.

Reconsideration of the archaeological evidence suggests an Early Mesolithic date for this assemblage and the closest comparable English Maglemosian assemblages appear to belong within the eighth millennium b.c. A similar age is therefore suggested for this occupation at Daylight Rock, while fragments of typologically apparently identical assemblages can be recognised from Valley Field on Caldey (Lacaille and Grimes (1961) p. 60, also Grimes, W. F. pers. comm.), and from Palmerston Farm near Haverfordwest (J. Rees, pers. comm.).

Archaeological material has been collected for just under a century (Leach (1933) p. 229) from the promontory of Old Red Sandstone known as the Nab Head. With the exception of a spindle-whorl (Grimes (1951), pp. 12-14) and a flake recognised by C. H. Houlder as from a polished axe of group VIII rock, provenanced St. Brides rather than more specifically as from the Nab Head (Tenby Museum), all the material recovered from here would appear to be Mesolithic.

Most important among these finds is what appears to be the only carving in the round known from a Mesolithic context in the British Isles (Plates I-III). Worked in black shale this object which is 42 mm in height is interpretable either as a Venus figurine or as a phallus (Gordon Williams (1926) pl. 5, Breuil (1953) and Geidion (1963) p. 234, and fig. 166). Its back is left naturally flat and slightly rough to the touch, with cutting and shaping confined to the front and sides of the body and thighs. If it is a 'Venus figurine', there is no indication of a head or of the arms, while the lower legs, if any, appear to be absorbed up into the thigh region. Much of the surface of this object is glossed as if through persistent handling, and when found the object was covered with a "soapy substance".

A second shale object from the site (Pl. IV), an ovoid pebble, 57 mm high, and with a flattened-oval cross-section, has a sharply defined incision just over 7 mm long at its base, and it would be attractive to interpret this object as a second more "elementary" Venus representation, with the vulva indicated by the incision. The object is rather weathered and any further traces of shaping will thus have become obscured (originally illustrated Gordon Williams (1926) pl 3.).

With the shale figurine (Ibid pp. 88 and 94) were nine out of a present total of 523 shale beads—the number traced to date—the bulk collected by Gordon Williams after his excavations in 1925 (Fig. 18, no. 2-10). Ranging from circular to oval in outline through celtiform to triangular or even rectangular, the beads, which average 2 mm in thickness, are perforated by means of a single crater drilled from one face. That some at least of these beads were manufactured on the site is, as we have suggested, indicated by the presence of a single partly perforated shale disc (Fig. 4.18, no. 10) and a number of unworked discs (Nos. 11-13). Strung together and without knots between to act as spacers, these beads would together make up just under three and a half feet of necklace. In size and material these beads are identical to 27 from Star Carr (Clark (1954) Pl. XX A and F) and single specimens from Freshwater East (see above) and Limney Burrows (Fig. 4.5, left hand side). With this last were found both the microliths illustrated and a group of Neolithic arrowheads.

As far as is known, all these beads occurred as surface finds and there is no evidence as to whether they formed parts of a single object of adornment, such as a necklace, derived from a number of necklaces, or from smaller objects such as armlets or anklets. The number of finished beads abandoned on the site appears too great to be accidental, and one might perhaps justifiably speculate if their presence (taken with that of the figurine(s), and such other early finds from the Nab as stone rings and a pecked stone adze, this last also without obvious parallel in the British Mesolithic) is not perhaps suggestive of deliberate deposition, perhaps originally with burials. That beads at least have been found over a period of almost a century, as erosion proceeded, might be taken as indicative of a number of groups of jewelry.

Finds of objects of adornment are, in fact, rare in the Mesolithic. Perforated stone beads appear to be confined to the early stages of the period with finds outside Wales coming only from Star Carr (see above), Thatcham site VI (in preparation) and perhaps Colne Valley in Essex (Layard (1927) Fig. 4, no. 5: now lost). From the later Mesolithic we have instead more weighty pendants from Hawkcombe Head on Exmoor and Culverwell on Portland. At Star Carr these stones beads are supplemented by perforated amber lumps and at Culverwell by beads made of sea-shells.

The statuette and the first two dozen of the beads from the Nab Head were recovered from the site by Gordon Williams during his explorations of 1925 (1925 a-c and 1926). Subsequent to this date, Gordon Williams clearly revisited the site and recovered much more material. The collections which survive from the site, however, present many problems. Firstly, it is clear that what was excavated in 1925 is now largely missing or unrecognizeable, while one part of the material in the National Museum and published as deriving from this site (Wainwright (1963) Fig. 6, Nos. 4, 9, 15-19, 28 and 37) may incorporate artefacts from other Pembrokeshire findspots—Stackpole, Castlemartin and Brownslade Burrows where Gordon Williams had also collected—and it cannot, therefore, be made use of (documentation in the National Museum). Large collections of material in the Pembrokeshire County Museum are believed also to derive from Gordon William's work at the Nab Head but now lack the documentation to make the identification certain.

Flint artifacts, therefore, which can certainly be recognized as coming from the Nab Head are only those which can be identified as among the illustrations published in the 1926 report or are among the drawings and photographs in the Museum register at Abergwili (Carmarthen). To these it is possible to add the separately labelled core adzes and beads found by Gordon Williams and also now at Abergwili. Small collections of microliths and other tools made from the surface of the site by other workers are now in the National Museum and the Tenby, and Pembrokeshire County Museums while small groups of material exist also in the Ashmolean, and the Cambridge Museum of Archaeology and Ethnology. The material in the Pembrokeshire County Museum includes that excavated by Wainwright (Wainwright 1970) while small well provenanced collections exist also in private hands (Rees, 1973).

Among microlithic material which thus survives (Fig. 4.15), it is possible to recognize a group of obliquely blunted points (Nos. 1-10), a broad isosceles triangle (No. 11), a group of large scalene triangles (Nos. 12-17) and a pair of bitruncated points (Nos. 18 and 19). All these together with the convex backed microliths (Nos. 20-24) have precise parallels among the assemblage at Daylight Rock, and argue for an early Mesolithic occupation of the site. Directly comparable also with Daylight Rock is the large group of "Mèches de Foret" recovered (Nos. 29-46 and Fig. 18, Nos. 14-18) and the greater part of the non-microlithic tool kit with its convex-edged rather than denticulated scrapers, and core adzes (see below). It would not appear totally unreasonable to regard the great bulk of the material recovered from this site as comparable to Daylight Rock and hence perhaps, if we extend the argument, as early as the eighth millennium b.c.

Outside the range of anything recovered at Daylight Rock, however, are a small number of microliths: No. 25 a narrow rod-like piece, No. 26 the forepart of a scalene micro-triangle and Nos. 27 and 28 "four-sided forms". While all these shapes are characteristic of later Mesolithic assemblages the four-sided forms do not appear much before about 5,000 b.c. and their presence at the Nab Head would seem to imply reoccupation of this rock promontory at a moment when eustatic rise of sea-level had effectively achieved its present-day level.

Location of Early sites and site economics

Of the sites which, it has been suggested, fall earliest within the local sequence, Burry Holms and Daylight Rock are on islands, Trwn Ddu (Aberffraw) is on the coast of the present Anglesey Island while Rhuddlan is only three miles inland. In considering the economic potential of these sites and the two other points where it has been suggested early material has been recovered—the Nab Head and Aberystwyth—it is clearly necessary to compensate for loss of dryland resources within their respective site territories due to eustatic rise of sea-level since the moment of their occupation. With the minimal data at our disposal such estimates must be of the crudest.

Nevertheless, if we accept the ages for Rhuddlan E and Trwn Ddu of close to 6,700 b.c. and for Rhuddlan M of about 6,500 b.c., it is clear that contemporary sea-level would have stood lower than—70 ft. (-21.5 m) O.D. closer to 6,700 b.c. and below—68 ft (- 20.7 m) closer to 6,500 b.c. (Shotton

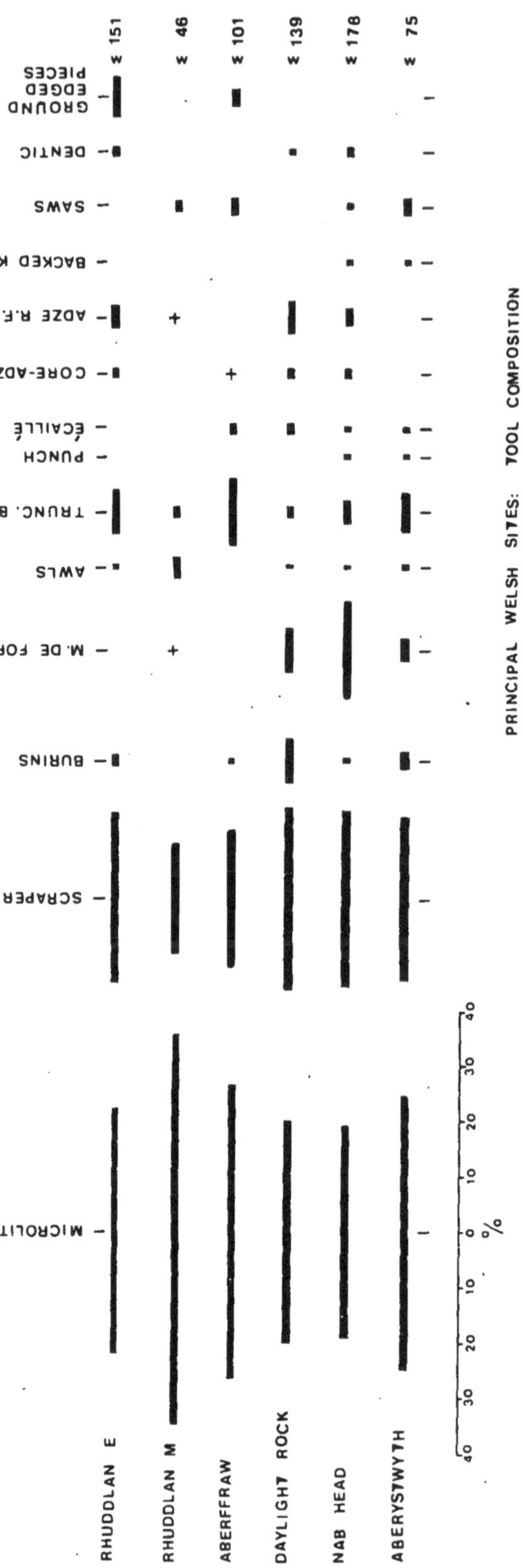

Fig. 4.19 Principal Welsh sites: tool composition

et al. (1974) p. 289 and Hawkins (1971) p. 121). Accepting these estimates as a useful working basis, the Rhuddlan sites lose their areas of fen and coast and Trwn Ddu, while in all probability retaining within its home range a length of shore, ceases to be a cliff-top site and Anglesey of course is no longer an island. If we convert these horizontal estimates into meat yields—employing a density of one red deer per 100 acres, as assumed for Star Carr (Clark 1972 p. 26)—it becomes clear that at a 6 mile (10 km) radius and with a 1:6 cull, the territories below 200 ft. O.D. around Rhuddlan and Trwn Ddu could each have provided 26 kg of meat (60% of deadweight) per diem over the whole year. These estimates obviously compare more than favourably with Clark's (1972) allowance of 21.7 kg of meat to fulfil the daily requirements of his four-family occupying group. In fact the yield of the Star Carr territory below 200 ft estimated at some 22.7 kg at a 1:6 cull (Clark (1972) p. 27) is obviously rather less than for these three Welsh sites.

No absolute age estimates are available for Burry Holms, Daylight Rock or the early artifacts from the Nab Head and Aberystwyth. The occupation at Daylight Rock, we suggested above, could fall within the 8th millennium. If, however, we make the probably more conservative assumption that these sites are broadly contemporary with Rhuddlan or Aberffraw, the dry land yield from below 200 ft. O.D., and at a 1:6 cull, for Burry Holms would be some 31 kg. per diem; for Daylight Rock 21 kg.; for the Nab Head some 21.5 kg and for Aberystwyth some 22.7 kg. Burry Holms and Caldey of course cease to be islands and in each case any length of contemporary coastline is pushed right to the edge of the respective site territory. In each case also it would appear that a population of Star Carr size could be supported over the whole year on ungulate resources alone.

It cannot of course be shown that any of these sites would be used over the whole year or by as many as four family groups. Indeed, if we take the area of flint debris at Daylight Rock and compare this with Weissner's (1974) regression of log of Kung! camp area viz. "... all of the space within a continuous border that encircles huts and major concentrations of bone, charcoal, vegetable remains etc...", on log of camp population, the resulting estimate suggests an exploiting population of about fifteen rather than twenty four. It remains also an assumption that there was no major redistribution of ungulate resources over the year (see below). The essential conclusion, however, is that none of these Early Mesolithic Welsh sites, although situated on or near the present-day coast, was, at the time of occupation, in any way less favourably situated for the successful cropping of ungulates than Star Carr.

Independent confirmation of the exploitation of these dry land resources would appear to derive from a comparison, presented on Fig. 4.19 of the total tool kits from Daylight Rock, the Nab Head and Trwn Ddu with those from Rhuddlan sites E and M whose site catchments take in only areas of dry land (see above). The total sample of tools from Burry Holms is, unfortunately, too small for statistical treatment but the non-microlithic tool kit includes single- and double-ended convex scrapers, finely serrated blades, a burin on natural truncation and several notched pieces. Survival of material from Aberystwyth is likely to be selective while existing collections are almost certainly mixtures of Early and Later Mesolithic artifacts. The material from Trwn Ddu is that collected previous to the recent excavations while for Rhuddlan only tools from undisturbed contexts have been incorporated.

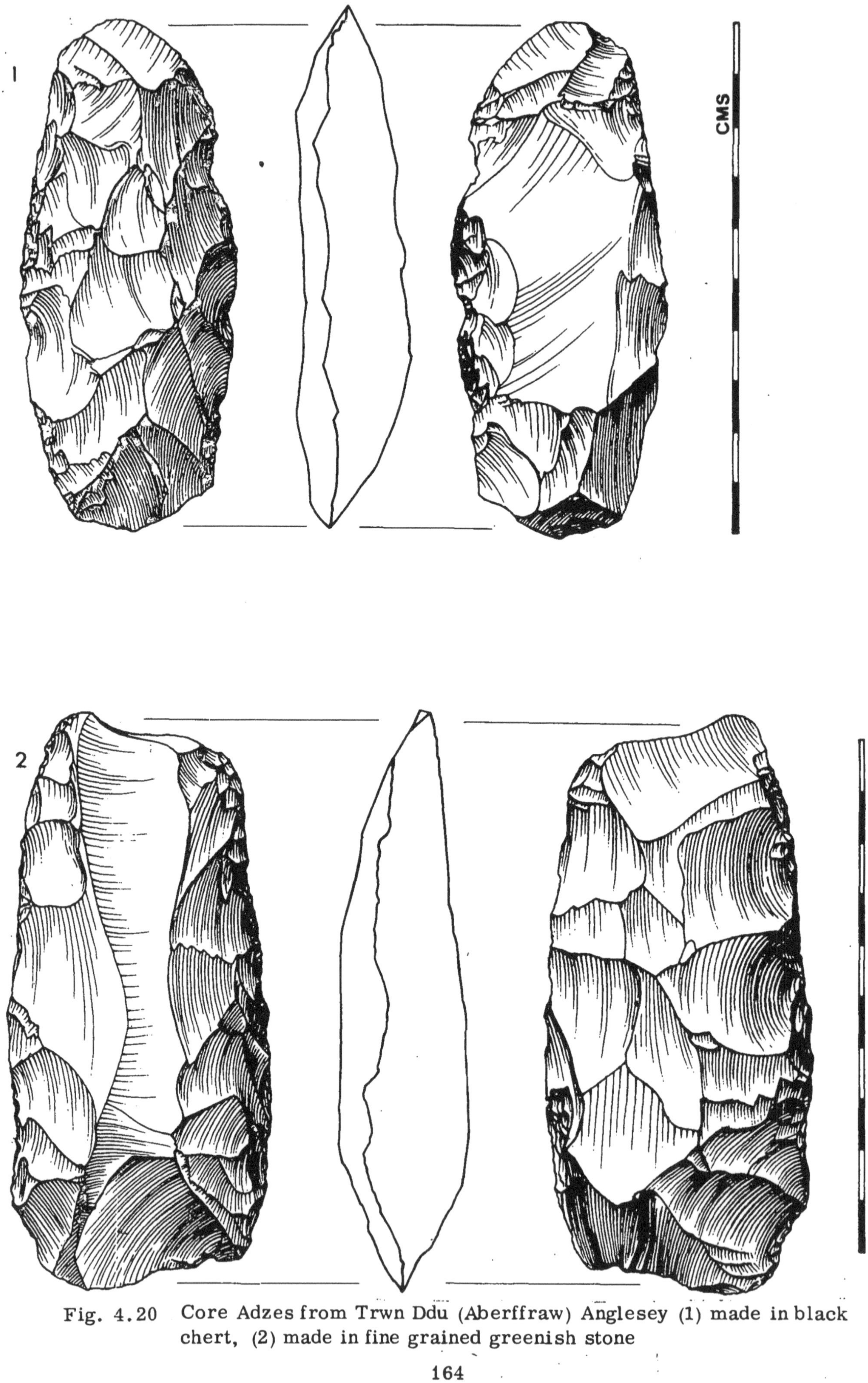

Fig. 4.20 Core Adzes from Trwn Ddu (Aberffraw) Anglesey (1) made in black chert, (2) made in fine grained greenish stone

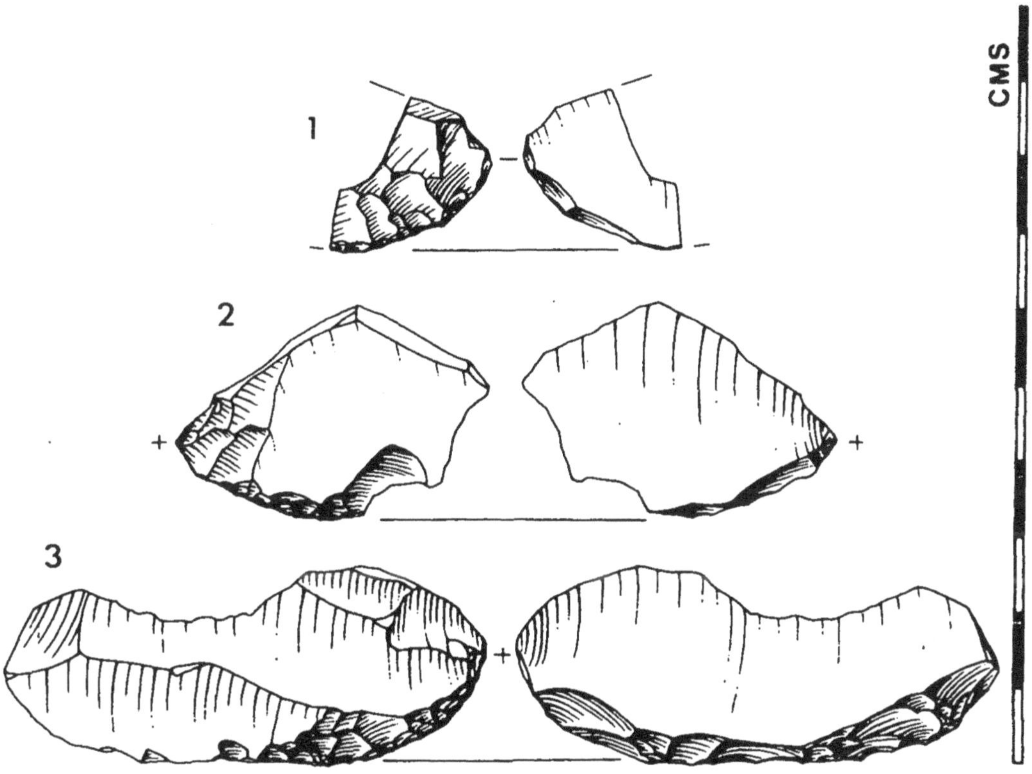

Fig. 4.21 Adze resharpening flakes from Daylight Rock (Caldey): all in stone

The figures show striking similarities both among themselves and to those for "Maglemosian" industries from northern and eastern England. On all the sites microliths make up close to half of the finished tools, while, in each case also, scrapers varying between 20% (Rhuddlan M) and 34% (the Nab Head) are the most numerous non-microlithic tool type represented (Fig. 4.16, nos. 1 to 9). Burins (Fig. 4.16, no. 15) are relatively few, but absent only from Rhuddlan M, with a total sample of only 46 tools. Representation of truncated blades (No. 11), backed knives (No. 12, Awls (No. 14) and "fabricators" (No. 16) is more variable, while retouched blades (Nos. 10 and 13) have been omitted from the diagram since it is probable that these will not have been deliberately retained in all the collections studied. Similarly, flakes and blades with their edges apparently "ground" smooth, a tool category represented on the majority of large Early Mesolithic sites, will have been recognized only by recent workers. The greatest variability observable is in the representation of "Mèches de Foret".

Four complete core adzes are recorded from the Nab Head (Fig. 4.17, No. 21) together with a possible butt fragment while there is a pair of adzes from Trwn Ddu (Fig. 4.20) and single examples from Daylight Rock (Lacaille and Grimes (1955) Fig. 4.21, no. 16) and Rhuddlan site E. These are supplemented in every case by groups of resharpening flakes (Fig.4.17, Nos. 3-6 from the Nab Head and Fig. 4.21, Nos. 1-3 from Daylight Rock) while resharpening flakes alone are present at Rhuddlan M. On each of these sites, fine-grained rock was used for core tools, a local adaptation in an area where flint of a sufficient size was not regularly available. At the Nab Head and Daylight Rock this takes the form of a "partly mineralized siltstone" while at the Nab head this is supplemented by use of a "Welded siliceous tuff" (C. H. Houlder: pers. comm. 27. 6.75). A resharpening flake from the Nab Head (Fig. 4.17, No. 5) is made in a chert visually similar to the Cretaceous cherts of Somerset, and isolated waste flakes of this material are recorded from the site. Flakes of stone from Aberystwyth and Valley Field on Caldey appear too thick to be derived from the thinning or resharpening of core adzes.

Summarized, the table suggests that there is very little difference quantitatively between the flint equipment found on these sites, while the range of equipment is equivalent to that found on Early Mesolithic sites elsewhere in Britain. The crude proportions of one tool to another are also very similar to those encountered in Maglemosian assemblages outside Wales. The Aberystwyth site with its assemblage probably contains Later as well as Early material and is included as a sixth 'substantial' assemblage from Lowland Wales.

At the Nab Head, the flint equipment appears to be supplemented by a "pecked-stone axe" (Fig. 4.18, no. 1 (Leach: 1933), without parallel within the British Mesolithic, but with its splayed blade possessing an outline closely similar to a ground and polished axe of Quartz-Dolerite, probably of N. Pembrokeshire origin from Caron-uwch-clawdd (Cardigan:Nat. Mus: 47. 164/201). Apparently found with it were a pair of stone rings with pecked perforations (Fig. 4.17, no. 1) but the original archaeological context of these finds is not now recoverable.

Bevelled and abraded pebbles which are usually connected with some aspect of coastal exploitation (see below) are recorded from both the Nab Head

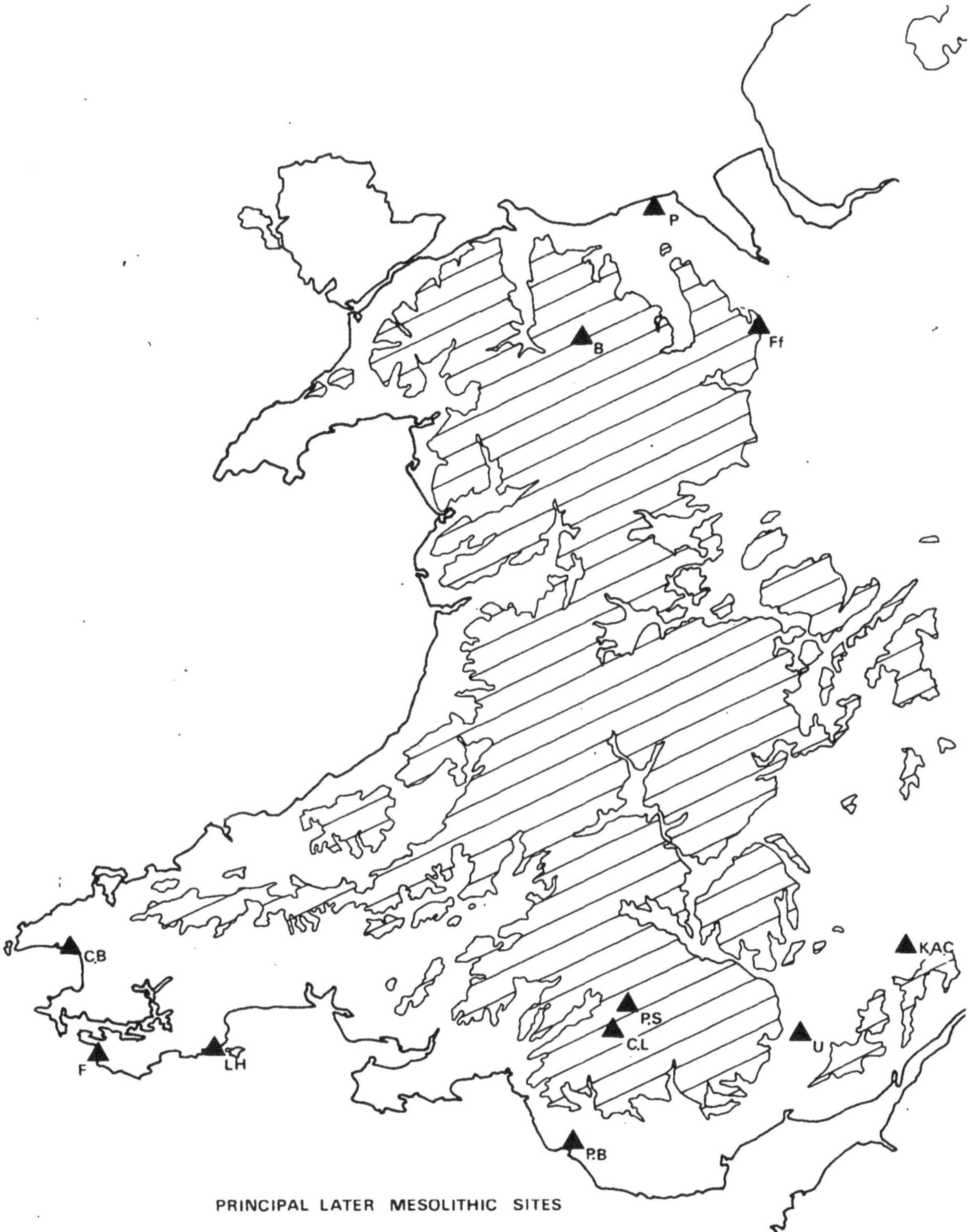

Fig. 4.22 Principal Later Mesolithic Sites in Wales

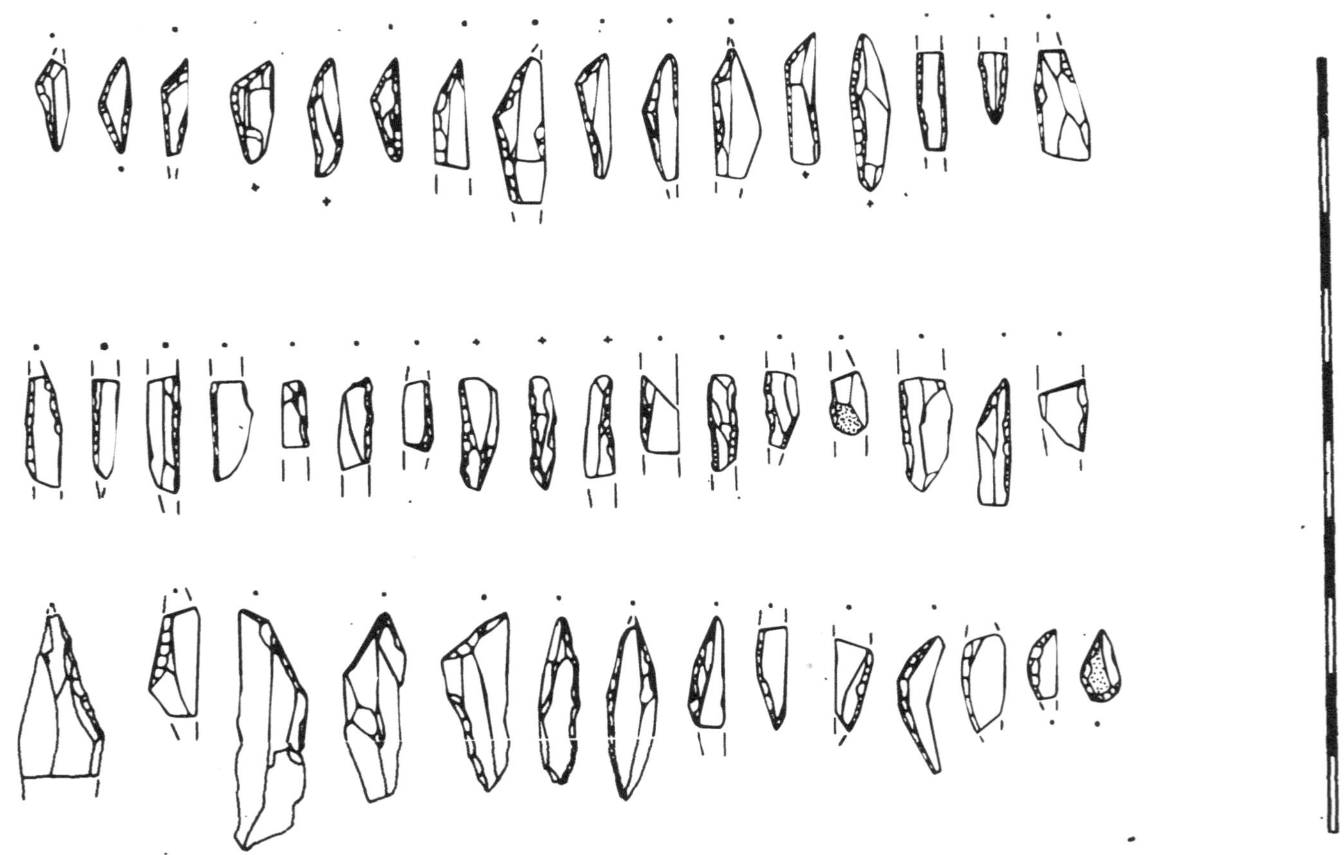

Fig. 4.23 Microliths from group of sites around Craig-y-Llyn (Glamorgan)

(Gordon Williams (1926) p. 88: Abergwili Museum) and Aberystwyth, (Thomas and Dudlyke (1925) pp. 85-86: now lost). Before, however, attaching these to the Early Mesolithic equipment from these sites, it should be remembered that microliths indicative of use of these localities considerably later in the Mesolithic and with a sea-level nearer that of the present-day, have been recovered from both. These artifacts cannot, therefore, be introduced into any argument for coastal exploitation along the western margin of the Maglemosian technoterritory. As with Star Carr, the sea-shore may have been only the source of raw materials and objects of adornment—amber at Star Carr; shale at the Nab Head.

The Later Mesolithic: Chronological Considerations

Of sites characterized by scalene micro-triangles (class 7) and narrow straight rods (class 5) which one would, therefore, describe as Later Mesolithic, the most valuable excavated samples derive from Prestatyn (Flints, see below), Pen-y-Bont (Ogmore, Glamorgan: Webley (1976), p. 35) and the Brenig site 53 (45) in Denbighshire. Other excavated groups, too small for statistical analysis from the Roman sites at Usk and Ffrith (Denbighshire), are supplemented by surface collections from a series of sites on ploughed land along the south side of the St. David's peninsula and from the Glamorgan uplands above 1,500 ft. O.D. around Craig-y-Llyn (Savory, 1961) and just above 1,100 ft. O.D. at Pant Sychbant (Brecon; Burke, 1966). In total some 66 Later Mesolithic findspots could be recognized from Wales (including Anglesey) and the Marches, the positions of the more important sites being recoverable from Fig. 4.22.

The forms that these new microliths take can be grasped from the top two lines of Fig. 4.23 of microliths from the large and scattered series of surface sites around Craig-y-Llyn. By comparison the first five microliths of line three are distinctly more "archaic" shapes and if these do not belong with the rods and microtriangles of the upper lines might suggest some Early Mesolithic activity on the site. Numerically, the Later Mesolithic microlith assemblages found in Wales fall into the range of variation displayed by "microtriangle" assemblages in southern Britain, and there is as yet no suggestion of any sort of local grouping amongst them such as is detectable among the microlithic industries of the South West Peninsula. In southern Wales (most notably in Pembrokeshire), the associated non-microlithic tool kits include many denticulated scrapers, and in this resemble industries from Devon and Cornwall while in northern Wales scrapers tend to be small and convex and at Prestatyn at least to be accompanied by steeply retouched flakes. Common to both areas are truncated pieces, burins and awls: core adzes appear to be unrepresented.

If attempting to date the earliest microlithic equipment in Wales presents problems, any estimate of just when this later "micro-triangle" technology appeared runs into equivalent difficulties. Only two determinations from Wales are actually directly associated with "microtriangle" material: both are on charcoal from the infill of a pit feature (Pit 19) below a barrow, Brenig 53 (45) at 1,200 ft. in Denbighshire. The actual dates are 5,240 b.c. ± 100 and 5,350 b.c. ± 100. Worked flint recovered from this pit included, besides

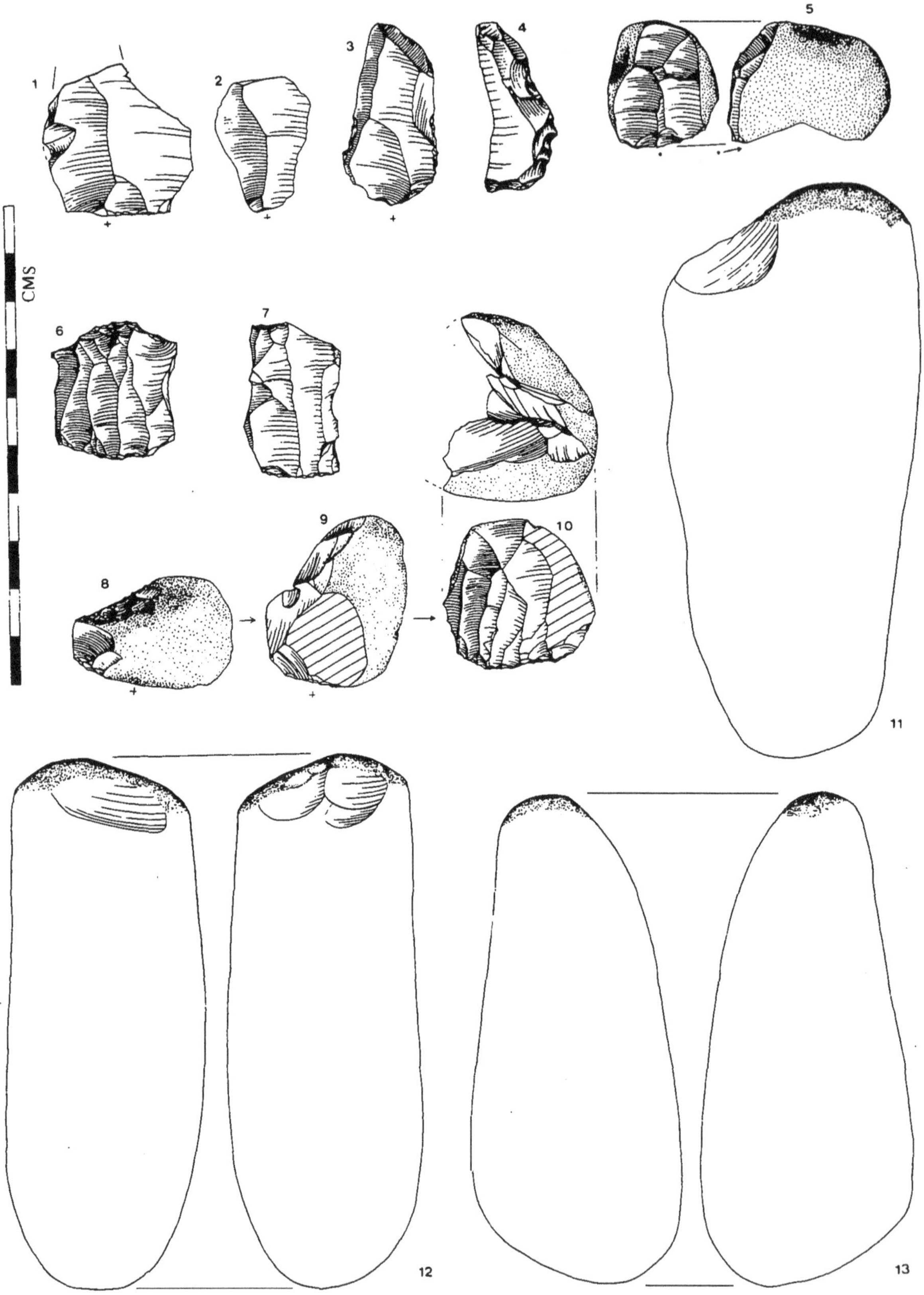

Fig. 4.24 Worked flints and bevelled pebbles from sites at Frainslake, Pembrokeshire

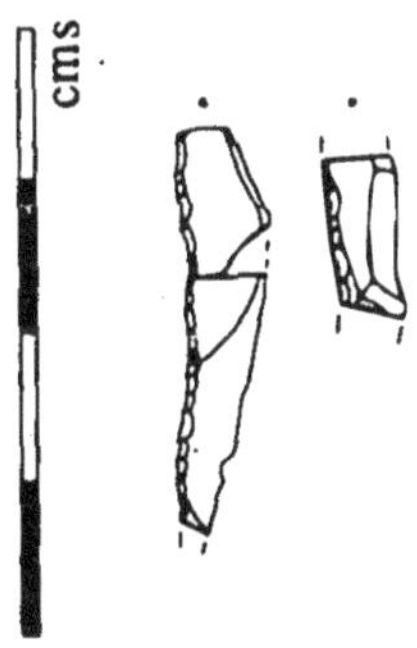

Fig. 4.25 Micoliths found with pig skeleton at Lydstep Haven (Pembrokeshire): from a lantern slide in Tenby Museum

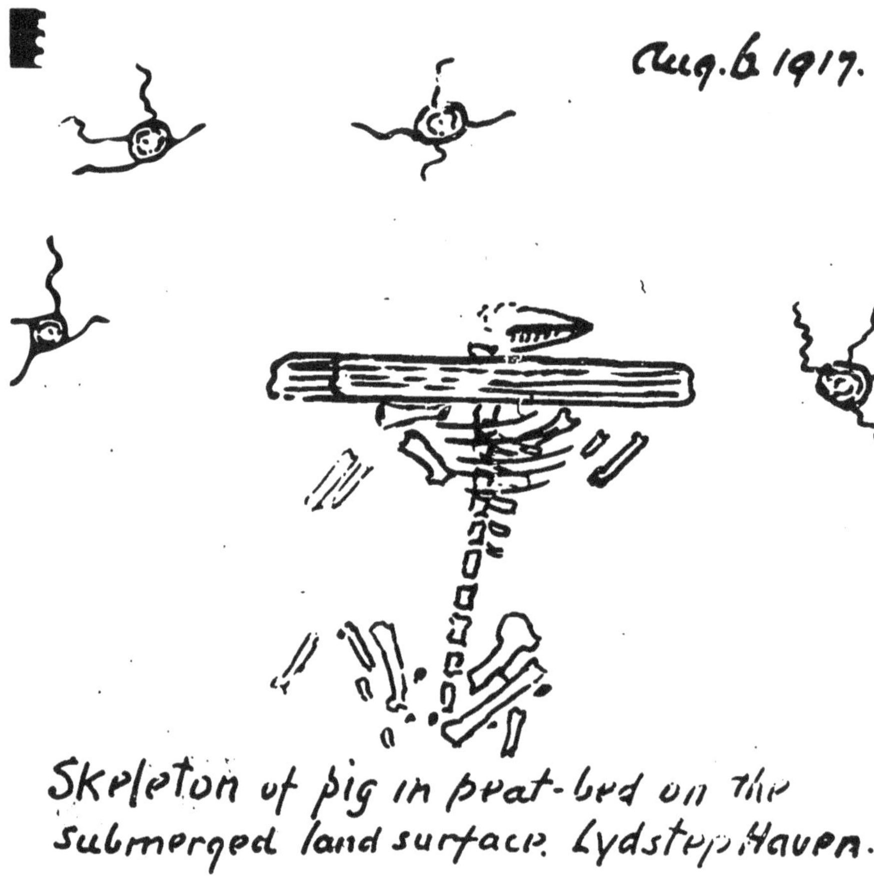

Fig. 4.26 Plan of find of pig-skeleton at Lydstep Haven (Pembrokeshire), reproduced from a lantern slide in Tenby Museum.

Plate 4.V Pig skeleton from Lydstep Haven: skull to one side of fallen tree

waste flakes and blades, only a single scalene micro-triangle, and no way exists of extrapolating this age-determination to the very much larger sample of flint work recovered from other points beneath the barrow. This latter flint work included only narrow-blade forms of microlith, scalene micro-triangles and rods blunted down one or both sides, and there is a striking absence of such obvious earlier forms as obliquely blunted points and large isosceles or scalene triangles. If we can suggest that one element at least of the "micro-triangle" technology was present in North Wales by a little before 5,200 radiocarbon years b.c., no way exists of estimating the probability that elements of this technology could have been innovated as early as 6,500 b.c. as in northern England or 6,000 b.c. as in southern England (Jacobi (1976) Fig. 7). Indeed, that there was a human presence at all in north Wales earlier in the 6th millennium b.c. is suggested by just a single radiocarbon date, unassociated with any artifacts of Har. 656 = 5,700 ± 80 b.c. on charcoal from a shallow depression below Brenig 40 (Musson (1975) p.18).

In short, we cannot argue realistically one way or the other for the introduction of a Later Mesolithic technology contemporary with its innovation elsewhere in Britain, and it is even difficult, except on grounds of probability, to argue for the (not necessarily continuous) presence of a human population in Wales at a time contemporary with the early stages of the Later Mesolithic elsewhere before the 6th millennium b.c. Indeed, evidence which can be taken as documenting continuing human activity in Wales later than 5,000 b.c. and as confirming the detailed evidence of flint typology in indicating a human presence in South Wales at least later than this date (see below), appears to be given by only two finds both from the "submerged forests" of the South Pembrokeshire coast—at Frainslake (Gordon Williams, 1926) and at Lydstep Haven (Leach (1918) pp. 50-51).

Both these finds are from within wood-peat exposed on the modern foreshore at low tide. A single determination of Q530 = 4,010 b.c. ± 120 (Godwin and Willis (1964) p. 127) from Freshwater West, only a few hundred metres from Frainslake, accords well with other estimates for peat at the same ordnance datum around Cardigan Bay (NPL-113 = 4,020 b.c. ± 90 (Clarach, Cardigan: Taylor, 1973; Q380 = 4,076 b.c. ± 135 and Q382 = 3,948 b.c. ± 135 (Borth. Cardigan: Godwin and Willis, 1961) in suggesting an age close to 4,000 b.c. for the peat at Frainslake. That the peat deposit at Lydstep Haven may well be of a broadly similar age is necessarily an assumption based upon its similar stratigraphical position and obviously requires independent confirmation.

At Frainslake, Gordon Williams (1926, pp. 106-108) observed a pair of closely adjacent flint scatters (sites C and D) in peat covered by shingle. While one (site D) consisted only of a "... concentration... of flint and pebble chip..." storm action on the other site revealed what Gordon Williams described as "... a windscreen of gorse, birch, hazel and alder... set in the peat ... (with) ... on the North side of this shelter which ran in a gentle curve for $4\frac{1}{2}$ yards... an area bearing much charcoal, flint chip... a large rabattu point... etc.". Of 56 worked pieces which survive in the Museum at Abergwili (A76-1749 and 1824) from these two scatters all except two are worked on beach pebble flint: The exceptions are single flakes of a coarse mottled cretaceous chert and of a greenish stone. Forty two pieces are unretouched flakes (Fig. 4.24, nos. 1 and 2) while there is one fragment of a split pebble. Two flakes show signs of utilization (no. 3) and one flake has a shallow notch (no.8). There are 7 cores (nos. 5-7 and 10), onto one of which refit the notched flake and a second previous removal (nos. 8, 9 and 10), and also a pair of core fragments. Finally, there is a broken scraper (no.4). The "... large rabattu point..." from site C, quite conceivably, given the terminology of the time, a microlith, has, however, disappeared. With this flint material was associated a small group of elongated pebbles (6 from site C and 2 or 3 from site D) of which 6 survive (Fig. 4.24, nos. 11-13), all showing characteristic bevelling at one or both ends and being between 110 and 130 mm long. The "bone implement" from site D (ibid. p. 108) can be identified as the heavily rolled proximal fragment of the radius from a juvenile Bos whose distal end might be taken as showing some signs of bevelling as if from use as a "limpet scoop". The rolled condition of this bone fragment is so marked, however, that it could perhaps better derive from the modern beach shingle than from the peat.

The second find, that at Lydstep Haven, was of a pair of broken microliths, both narrow "rods" blunted along one edge (Fig. 4.25: not three microliths as stated in Wainwright (1963) p. 105). Found in the peat immediately below were the neck vertebrae of a pig (Leach (1918) p. 51, also pp. 55 and 58). Above the pig and sealing both the microliths and the neck vertebrae was a tree trunk (Fig. 4.26: from a lantern slide in Tenby Museum). The "... unnatural and sprawling position..." of the skeleton was attributed by Leach (Ibid. pp. 50 and 51) to the animals having been crushed to death by the falling trunk, this lying "... across the neck and shoulders, the skull being to the west (Pl. 4.V source as above), the neck bones beneath, and ... the rest of the skeleton to the east..." If this was indeed so and the trunk is not simply a casual component of a gradually accumulating wood peat, it would appear to follow that the microliths may, if the arrowhead carrying them was not yet a third casual component of this peat, have been lodged in the back of the animal's neck at the moment of death. In short, like the Poulton Elk (see previous chapter) or the poorly documented find of an aurochs in Ham Marsh, near Newbury (Berks: Palmer (1878) p. 133: Peake (1931) p. 45) with a "small sharp flake of flint...", locally believed to have been a microlith, embedded in the frontal sinus, the pig from Lydstep might have escaped its hunters whilst still carrying elements of a projectile within it.

Recent work suggests that a number of microlith shapes,—"Four-sided forms" (or "micro-rhomboids") and "micro-tranchet" forms—are absent from flint assemblages dating to before about 5,000 radiocarbon years b.c. A plot (Fig. 27) of the small number of findspots of such microliths from Wales would seem to confirm the proposition suggested by the finds from Lydstep and from Frainslake that settlement, at least in the south-west of the Principality, extended into some part of the last two millennia of the Mesolithic. It may be suspected, however, that when sufficient collections of microlithic material become available from central and northern Wales the apparent area of this settlement will be correspondingly extended.

Evan's recent identification at Cwm Nash (Glamorgan) of a tufa which started to accumulate in the Pre-Boreal, would appear to cast considerable doubt on any earlier assumptions, seemingly supported by isolated radiocarbon dates, that such deposits, and hence any artifacts incorporated into them, belonged exclusively to the Atlantic—i.e. to after about 5,000 b.c. It is thus impossible to confirm suggestions of an Atlantic date for the well known Later Mesolithic assemblage from Bryn Newydd (Prestatyn: Clark (1938) Fig. I and Wainwright (1963) Fig. 10), with its numerous micro-triangles. Here broken hazel nut shells (Clark (1939) p. 201) were claimed to have spread from the "...half dozen square yards..." (Smith, 1926) of the chipping site into the tufa immediately adjacent to the ridge of sand and clay upon which the site was situated, and clearly a fresh study and dating of tufa profiles here will be necessary before the site can again be introduced as evidence for human exploitation of this area of north Wales later than 5,000 radiocarbon years b.c.

While there is no evidence as to technological developments during the 4th millennium b.c. there is certainly no reason to suspect that microlithic industries were replaced (or supplemented) by a technology concentrated on the regular production from preformed cores of blades, approximating in form to "Bann" flakes and corresponding to the Irish "Late-Larnian". While "Bann" flakes are certainly recorded from the Isle of Man (Clark (1935) Fig. 6) and presumably indicate voyaging during the Later Mesolithic none of the pieces published from Wales (Lynch (1970), Livens (1972)) appears capable of fulfilling recent, more stringent, definitions of this tool type (Liversage (1961), Woodman (1974b), p. 254). The same comment could be offered for the few similar pieces recorded in the past from the Cheshire coast.

There are two claims, however, for "microtriangle" equipment of Later Mesolithic type apparently contemporary with the local Neolithic. At Pen-y-Bont, Ogmore, a series of microliths predominantly of Later types is associated with "Peterborough" ware in the enriched B horizon of a podzol developed on the terrace deposits of the River Ogmore and sealed below dune sand (Webley (1973) pers. comm.). A date of BM-1112 = 2,709 b.c. ± 52 (Webley (1976) p. 35) on a sample of charcoal is taken to date this "earliest occupation". Before any conclusions can be drawn, however, as to the significance of this date, the real association of these objects—microliths, pottery and charcoal—beyond burial by natural means to a common depth in the lower part of a once less acid soil, needs to be demonstrated. Such burial may have been ceasing towards the end of the 3rd millenium b.c. as soil acidity increased, since sherds of "evolved" Neolithic wares associated with a date of Har. 1140 = 2,370 b.c. ± 80 are confined to the top or surface of the A horizon (Webley (1973) pers. comm.). Elsewhere in south Wales, replacement of microlithic equipment by the more "normal" pressure-flaked projectiles of Neolithic type at the opening of the 4th millenium b.c., appears confirmed by an estimate of NPL-132 = 3,050 b.c. ± 95 from Coygan Camp (Carmarthenshire: Wainwright: 1967).

In North Wales, while dates slightly before 5,000 b.c. were obtained from Pit 19 below the round barrow Brenig 53 (45), other features belonging to the pre-barrow surface gave a series of rather more recent estimates: 3,170 b.c. ± 100 (Pit 28), 2,850 b.c. ± 90 (Pit 25), 2,830 b.c. ± 100 (Posthole 100) and 2,620 b.c. ± 90 (Pit 20: D. Allen (1976) pers. comm.). Quite clearly, this range of dates overlaps those for the earliest Neolithic domestic occupation at Llandegai, near Bangor (25 miles north-west as the crow flies: Houlder 1976)—NPL-223 = 3,290 b.c. ± 150—and for features within the Northern Henge—NPL 220 = 2,790 b.c. ± 150 and NPL 224 = 2,530 b.c. ± 145.

With the exception of a microlith from the 6th millenium from Pit 19 and an isolated angle-burin from Post-hole 100, the worked flint and chert from within the dated features is slight and without typological significance. While certainly no Post-Mesolithic artifacts were recovered from the Pre-barrow surface a leaf arrowhead incorporated into the mound material (Grealey (1974) p. 100) is strongly suggestive of local derivation and would fit well with the sequence of dates obtained from the pits.

If, however, the premise is granted that use of ground at this altitude will have been only "part-annual" then, before a choice can be made between (1) associating these late dates with the microlithic material and using them as evidence for the survival of a microlithic technology in Northern Wales right into the middle of the 3rd millennium b.c. or (2) regarding the majority of the pits as Neolithic features cut into a former Mesolithic occupation site, it remains to be established what was the technology and economy of precisely contemporary sites in low lying situations which could be interpreted as the base sites of the groups using the upland around Brenig. Immediately relevant of course would be the location and dating of settlements within the Vale of Clwyd to which Brenig is naturally linked by the Valley of the Afon Ystrad.

Summary

In lowland Wales it is possible to recognize three lithic technologies, the first two of which we have already discussed in detail. Firstly, an Early technology characterised by simple "non-geometric" microliths and many single and double ended convex-edged scrapers. With these are associated burins, finely serrated blades, ground-edged pieces, "punches" and core adzes—the last made in stone more often than in flint. A second "Later microlithic" technology is dominated by narrow forms of microlith, principally scalene "micro-triangles" and "rod"-like pieces blunted down one edge. In southern Wales these are associated with many deeply denticulated-instead of convex—scrapers, notched pieces, truncated pieces and narrow burins. Core adzes are absent and in their place appear to be various forms of chopper or chopping tool. In North Wales we find instead small convex edged scrapers and steeply retouched pieces. Both technologies are based on the production of blades or elongated flakes from regular conical, columnar, and in the later technology, discoidal cores. On many "coastal" sites in Pembrokeshire, patinated industries with "micro-triangles" and/or denticulated scrapers are mixed with an unpatinated remainder. For the latter, the technique of primary working consists of the vertical "splitting" of beach pebbles between a hammerstone and a fixed anvil, which produces characteristic thin outer cortical flakes, the centre of the pebble being reduced to a splintered wedge of flint—a "scaled-piece" or outil écaillé". This same phenomenon is regularly encountered along the Cumberland coast and in both areas diagnostic Neolithic artifacts—scale-flaked tools, pressure flaked points and polished axe fragments—when present are also unpatinated. In Cumberland an "outil écaillé" is dated to after the Elm decline (Pennington, 1965) while they appear most numerous on Later Neolithic occupation sites. Significantly also, one of the very many "outils écaillés" from Freshwater West (Wainwright, 1959) is made on a fragment of polished flint axe (National Museum Acc. No. 74.154/252). Evidence for use of this technique is, by contrast, absent from the large sample of flint work available from the midden at Westward Ho! and from the small group of waste recovered from Frainslake. There would be reason, therefore, to suggest that at earliest this third technology which regularly produced such "outils écaillés" might belong to the very end of the Mesolithic, but more convincingly it could be argued that this technology is to be associated with Neolithic or later groups.

The Lowlands and the Coasts of Wales: Vegetation and Resources

Previous writers on the Welsh Mesolithic have emphasized the "...greater attractiveness...of (the) coastal regions ..." (Ap Simon (1976) p. 41) as compared to the inland and upland areas of Wales for human settlement and certainly it is possible to recognize over 200 findspots on the coastal lowland between the Dee and the Taff, some 80% of which are either in Pembrokeshire or in the Gower west of the Neath. "Dependence on the seashore both for food and raw materials..." (Wainwright (1963) p. 126) has been given as the most probable explanation of this coastal distribution.

Alternatively, the distribution could be seen to correlate closely with areas of relatively high winter temperatures or, more precisely, with areas of maximum accumulated annual temperatures (Webley (1976) p.20 and p.21, Fig. I), the greater number of day-degrees per annum identifying areas of highest biotic potential—both marine and dry land. Webley (Ibid p.22) has gone on to reason that combination of these high accumulated-temperatures with a well distributed precipitation (giving only minimum loss of growing days due to moisture stress) will have made lowland south-western Britain—the South-West Peninsula as well as Wales—more attractive for grazing for a greater part of the year than south-eastern England. To Webley's observations it is important to add a third variable, the extension of forest cover over Britain as a whole.

Crude comparison of pollen diagrams for this region with those from Eastern Britain suggest, for Wales, the probability of a relatively more light-admitting (deciduous) canopy during the Mesolithic, more particularly after about 5,000 b.c. when diagrams from East Anglia, the Midlands, and from within the Weald have been interpreted (Birks et al. (1975), Girling and Greig 1975) as indicating the development of lime woodland even on what are now areas of acid sand. The heavy shading to be expected below such a lime canopy has frequently been stressed (cf. Iversen (1973) pp. 65-67). Given the critical role of canopy type in determining the production and quality of browse, the development of herbaceous vegetation, and hence ultimately in controlling ungulate densities (Mellars 1975), it might be possible to argue that in many areas of eastern England the meat resources available to Mesolithic groups will have been relatively less abundant than in lowland western

Britain—thus giving added confirmation to Webley's original conclusion that the Welsh coastal plain—most notably West Pembroke, West Glamorgan and Anglesey—was an area of high biotic potential.

The remarkable concentration of findspots in Pembrokeshire, if not a direct function of intense archaeological activity, could reflect not only selection of the area of highest local winter temperatures, but also that area where, with highest winter sea-surface temperatures, year to year survival of shellfish beds could be expected with minimum destruction by frosts. Such a concentration correlates too with the highest local values achieved for hazel pollen both in the Earliest Flandrian (Moore, P. D. pers. comm.) and in the Atlantic (Birks et al. (1975) p. 96, Fig. 7), and the importance of hazel nuts at least as a subsidiary winter food would probably now be generally accepted. (Mellars (1976) p.376).

Contemporary Fauna, and Evidence for Ungulate Exploitation

Evidence of the fauna contemporary with the Early Post-glacial settlement of Wales and hence potentially exploitable by Mesolithic groups, together with the evidence of the animals actually exploited, derives from two distinct categories of findspots—(i) foreshore peats, the majority of Atlantic age, exposed by marine erosion around the coasts of Cardigan and Pembrokeshire and (ii) the contents of a small number of stratified cave sites.

From peat exposed near the centre of the bay at Lydstep Haven we have the skeleton of a pig closely associated with a pair of microliths (see above) while a single bone of pig is recorded from a flint scatter sealed by peat deposits at Freshwater West (site B5: Leach (1918) p. 55). A tooth of roe deer is also recorded from the same site. Partial skeletons of Bos primigenius are known, but without archaeological associations from Borth (Cardigan): Taylor (1973) pp. 286-289) and Amroth (Pembs: Nat. Mus. Wales Acc No. 31, 363). Probably also without archaeological associations are red deer antlers and the jaw of a brown bear recovered in about 1867 from peat exposed in Whitesand Bay (Hicks (1887) p. 10: but see also Leach (1918) p. 52 and footnote).

Fauna, apparently associated with Mesolithic type artifacts, has been recovered from three caves—Potter's Cave on Caldey, Ogof Garreg Hir in South Pembrokeshire and to the east of our area King Arthur's Cave near Ross-on-Wye. From Potter's Cave Lacaille and Grimes (1955, pp. 125-126) recorded from what they term the basal "reddish-brown loam" (?equivalent to layer 5 of Davies (1978A) p. 10) a fauna of Bos, pig and fox, together with a piece of shed deer antler (not further identified). Associated with this fauna was a small group of artefacts (unretouched blades and cores: Lacaille and Grimes (1955) pp. 126-129 and Fig. 19). That the fauna does not belong later in the Post-glacial than the Mesolithic is suggested both by the absence of ovicaprid bones and by the stratigraphy of the cave itself within which the only evidence for Neolithic occupation appears rather higher in the deposits, in the upper portion of the double band of stalagmite which overlies the reddish loam (Ibid, p. 129 and Nédervelde and Davies (1978) p. 2). Additional to these remains, interpretable as "game animals", is part of the mandible of a dog, the animal being described as"... about the size of a fox terrier..."

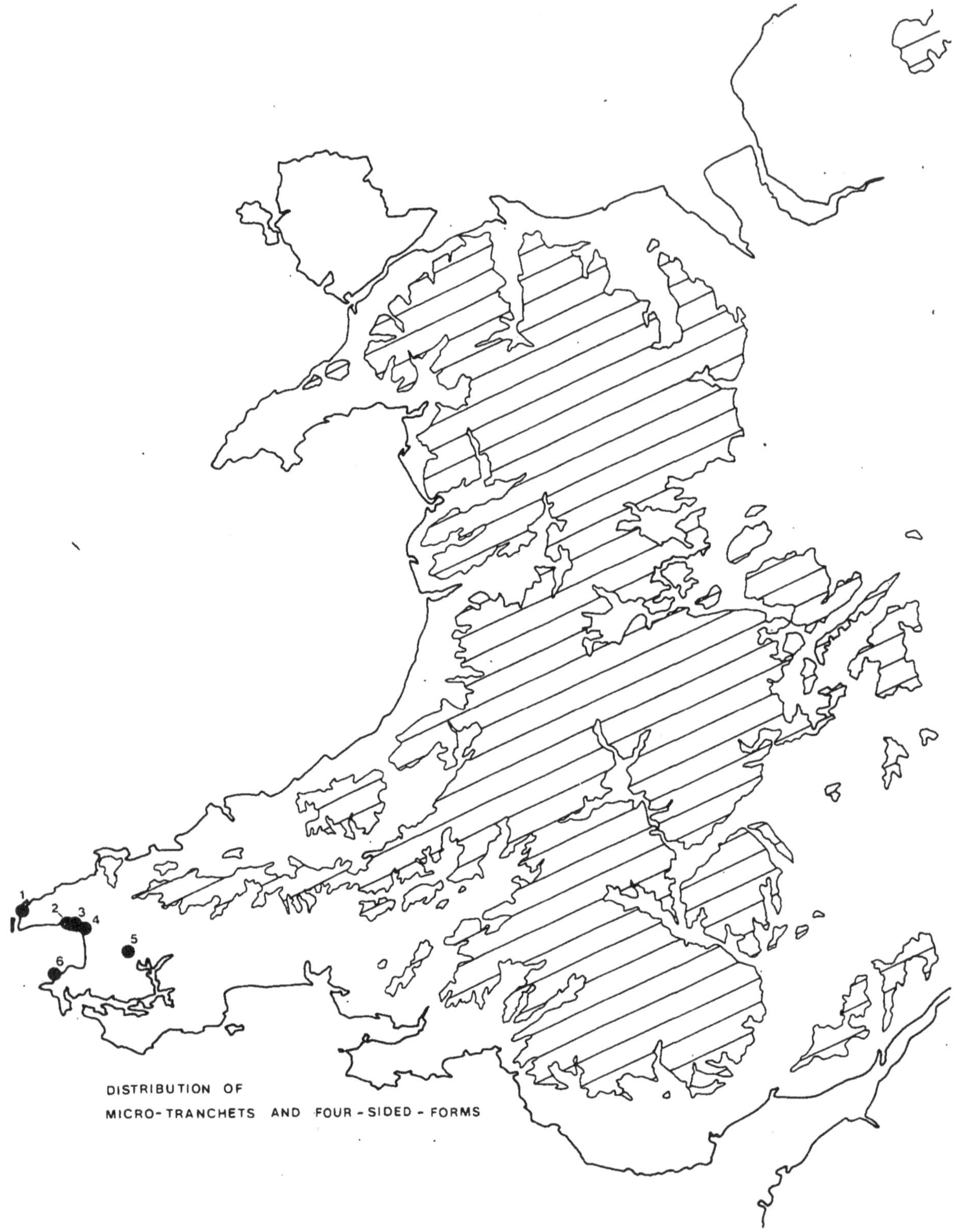

Fig. 4.27 Map of Late forms of Microlith within Wales

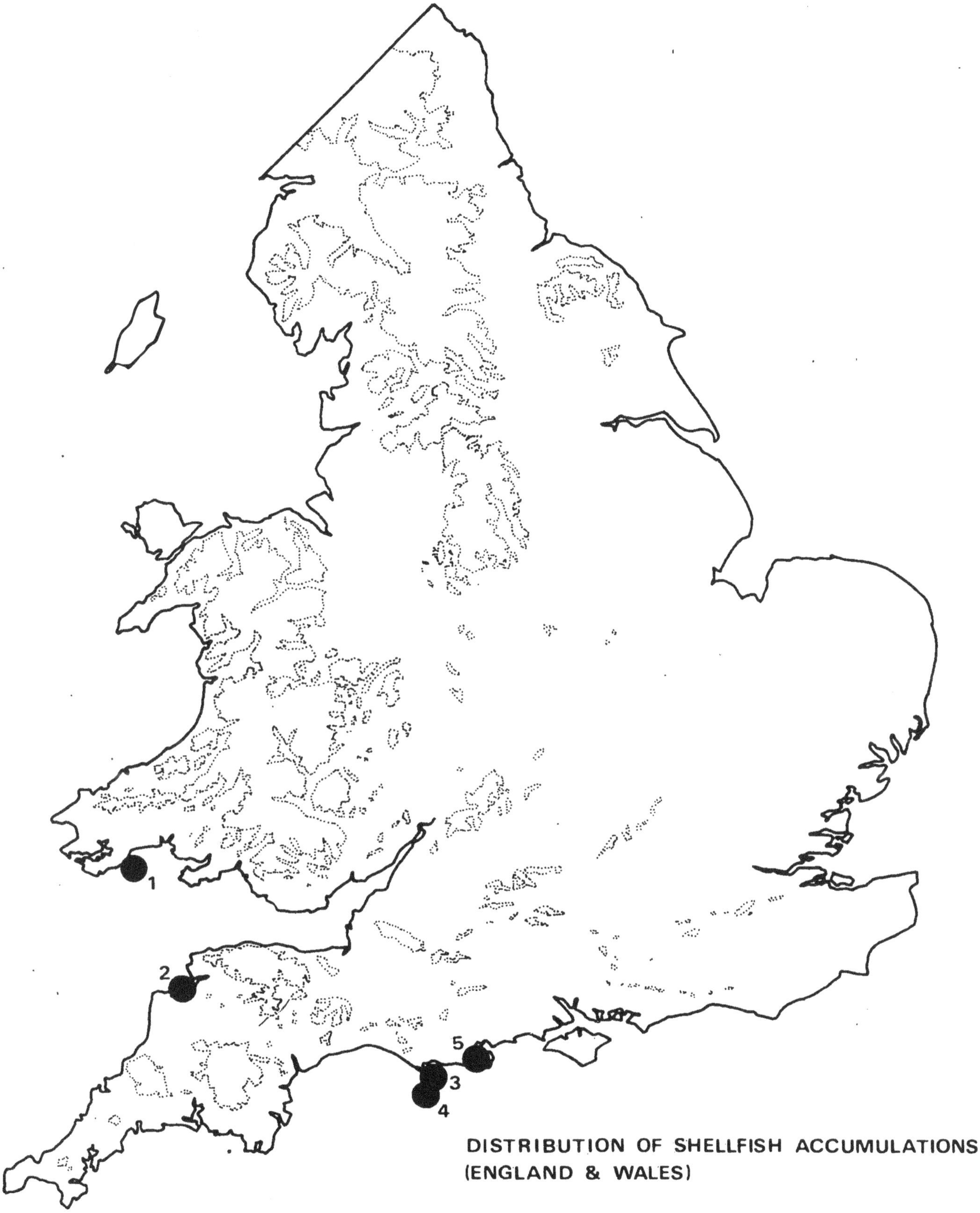

Fig. 4.28 Map of Mesolithic shellfish accumulations in England and Wales

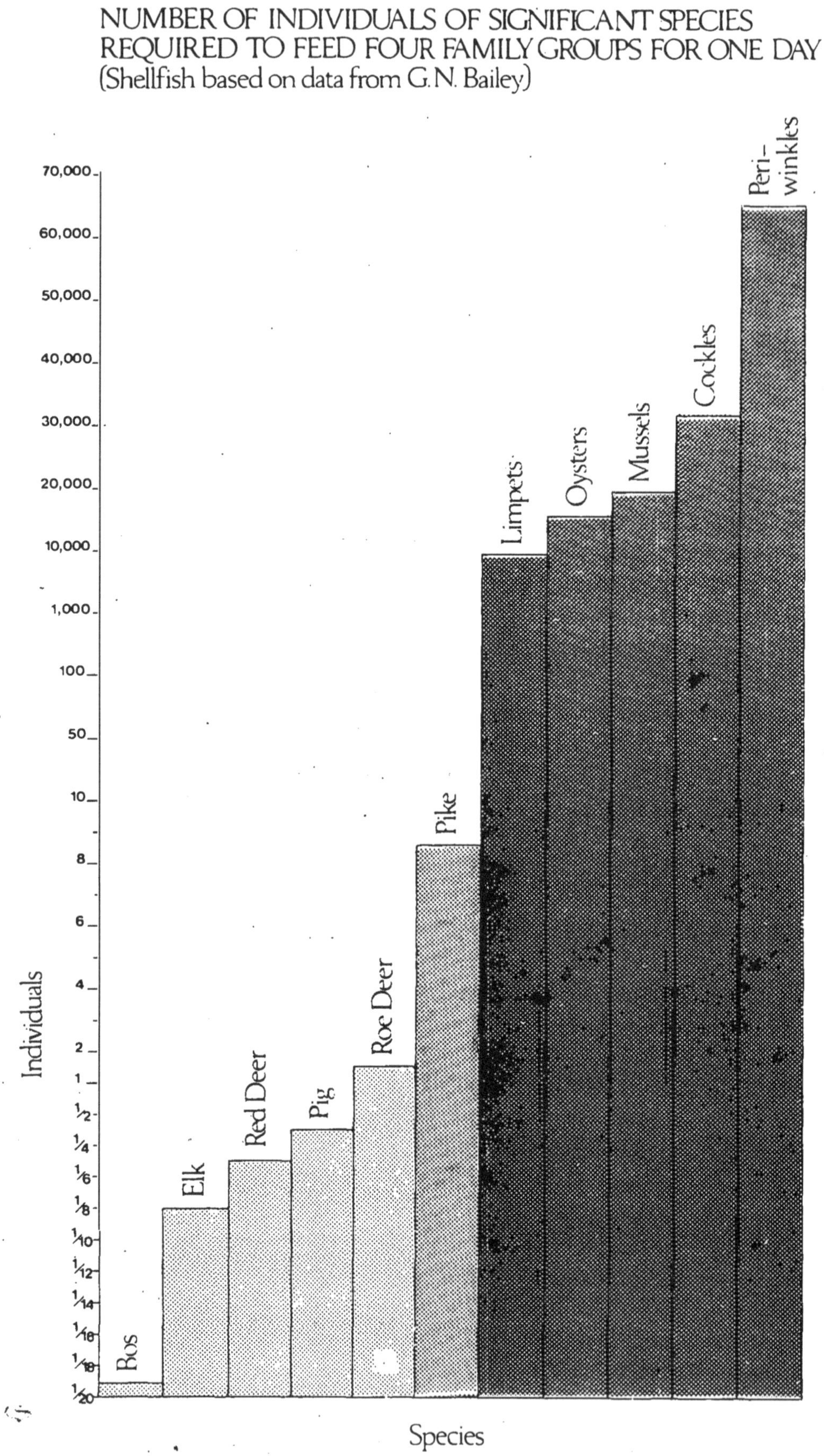

Fig. 4.29 Number of individuals of significant species required to feed four family groups for one day (Shellfish data based on data from Bailey, G. N.)

(Ibid, p. 126 and p. 157). This occurrence joins only two other Mesolithic records of dog, both belonging to the 8th millennium b.c. and from Star Carr and Thatcham site V respectively. The apparently associated flint waste from Potter's cave could also, subjectively, be seen as belonging within an Early rather than a Later Mesolithic context, and this observation only emphasizes the absence of records for dog from British Later Mesolithic faunal assemblages. Less significance attaches to the record of dog from the Red Loam at Nanna's Cave since this deposit contained also a horn core of goat (Ibid pp. 104-105) thus indicating an admixture of Neolithic or Later material with the Mesolithic artifacts from this horizon.

From Ogof Garreg Hir (Davies, 1978B) a microlith of later Mesolithic type, an elegant blade and a pointed bone (? an awl) were recovered from a cliff tunnel, the contents of which had been heavily mixed by wave action. The certainly identified fauna from this tunnel comprised red and roe deer, fox, wolf and hare—also bird bones—and these are considered to be from a "Mesolithic midden" (Ibid). However, the "possible" identification of astragali of sheep/goat must raise the possibility that post-Mesolithic material had also entered the cave. The absence of marine molluscs from both Ogof Garreg Hir and Potter's Cave could be taken to imply use of these sites at a time when the coastline lay further away from the caves than at present.

Exploitation of deer—presumably red—pig and Bos—is confirmed by the evidence from the Later Mesolithic "First Hearth" on the platform in front of King Arthur's Cave (Taylor (1927) p. 68) while at the same site beaver was taken for its meat and pelts and wild cat or pine marten were presumably trapped.

It remains speculative whether elk survived to be exploited by the earliest Mesolithic groups in Wales, while the bones of horse recorded by Taylor from the "First Hearth" could either represent a real component of the Post-glacial fauna, or, like the bones of reindeer from the base of the layer, material residual from the final Devensian deposits in the cave.

If the evidence for the dry-land resources exploited by Mesolithic groups within Wales is both scattered and distinctly scanty, it shows no features which set it apart from similar data for England (Grigson, 1978). While faunal material can thus be produced from only a small number of sites in Wales, the presence on nearly all sites of Mesolithic date of flint projectile points and processing tools indicates these as points from which ungulate exploitation took place.

Shellfish Exploitation: Wales and England

In England and Wales there are only three accumulations of shellfish which can be shown to be Mesolithic. Firstly, there is Culverwell on Portland (Palmer, 1976) where we have an accumulation of winkles, topshells and limpets with, apparently, no fish bones and little animal bone. Here the midden could be seen as formed by groups visiting the Island to exploit the substantial natural outcrops of chert both for the purposes of immediate tool renewal and for trading—the southern tip of Portland being obviously unfavourably situated for the successful cropping of land animals. At the second site,

Westward Ho! in north Devon, the principal mollusc to be collected was the oyster but the midden, with a site catchment which takes in a wide area of lowland hunting, contains also much animal bone, mainly Bos and red deer. In Wales, only the stalagmite deposit in Nanna's Cave, Caldey, described by Leach (1916, p. 163: see also Leach 1917) as "...crowded with shells..." of estuarine and rock mollusca, could represent a midden accumulation of Mesolithic age. This dating, however, rests on the recovery from the deposit of a single microlith (1916: p. 165, fig. 5, no.1) and 18 pieces of waste and clearly requires independent confirmation. It could, however, be argued that the stratigraphy of this midden, within a deposit of stalagmite, would imply its accumulation during the Atlantic period at earliest. If this is indeed so—like Frainslake and Lydstep Haven—its probable age would be somewhat later than 5,000 b.c. (For discussion of dates of stalagmite formation see Van Nédervelde and Davies (1978) p. 3).

Only two other sites could perhaps be added to our list of Mesolithic midden accumulations (Fig. 4.28): Portland site I (Palmer, 1969), if the shell debris observed was really associated with the Mesolithic artifacts recovered, and Blashenwell, near Corfe Castle, a site four miles inland from its contemporary coastline to which shellfish in unknown quantities were being brought. Here, as at Westward Ho!, it may be suspected that land animals made up the greater part of the food consumed.

In their report on Daylight Rock, Lacaille and Grimes (1955, pp. 134 and 136) would appear to associate with the mesolithic occupation a localized deposit, rather to the south-east of the main artifact spread, which contained, besides patinated flints and charcoal "... an admixture of broken bones (unidentified) and the friable vestiges of marine and terrestrial molluscan shells..." Its contents "... point(ed) to the proximity of a generous seashore..." (ibid p. 136). Identification (by Dr. J. Chatfield) of Helix aspersa (mull) in samples of this midden material together with Discus rotundatus (mull) Clausilia bidentata (Stron) and Patella sp. would however suggest that this accumulation is at earliest Roman and cannot therefore be introduced into any discussion of the prehistoric use of the site.

Radiocarbon dates for shellfish accumulations at Culverwell and Westward Ho! indicate that such collection had begun in South Western Britain just before 5,000 b.c., a date approximately contemporary with those for the earliest surviving Danish shell accumulations and slightly earlier than a determination from Hoédic in Morbihan (Giot, 1966). While in Denmark, for reasons of local geology, the record for coastal settlements—with or without evidence of marine exploitation—begins only at this date, the identification from the late 7th millennium b.c. inland site of Svaerdborg I (Zealand: Degerbøl, (1933) p. 376) of a grey seal jaw has been taken to suggest the possibility of part annual occupation of coastal sites at this rather earlier date. Perhaps amenable to a similar interpretation are records of chert deriving from the Isle of Portland—then a rock promontory—from eighth millennium b.c. sites in East Hampshire and West Surrey (Oakhanger VII(2) and Frensham South).

Evidence from England and Wales for Mesolithic shellfish collection appears then peculiarly slight with a maximum of five sites. It might be suggested that shellfish exploitation may be expected to have made a significant contribution to overall diet only in areas of sheltered soft-shore deposits, for example the Wash, Thames Estuary (Jacobi 1978b) and Burry inlets, areas of present-day commercial exploitation and where marked changes in local environment appear improbable and where shellfish could be gathered quicker and more efficiently, even with the use of boats, than along areas of rock coast where not only would shellfish require individual collection by hand but where, certainly, along the west coast of Britain, the narrowness of rockcut beaches combined with exposure to Atlantic gales would lead to the formation only locally of rich shellfish beds. Along both forms of shoreline it may be suspected that marine erosion and transgression coupled with alluvial build-up will have respectively destroyed or masked evidence for shellfish processing and consumption, activities which it may be assumed took place most frequently on the contemporary beaches or just behind and as close to the points of original collection as practicable. Thus, discovery of such middens as Westward Ho! or Polmontill (Lacaille (1954) pp. 167-169) connected with the exploitation of soft-shore bivalves must be regarded as exceptional while the survival of the the middens at Portland or inside Nanna's Cave is simply due to the carrying of shellfish to processing sites well above the local inundation level achieved by the Post-glacial eustatic rise in sea-level, the steeply shelving rocky beaches fronting both sites being perhaps themselves too narrow for the establishment of successful processing and living sites. Survival of the rather greater number of later prehistoric middens along the coast of South Wales and the South-West Peninsula must in large part be due to their incorporation into deposits of calcareous dune sand, the formation of which began along the west coast not before the end of the Atlantic (Evans (1972) pp. 291-296) and the opening of the Neolithic.

While the suggestion has often been made that shellfish collection, particularly that of limpets, will have been important to Mesolithic groups in this part of Britain, Fig. 4.29 suggests the number of individual shellfish which would have to be collected from a rockshore coast (the principal shoreline type within the area under consideration) to supply sufficient calories for four families for a single day. These totals are compared on this staircase diagram with the number of ungulates which would have to be taken to achieve the identical contribution. Consideration of this diagram suggests not only that, with the quantities of molluscs required seemingly impressive, middens could very rapidly be built up, but also suggests that few beds, even those with a maximum European density of about 1000 limpets per square metre—a figure likely to be approached only very rarely along the British west coast—could sustain exploitation over more than a few years at the rate implied—the total cropping of roughly 10 sq. metres per diem. While initial returns for exploitation of an optimum bed might be suspected to be rapid, perhaps requiring as little as 2½ hours work by each family group in a day, the distance to be covered between settlement and food source will be increased as the nearer beds become overexploited. Yields may be expected to drop sharply if beds become more difficult of access, the shoreline narrows or if exposure to storms increases, factors all leading to a rapid decrease in shellfish density. In short, input of effort may be expected to rise rapidly to a point beyond which exploitation of rock molluscs may no longer repay the effort invested.

DISTRIBUTION OF BEVELLED PEBBLES
CONFIRMED RECORD
UNCONFIRMED

4.30 (opposite) 'Map of finds of bevelled pebbles in South Wales and S.W. England''.

Key:

1. Portland (S. Palmer: Pers Comm.), Dorset.
2. Pedn. men on mere, Cornwall.
3. Roskestal cliff (Morsden (1919) p. 63), Cornwall.
4. Stamps, Cornwall.
5. Greeb, Cornwall.
6. Gwithian site BZ, Cornwall.
7. Gwithian site GU, Cornwall.
8. Godrevy Barrow GB, Cornwall.
9. Gwithian site CM, Cornwall.
10. Gwithian site HU, Cornwall.
11. Gwithian site ND, Cornwall.
12. Gwithian site NE, Cornwall.
13. Elmscott Gutter, Devon.
14. Croyde Hoe, Devon.
15. Boggy Point, Devon.
16. Spreacombe, Devon.
17. Burry Holms Island, Glamorgan.
18. Drinkim Field, Caldey, Pembrokeshire.
19. Stockpole Warren (Gordon Williams (1926) p. 106) Pembrokeshire.
20. St. Govan's Head, Pembrokeshire.
21. Longstone Warren, (Gordon Williams, (1926)p. 106) Pembrokeshire.
22. Linney Burrows, Pembrokeshire.
23. Frainslake site C, Pembrokeshire.
24. Frainslake site D, Pembrokeshire.
25. Mill Bay Bottom (Brunt), Pembrokeshire.
26. Little Castle Point, Pembrokeshire.
27. Short Point Gulley, Pembrokeshire.
28. Ramaskell, Pembrokeshire.
29. Langlands Farm, Dale, Pembrokeshire.
30. The Nab Head, Pembrokeshire.
31. Cwm Bach Field I, Pembrokeshire.
32. Cwm Bach Field II, Pembrokeshire.
33. Pointzcastle Field II, Pembrokeshire.
34. Llanunwas, Pembrokeshire.
35. Carfai Field I/II, Pembrokeshire.
36. Carfai Field III, Pembrokeshire.
37. Porthllisky Field I, Pembrokeshire.
38. Porthllisky Fields II/III, Pembrokeshire.
39. Aberystwyth, Cards. (Thomas & Dudlyke (1925) p. 85).

The probability is thus suggested that the long-term strategy of any group which contains sufficient active hunting personnel and sufficient dependants to ensure efficient consumption of game taken, may well be to select to exploit land resources and the point should be remembered that the recognition and dating of Mesolithic findspots in coastal situations depends entirely upon the recognition of a range of artifact classes-scrapers, burins, knives, truncated pieces and projectile points (microliths)—which it appears reasonable to associate with the hunting and processing of land mammals. That is, all these tool forms can be found on inland sites without obvious alternative food sources, except perhaps plant foods, and apparent confirmation of this proposition derives from Westward Ho! where a complete range of Mesolithic tool types is associated with a fauna of Bos, red and roe deer also pig (Ellis 1867: Hall 1879). Indeed, to demonstrate an "...enforced dependence..." (Wainwright (1963) p. 126) upon coastal exploitation for a Mesolithic group, and for anything but the short-term in a mainland situation, it is necessary to demonstrate its inability to take sufficient other resources—i.e. that the "dependence" was indeed "...enforced..." as could perhaps be argued for such an Island as Oransay if access to the larger island of Jura or to the mainland was denied.

Other Evidence for Coastal Exploitation

If midden accumulations can be seen to be few, artifact evidence has in the past been used to supplement arguments supporting shellfish exploitation. Thus, elongated, flattened pebbles of soft stone, "ground" to a convex bevel at one or both ends, have been interpreted as "scoops" for scraping the meaty part of shellfish from their shells. Less common are heavily battered and splintered pebbles which are considered to be the "hammers" used to dislodge limpets from rocks. While the bevelled pebbles (Fig. 4.24) are found only in coastal contexts and are generally interpreted as scoops, the "bevels" on the majority of these pebbles, however, are too wide to be practical and are, as was earlier pointed out (Cantrill, 1915) markedly wider than those of similarly bevelled pebbles found in Obanian middens. In overall size, however, and particularly in the widths of bevel observed, these "pebble tools" resemble closely the "sponge-finger stones" recovered from Beaker contexts and recently linked to leather working activities (Smith and Simpson (1966) pp. 134-5). Thus, while in the past an association with the processing of shellfish has often been assumed for these tools (cf. Wainwright (1963) pp. 117 and 126), their real function remains problematic, representing either some activity which is really confined to the coast or the translation into easily available stone, perhaps during the summer months, of a tool made elsewhere in antler.

Three lines of evidence support the dating of these "bevelled pebbles" as Mesolithic. Firstly, their recovery by excavation from specifically Mesolithic contexts at Elmscott Gutter in N. Devon (Gardner (1957) pp. 167-172), at Frainslake sites C and D in Pembrokeshire where the finds appeared stratified into a peat most probably of Atlantic date (see above) and the record and identification of examples from Gordon William's excavation at the Nab Head. Secondly, in every case when these pebbles have been recovered, and where accessory archaeological material survives so that their original context can thus be investigated, this material includes an often sizeable Mesolithic, and always where datable, a Later Mesolithic component. Thirdly, in no case have "bevelled pebbles" been recovered from an unmixed excavated context in association with only later prehistoric tool types or where an écaillé technology alone is present.

That these broad "bevelled pebbles" are not simply to be associated with the processing of shellfish may be suggested by their absence from the midden accumulations at Westward Ho! and Nanna's Cave, and by the presence of only a single possible example in the midden at Culverwell (Mrs. Palmer: pers. comm.). Their total distribution (Fig. 4.30) in England, from Baggy Point down to the Land's End and then again at Culverwell and Element Head in Northumberland and in Wales from St. David's Head round to Burry Holms Island, is obviously more limited than that of shellfish beds around our coasts—unless their present-day distribution is simply highlighting those portions of the the coast where local beach topography will have forced any processing sites where these pebbles were in use up into cliff-top situations. If so, it could be argued that their original distribution had once been more continuous, the intervening lost processing sites being situated on contemporary beaches now destroyed by marine action. Equally, however, their distribution could be interpreted as correlating closely with the breeding colonies of another somewhat richer potential coastal resource—grey seals. In the South-West Peninsula these pebbles are most numerous near the larger "rookeries" existing either directly on the mainland or on immediately offshore rocks and islands, and it would be tempting, working on our analogy with the "sponge finger stones" of the later prehistoric period, to associate these bevelled pebbles with the cleaning of seal skins, perhaps in preparation for boat building. Beyond their skins of course the fully grown grey seal has a meat weight equivalent to a red deer, meat which, rich in fat and oil, is a particularly desirable winter food.

In summary, survival from England and Wales of evidence for shellfish collection is poor but would appear to confirm the exploitation of both rock molluscs and soft-shore bivalves. This evidence begins to survive from about 5,000 radiocarbon years b.c. . Probability suggests that individual shellfish beds would become rapidly overexploited and that there would be a rapid fall-off in returns for effort invested. Bevelled pebbles from coastal sites appear functionally unsuited to the processing of shellfish, and various arguments could be advanced to explain their discontinuous distribution, one of which is that they are to be connected with the processing of seal skins. The presence of a standard range of flint artifacts—Later microlith shapes or denticulated scrapers, indicating occupation when beaches around most of the Welsh coast would have been only a few hundred metres away—appears to confirm the exploitation of land mammals, and probability suggests that this food source would perhaps have given the most consistent returns for labour expended. There is no evidence as to whether fishing was practised by groups on the coast of Wales or western England, and it is impossible now to confirm as "Mesolithic" the record of skate from Westward Ho! (Royal Albert Museum: Exeter "Register of Fossils"). Patterns of fishing appear in fact to have varied greatly around the coast of Britain. Thus, no fishing took place from Portland while deep sea fishery took place from Morton Tayport in Fife.

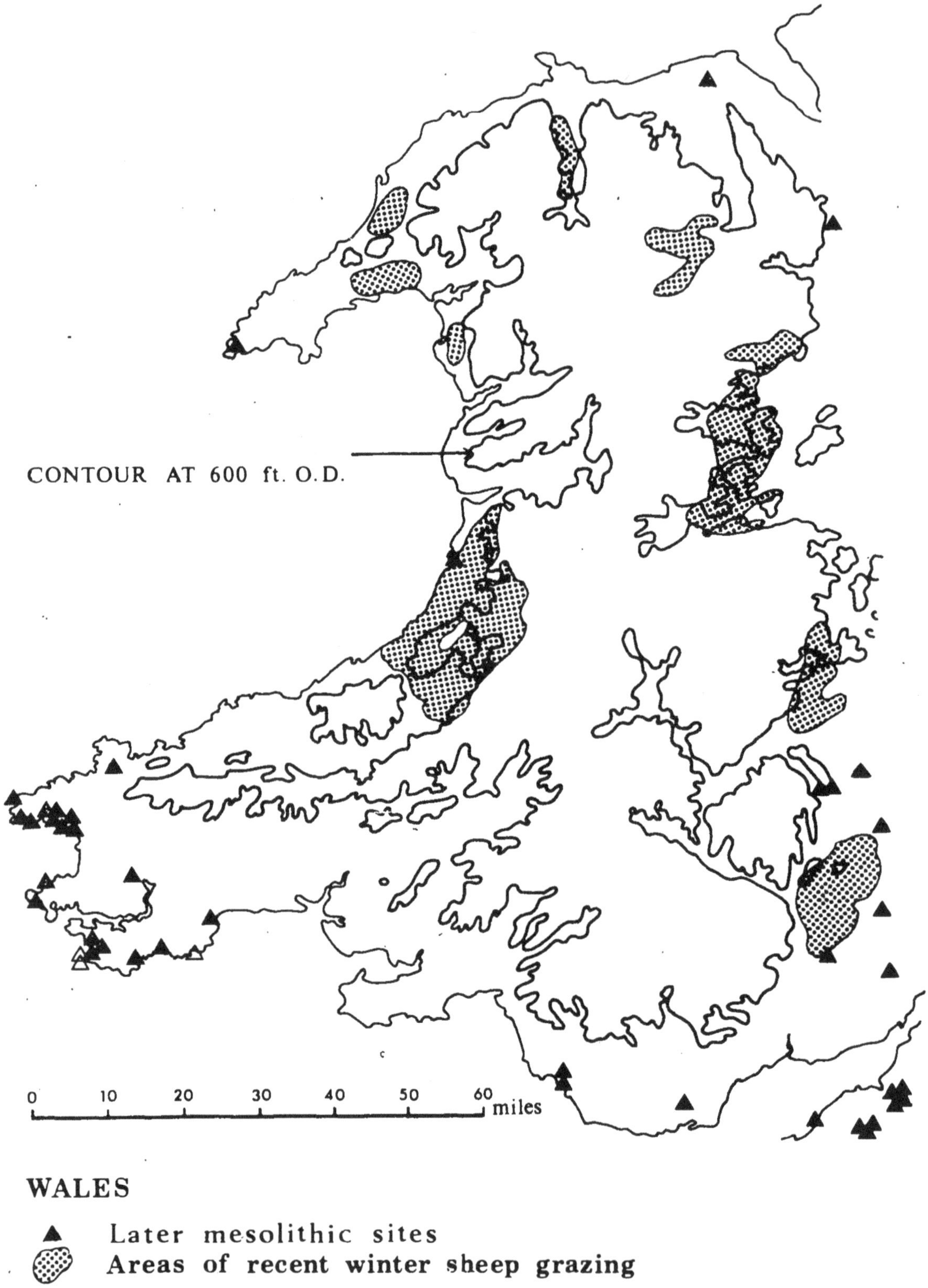

Fig. 4.31 Welsh Later Mesolithic sites found at altitudes lower than 600 ft plotted against areas of recent winter grazing

In the Obanian of Western Scotland, by contrast, potentially rich, deep-water fishing was ignored in preference to the heavy exploitation of the infralittoral zone.

Coastal exploitation has always been regarded as a Post-glacial phenomenon (Clark 1968). While in North Spain collection of shellfish can be shown to extent back to the 10th millennium b.c. (Clark, 1971), in Northern Europe evidence for this activity certainly does not survive from before about 5,000 b.c.. Analysis of antler material from Late-glacial sites in both western and northern Britain is conclusive in suggesting the presence of reindeer herds in these regions for calving and during the summer months. In other words reindeer will have been available for only part-annual exploitation, and may be expected to have wintered eastwards on the Danish and North German lowlands. While the "shortfall" will have been in part made up by the exploitation of the other ungulate species,—which regularly occur in Late-glacial occupation deposits,—notably horse and various species of deer—it must remain open to speculation whether these groups will not also have exploited the river and marine resources of the west coast, notably populations of seal. Indeed, recent work would seem to suggest (Burch, 1972) that the most successful groups in a Late-glacial situation—success expressed in terms of population density per km^2—would be those who made use of coastal and river resources for at least part of the year. Fishing of river and coastal reaches appears confirmed by the evidence of fish bones from Pin Hole Cave (Armstrong 1929) and Peakland sites (Bramwell, pers. comm.), while outside Britain exploitation of seals by Magdalenian groups is indicated both by seal teeth and bones brought back to sites in the Perigord and by the extremely accurate art representations of these animals which are suggestive of more than just casual encounters (Bahn 1977). In the apparent absence of a land connection to Ireland by the later part of the Late-glacial, it may be suspected that breeding populations of seals will have become distributed along the then west coast of Wales and south-west England, and that winter shortfalls of reindeer may have been countered by exploitation of these populations. Of the extractive patterns which go to make up the coastal economics of the Mesolithic, fishing and sealing could in all probability be argued to be strategies with origins in the Late-glacial. On the other hand, shellfish collection on its own may in Northern Europe, and here one is arguing from negative evidence, prove to be a specifically Post-glacial phenomenon.

The Lowland and the Upland of Wales

In concentrating so heavily upon the coastal lowland of Wales we have ignored the fact that very many flint scatters clearly of Mesolithic date have been recognized as a result of recent field work in areas of lowland without any coastal fringe, most notably in Herefordshire (Pye: 1959) and Radnorshire (Dunn: 1969). This work, selectively expressed in the form of a plot of Later Mesolithic findspots from below 600 ft. O.D., shows them to extent round the edges of much of the Welsh upland, empty areas within the "Marches" highlighting an absence of recent research (Fig. 4.31). Again, however, it is noticeable that the rather greater number of findspots in Herefordshire directly reflects, as along the coast, the areas of maximum accumulated annual temperatures.

In the absence of complete ground plans or samples of faunal material from these lowland sites, most representing little more than surface scatters of diagnostic artifacts, statements on the role of any of these sites must remain speculative. Indeed, there is only one British site exploiting an area of low ground immediately adjacent to an upland massif for which we possess any such relevant information. This site is, of course, Star Carr in the Vale of Pickering, N. Yorkshire, and here analysis of the fauna, once one has extended consideration beyond the substantial number of fully developed red deer antlers imported in winter onto the site for industrial purposes, suggests use of the site by part at least of the occupying group at least as late as May or June at least once. This is suggested by the roe deer antlers recovered which were all unshed— Roe deer antler losing its velvet in April (Clark (1954) p. 94)—as well as by the presence of newborn calves of both elk and red deer—born in May or June (Noe Nygaard (1975) pp. 14-15). There is no reason of course why the roe deer could not have been taken even later into the summer, the antlers being shed only in October or November. No seasonal evidence, however, has been produced for the *Bos* and pig remains present, and it should be remembered that, in meat weight terms, *Bos* was the most significant single contributor. It could, therefore, be suggested that, for one season at least, Star Carr was used into the summer months. The probability is thus suggested that use of Welsh sites around the foot of the central upland may similarly have extended into the summer at least in some seasons, and it is important to remember that available to the later Mesolithic groups of lowland Wales will have been two forms of resource not present around or within the lake basin at Star Carr—hazel nuts (Clark (1954) p. 60) and fish (Wheeler, 1977), and that the annual organization of Mesolithic groups in Wales may be expected to have taken advantage of these additional non-ungulate resources.

Identification at altitudes over 1200 ft. O.D. on the Clevelands, in situations most unfavourable to winter occupation, of flint scatters with a microlith component identical to that of Star Carr (Jacobi, 1977), suggests that groups at least closely related technologically chose to make use of this upland, presumably during the summer months. The very large number of Mesolithic sites of various technologies recognized from similar altitudes on many parts of the Clevelands, as well as the Middle and Southern Pennines suggests that regular exploitation of ground at these altitudes, presumably for ungulates, was an economically advantageous strategy. The very much greater number of Later Mesolithic sites than Early sites on these upland moors suggests too that even with the additional resources of hazel nuts and fish, upland hunting remained for many groups an extremely attractive economic proposition. Recent work has suggested (Jacobi, *et al*., 1976) the scale and significance of ecological changes brought about by the repeated use of fire as a hunting aid on these upland moors, while the recovery of the Horn sheaths of *Bos* from the Clevelands as well as the Northern Pennines suggests the possibility of a wider pattern of hunting than is usually assumed (Clark 1972, Jacobi 1977).

In summary, half a century of constant survey of the high ground of northern England has revealed many flint scatters at altitudes over 1200 ft. covering all phases of the Mesolithic. Their number suggests that exploitation of these moors remained an attractive strategy over the whole period. Evidence from Star Carr, the only lowland site with economic data, does not, however, deny the hypothesis that groups may also have continued to exploit the lower ground during part at least of the summer.

While in 1963 Wainwright was able to list only three finds from a "... highland massif..." that "... discouraged occupation..." of a present total of 296 Mesolithic find spots for Wales 49—or 1 in 6—lie over 1,000 ft. on these uplands which make up some 40% of the country. The bulk—33—are in Glamorgan where the archaeology of the Uplands north of the Rhondda has been the subject of an unique cooperation, initiated by H. N. Savory, between the National Museum, the Forestry Commission and amateur workers.

Best known of these sites, and that which has produced the largest sample of microlithic material is Craig-y-Llyn (Savory, 1961), while smaller sites producing microliths have been identified, near Ffos ton Cenglau, Cwm Fforch, Carn Foesen and from the ridge between Treherbert and West Cwmparc. (Savory, 1961-1968). This "block" of sites, clearly suggestive of the significant exploitation of at least one area of this upland, inevitably throws into prominence earlier but isolated finds from the Two Tumps Barrow on the Kerry Hills (1,666 ft. O. D. Daniels (1927) Fig. 5, no. c) and Pant Sychbant (Brecon: 1,100 ft. O.D.) while in Denbighshire finds from the Brenig (see above) and the Aled Reservoirs (1,250 ft. O.D.: unpublished) indicate exploitation of ground which rises locally to over 1,700 ft.

While the majority of microliths collected are quite clearly Later Mesolithic, the consistent representation of early shapes—obliquely blunted points or isosceles triangles—as for example at Cwm Fforch or on a group of sites just below 1,550 ft. O.D. on Cefn Hill on the North Eastern ridge of the Black Mountains (Robinson 1949)—inevitably raises the probability that exploitation of upland Wales began in the Early Mesolithic, a proposition which gains support from the 8th millennium dates obtained from comparable upland areas in Northern England (Switsur and Jacobi, 1975, and Jacobi, 1977). It remains, however, to identify any unmixed assemblage of Early material.

While the number of find spots from this high ground is infinitely fewer than from areas above 1,000 ft. on the Pennines or the Clevelands, there is nothing in the environmental evidence from this upland, as we have it preserved in the form of pollen diagrams (Godwin (1955), Seddon (1962), Moore and Chater 1969A and B, Moore (1972A) and Taylor, 1973) to suggest conditions any more "severe" than on the Yorkshire uplands, or even a woodland necessarily less productive of browse than that of lower altitudes. In North Wales it could be argued that the heavy representation of pine pollen above 900 ft. during the Boreal implies the likelihood of heavier shading while diagrams from Cardiganshire display a rather gentler but persistent increase in pine pollen inland (Moore, 1972b) and upland, as compared to Oak. In these latter diagrams, however, even at the time of the "marked expansion" of pine pollen towards the end of the Boreal, particularly at altitudes over 1,250 ft., recalculation of the counts to compensate for the heavy overproduction of pollen by pine (Faegri and Iversen: 1975) suggests the presence of greatly under-represented deciduous (particularly the sessile oak) component. The potential carrying capacity of such mixed deciduous—coniferous woodland has been recently discussed by Mellars (1975: Table I), and pollen diagrams for the succeeding Atlantic emphasize the importance of deciduous woodland at all altitudes, particularly oak in southern and western Wales (Moore (1977) pp. 79-80), while increased light penetration of the upper canopy with altitude may independently be suggested by a consistent rise in the representation of both birch and hazel pollen with height above sea-level.

The high tree pollen values in these upland pollen diagrams accord with scattered records of sub-fossil tree stumps in suggesting tree growth up to over 2,000 ft. (Woodhead and Hodgson: 1935; Crampton (1966) p. 50 and Taylor, 1975). Similarly stumps of oak trees at 2,000 ft. on Gader Mtn. (Brecon: Wheeler (1925) p. 58), a tree at the present-day confined to altitudes below 780 ft. O.D. (Jones: 1959) confirms the high representation of Oak in the upland diagrams considered.

An important factor in the recent economic history of Wales was the seasonal movement of sheep from their upland summer pastures to lowland wintering areas (Davies 1935), a movement making optimum use of the grazing potentials of both zones. Annual growth on the uplands will, as the distribution of floral isophenes indicates (Taylor, 1976), begin markedly later than in the lowlands, but despite severely delayed spring growth, summer growth yields can (as Taylor has pointed out (1976: p. 286) outstrip those for lower ground while, in many years, spells of fine weather will allow periods of rich growth during the autumn months—a phenomenon which must explain the occupation of upland Mesolithic sites into this season (see for example the quantities of hazel nuts recovered from the Brenig and from Pennine sites such as Saltersbrook and Ickhornshaw Moor). Summer nights on these upper slopes will be warmer than at lower altitudes due to temperature inversion and sun-shading effects and days will have longer hours of sunshine, receiving the sun earlier in the morning and keeping it later at night. In addition, diurnal temperature range during the summer months will be less above 1,000 ft. than it is in the lowland belt. There will thus be two incentives for animal populations to use this upland. Early summer growth may be expected to have attracted herds away from heavily grazed winter ranges while the more favourable summer daily temperature range may be expected to have kept herds on the upland through the summer.

Sharp deterioration in daily weather with decreased temperatures and eventually increased snow and blizzard frequency, leading to an onset of snow-lie, render the uplands of little economic value for what is a prolonged winter season. Pollen diagrams confirm a sharp drop in mean winter temperature with altitude during the Boreal, while the record of <u>Salix herbacea</u> in the early Boreal at Elan Valley (1,250 ft. O.D.: Moore and Chater (1969) p. 193) has been taken as indicative of a prolonged winter snowlie. The end of the growing season, and the onset of early snow may be expected to have caused animal populations to migrate to their winter grazing areas—areas more extensive than the artificial units mapped by Davies (1935) (Fig. 4.31) and in all probability extending peripherally right round the base of the Welsh upland.

It may be suspected that with decreased rainfall and higher summer temperatures during the later Boreal (Webley (1976) p. 22) plant growth on thin mountain soils will have been fore-shortened; while water surplus with a minimum of water stress/soil drought will have been a phenomenon of the Atlantic. Thus, as one comes forward in the Mesolithic so the regularity and the length of upland exploitation may have increased (Map Fig.4.32).

It remains, however, to link specific upland sites—interpreted as summer extractive camps—to specific winter base sites, and such a link could perhaps be convincingly argued for only Craig-y-Llyn and Pen-y-Bont (Ogmore). Here the microlithic industries are identical both quantitatively as well as qualitatively, while the bulk of the flints and cherts used at Craig-y-Llyn are of coastal origin. Particularly striking also is that the edge of the home range for Pen-y-Bont (74% of which at 6 miles radius lies below 250 ft. O.D.) comes almost into contact with that for Craig-y-Llyn—53% of whose home range lies by contrast above 1,000 ft. (Fig. 4.33). Effectively, each site is in a position to exploit, on what is presumed to be a seasonal basis, one end of the local topographical range.

If one accepts this proposition and follows through its implications it becomes attractive to interpret such early sites at Aberffraw 12 miles W.S.W. of Snowdonia and Rhuddlan in the Valley of the Clwyd, with 78% of its home range below 250 ft, as serving at least in part as winter base sites, a suggestion which would receive confirmation from the vast numbers of hazel nuts recovered. Aberystwyth, too, would be ideally placed as a winter site at the point where the valleys of the Ystwyth and the Rheidol leave an upland rising locally to 900 ft. for the extended coastal plain of Cardigan. If these sites do indeed represent one end of mobile economies of the early Mesolithic it remains, as we suggested above, to identify summer camp sites on areas of adjacent high grazing.

Preliminary data suggest that, as in the Pennines or Clevelands, sites appear to cluster either very high up—at altitudes over 1,200 ft.—or on the lowest lying ground, with a minimum of sites at intermediate altitudes. An explanation for this distribution must lie not simply in the bimodal distribution of red deer herds as a whole over the year but more specifically the distribution of stags and in summer barren females also. Clark has emphasized the advantages to any exploiting group of a cropping policy directed towards males who had achieved (or were close to achieving) full body weight, a policy largely exclusive of reproductive females and calves. Recent work on Rhum (Lowe: 1966) suggests that, of the constituent members of a red deer herd, it will be the males who graze the lowest parts of the winter range while, conversely, in summer it will be the males together with the non-reproductive females who will ascend highest to graze, well above the level occupied by the females and attendant young. To crop a herd to greatest long-term advantage, a human group can be expected to have cropped that—non-reproductive—part of the herd highest in summer and lowest in winter, a strategy which may go far to explain the extreme bimodal distribution of Mesolithic sites.

An upwards migration of red deer in summer is characteristic of the animal's movement but its population ceiling, as among other species, is dependent upon the maximum number of animals that can be supported through the winter months on the available, and normally lowland, grazings. Along the Welsh coast particularly, where such winter grazings will have immediately abutted the central mountains, their area will have been diminished during the

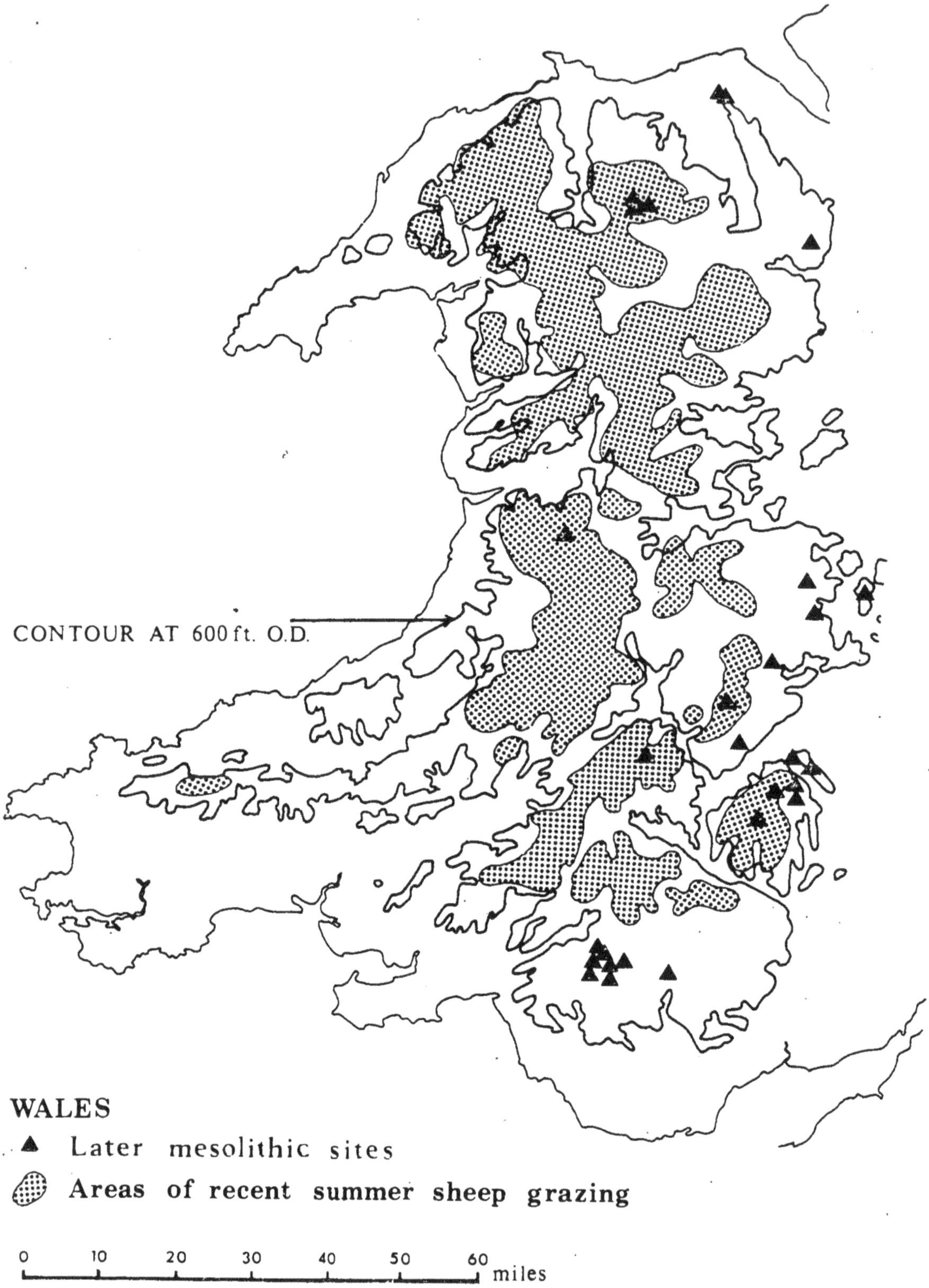

Fig. 4.32 Welsh Later Mesolithic sites found at altitudes greater than 600 ft plotted against areas of recent summer grazing

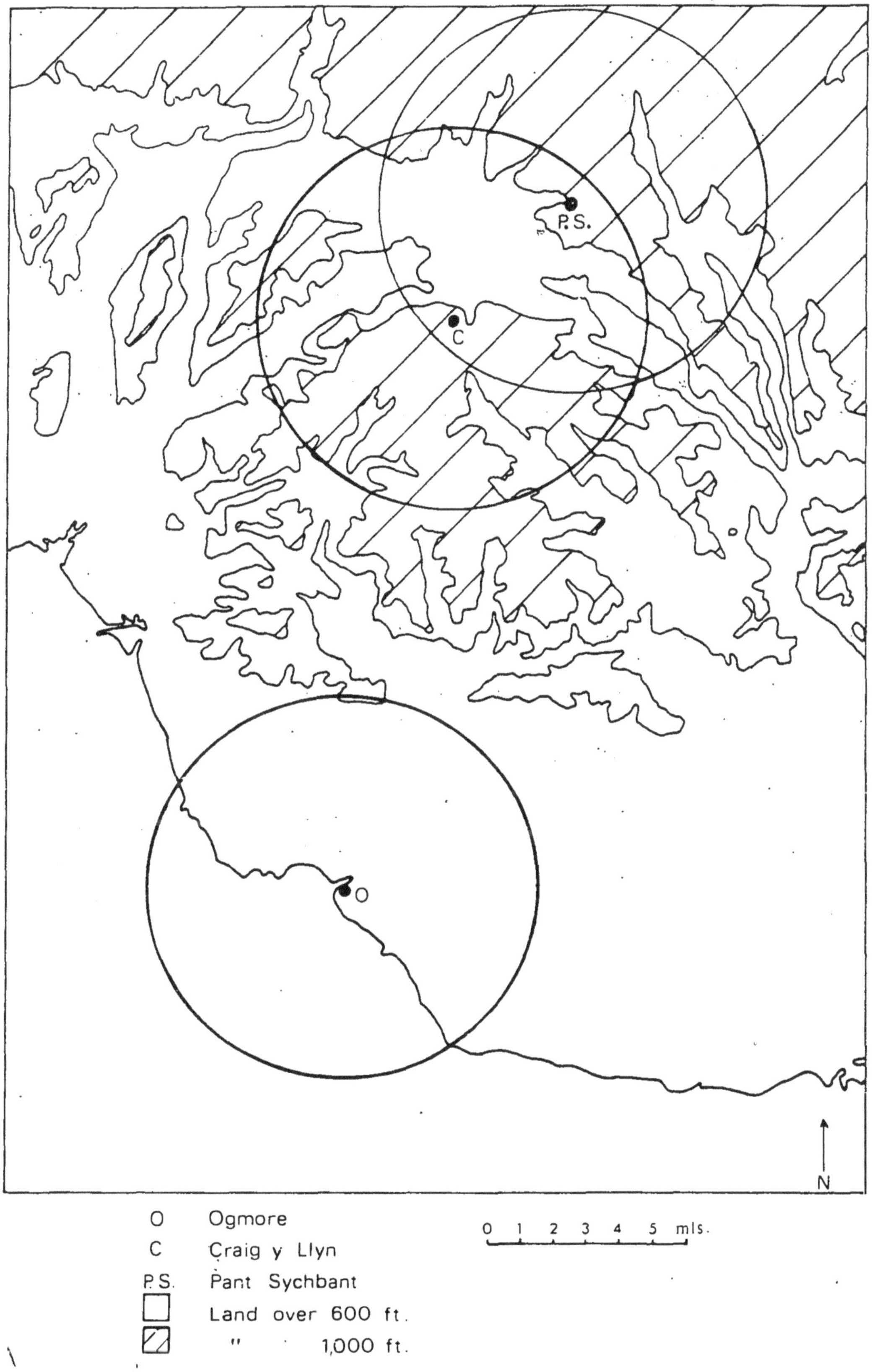

Fig. 4.33 Comparison of site territories at a six mile radius for Ogmore (Pen-y-bont), Craig-y-Llyn and Pant Sychbant (Glamorgan).

Post-glacial, at first sharply and then more gradually by eustatic rise in sea level. It must remain a speculation as to how much any increase in coastal exploitation in this period will have taken place to compensate for the resulting decrease in ungulate population.

There will also have been a clear correlation between areas of winter grazing and areas of prime arable soils, not only along the coast but also on the low ground and in the river valleys, east of the Welsh upland. Uptake of this lowland for arable cultivation in the Neolithic and the running of sheep over the upland will, as in Mediaeval Northern England (Pearsall, (1950) pp. 260-261) have ultimately destroyed the winter ranges essential to the maintenance of red deer populations and confined them to those areas of high ground not incorporated into sheep walks. With destruction of the winter ranges it may be supposed that destruction of the red deer herds themselves and the economies which exploited them will have been swift.

Acknowledgements

Finally, it remains to express my sincere thanks to the Curators and Keepers of Museums in Wales who went out of their way to make material available for study—particularly to Dr. H. N. Savory who helped, beyond showing the writer the collections in the National Museum, with suggestions for subsequent museum work. Professor Grimes, Melvyn Davies, David Allen, D. Webley and R. B. White made available much unpublished information. The artifacts on Figs. 4.21 and 4.25 are the work of Hazel Martingell who prepared all the maps and tables presented, as well as putting the finishing touches to and organizing the other pages of artifact material, and reading through and helping towards a final version of this text. I must thank also Stephen Green for commenting on various points in this script and helping me check many statements based on material held in the National Museum.

BIBLIOGRAPHY

ApSimon, A. (1976) "A View of the early prehistory of Wales" in G. C. Boon and J. M. Lewis Eds. "Welsh Antiquity", National Museum of Wales, pp. 230 Vide esp. pp. 37-53.

Armstrong, A. L. (1929) "Pin Hole Cave Excavations, Creswell Crags, Derbyshire: Discovery of an Engraved Drawing of a Masked Human Figure". P.P.S.E.A. vol. VI, pt. 1, pp. 27-29.

Bahn, P. G. (1977) "Seasonal Migration in South-West France During the Late-glacial Period". Journal of Archaeological Science, vol. 4, pp. 245-257.

Birks, H. J. B., Deacon, J. and Peglar, S. (1975) "Pollen Maps for the British Isles 5,000 years ago". Proc. Royal Soc. London. Series B, vol. 189, pp. 87-105.

Boon, J. C. and Lewis J. M. (eds.) 1976.

Breuil, H. (1953) "Statuette Bisexuée dans le Microlithique de Nab Head, St. Brides Pembrokeshire", Congrès Préhistorique de France: Compte rendue de la XIV e Session.

Brinch-Petersen, E. (1966) "Klosterlund—Sønder Hadsund—Bøllund: Les Trois Sites Principaux du Maglémosien ancien en Jutland..." Acta Archaeologia vol. XXXVII, pp. 77-185.

Burch, E. S. (1972) "The Caribou/Wild Reindeer as a Human Resource", American Antiquity vol. 37, no. 3, pp. 339-368.

Burke, T. W. (1966) "Excavations at Pant Sychbant, Penderyn (Breckn)", Bull. Board Celtic Studies, vol. XXII, pp. 78-87.

Cantrill, T.C. (1915) "Flint Chipping Floors in South West Pembrokeshire", Arch. Cambrensis, vol. LXX, pp. 157-210.

Clarke, G. A. (1971) "The Asturian of Cantabria: Subsistence Base and the Evidence for Post Pleistocene Climatic Shifts". American Anthropologist, vol. 73, pp. 1244-1257.

Clark, J. G. D. (1935) "The Prehistory of the Isle of Man", P.P.S. vol. I, pp. 70-92.

Clark, J. G. D. (1938) "Microlithic Industries from Tufa Deposits at Prestatyn Flintshire and Blashenwell, Dorset", P.P.S. vol. IV, pp. 330-334.

Clark, J. G. D. (1939) "A Further Note on the Tufa Deposit at Prestatyn, Flintshire", P.P.S. vol. V, pt, 1, pp. 201-202.

Clark, J. G. D. (1954) "Star Carr" C.U.P., pp. I-XXIII & 200.

Clark, J. G. D. (1968) "The Economic Impact of the Change from Late-glacial to Post-glacial conditions in Northern Europe". Compte Rendu VIII Congress of Anthropological and Ethnological Sciences, Tokyo, pp. 241-244.

Clark, J. G. D. (1972) "Star Carr: A Case Study in Bio archaeology", Addison-Wesley Modular Publications, pp. 42.

Clark, J. G. D. (1975) "The Earlier Stone Age Settlement of Scandinavia", C.U.P., pp. I-XXV , and 282.

Crampton, C. B. (1966) "Analysis of Pollen in Soils on the Peaks of South Wales", Scot. Geog. Mag. vol. LXXXVI, pp. 46-52.

Daniels, J. E. (1927) "Excavations on the Kerry Hills, Montgomeryshire", Arch. Cambrensis, vol. 82, pp. 147-160.

Davies, E. (1935) "Seasonal Movements of Sheep in Wales", J. Manchester Geogr. Soc., vol. XLV, pp. 24-40.

Davies, M. (1978A) "Extracts from Annual Excavation Reports: Caldey Island Caves 1973 to 1975 by Brother James Van Nédervelde," pp. 11 (privately printed).

Davies, M. (1978B) "Ogof Garreg Hir—Excavations of 7th March, 1977". Journal of Cambrian Caving Council no. 4 (1977-78), pp. 43-45.

Degerbøl, M. (1933) "Danmarks Pattedyri: Fortiden i Sammenligning med recente Former", Vidensk. Medd. fra Dansk naturh. Foren: Bd 96 pp. 357-641.

Dunn, C. J. (1969) "Recent Prehistoric Discoveries from a Ridge East of Radnor Forest and their significance", Trans. Radnor Soc. Vol. XXXIX, pp. 75-81.

Ellis, E. H. (1867) "On Some Mammalian bones and teeth found in the Submerged Forest at Northam", Rep. and Trans. Devon Assoc. Vol. 2, pp. 162-3.

Evans, J. G. (1972) "Land Snails in Archaeology". Seminar Press, pp. I-XII and 436.

Faegri, K. and Iversen, J. (1975) "Textbook of Pollen Analysis" 3rd Ed. Blackwell, Scient. Pubs. Oxford, pp. 295.

Gardner, K. (1957) "Mesolithic Survey—North Devon". Rep. and Trans. Devon Assoc. vol. LXXXIX pp. 160-174.

Geidion, S., (1963) "The Eternal Present", O. U. P. New York.

Giot, P. R. (1966) "Chronique des Datations—radiocarbone armoricaines", Annales de Bretagne, vol. LXXIII, p. 124.

Girling, M. and Greig, J. (1977) "Palaeoecological Investigations of a Site at Hampstead Heath, London", Nature, vol. 268, pp. 45-57.

Godwin, H. (1955) "Vegetational History of Cwm Idwal, a Welsh Plant Refuge", Svensk. Bot. Tidsk. vol. LIX, pp. 35-43.

Godwin, H. and Willis, E. H. (1961) "Cambridge University Natural radiocarbon measurements III" Radiocarbon Vol. III pp. 60-76.

Godwin, H. and Willis, E. H. (1964) "Cambridge University Natural Radiocarbon Measurements VI". Radiocarbon, Vol. VI, pp. 116-137.

Gordon Williams, J. P. (1925a) "The Nab Head Flint Factory: Unique Finds". Trans of the Carmarthenshire Antiquarian Soc. and Field Club, vol. XVIII, Part XLVI, p. 33.

Gordon Williams, J. P. (1925b) "Nab Point: Flint Industry". Trans, of the Carmarthenshire Antiquarian Soc. and Field Club, Vol. XVIII, part XLVI, p. 48.

Gordon Williams, J. P. (1925c) "Nabs Head Chipping Floor", Trans. of the Carmarthenshire Antiquarian Soc. and Field Club. Vol. XVIII, part XLVI, p. 80.

Gordon Williams, J. P. (1926) "The Nab Head Chipping Floor", Arch. Cambrensis, vol. LXXXI, pp. 86-110.

Grealey, S. (1974) "Brenig 45: Boneyn Arian" in Lynch, F. et al. , Eds. "Brenig Valley Excavations 1973: Interim Report", Denbighshire Historical Transactions, vol. 23, pp. 4-10.

Grimes, W. F. (1951) "Prehistory of Wales", Pub. by the National Museum of Wales, pp. I-XVII & 288.

Grigson, C. (1978) "The Late-glacial and Early Flandrian ungulates of England and Wales—an interim Review". in S. Limbrey and J. G. Evans, (Eds.) C.B.A. Research Report No. 21: "The Effect of Man on the Landscape: the Lowland Zone," pp. I-V and 153, vide esp. 46-56.

Hall, T. M. (1879) "Geology Notes" Barnstaple Anthenaeum Manuscript No. 14486. Probably the text of a lecture briefly reported in the Q. J. G. S. Vol. XXXV, p. 106. "The Submerged Forest of Barnstaple Bay".

Harding, J. R. (1950) "Prehistoric Sites on the North Cornish Coast between Newquay and Perranporth" Ant. J. Vol. XXX pp. 156-169.

Hawkins, A. B. (1971) "The Late Weichselian and Flandrian Transgression of South-West Britain", Quaternaria, vol. XIV, pp. 115-130.

Hicks, H. (1887) "On Some Recent Researches in Bone Caves in Wales". Proc. Geol. Assoc. vol. IX (1885), pp. 1-20.

Houlder, C. H. (1976) "Stone Axes and Henge monuments", in G. C. Boon and J. M. Lewis Eds. "op. cit. supra. pp. 55-62.

Houlder, C. H. (1977) "Man in the Welsh Quaternary", Cambria, vol. IV no. 1, pp. 98-104.

Ireland, J. and Lynch, F. (1973) "More Mesolithic Flints from Trwyn Du Aberffraw", Transactions of the Anglesey Antiquarian Society and Field Club, pp. 170-175.

Iversen, J. (1973) "The Development of Denmark's Nature since the Last Glacial". Danm. Geol. Unders. V. Raekke Nr. 7c pp. 126, Copenhagen.

Jacobi, R. M. (1973) "Aspects of the Mesolithic Age in Great Britain" in Kozlowski, S. K. Ed. "The Mesolithic in Europe" Warsaw.

Jacobi, R. M. (1975) "Aspects of the Mesolithic Archaeology of England and Wales" Unpublished Ph. D. Thesis. University of Cambridge.

Jacobi, R. M. (1976) "Britain Inside and Outside Mesolithic Europe" Proc. Prehist. Soc. Vol. XLII, pp. 67-84.

Jacobi, R. M. (1977) "The Settlement of Northern Britain in the 8th millennium b.c." for Mellars, P. A. Ed. "The Early Post-glacial settlement of Northern Europe", Duckworth, In Press.

Jacobi, R. M. (1978A) "Chronological markers in the British Later Mesolithic" In manuscript.

Jacobi, R. M. (1978B) "The Mesolithic of Essex" in Buckley, D. G. Ed. "The Archaeology of Essex to 1500 A.D." C. B. A. Research Report, Forthcoming.

Jacobi, R. M. and Switsur, V. R. (1976) "A Radiocarbon Chronology for the Early Post-glacial Archaeology of England and Wales" Presented at the Chicago Radiocarbon Conference (In Press).

Jacobi, R. M., Tallis, J. H. and Mellars, P. A. (1976) "The Southern Pennine Mesolithic and the Ecological Record". J. Arch. Science, vol. III, pp. 307-320.

Jones, E. W. (1959) "Biological Flora of the British Isles". Quercus. L.". J. Ecol., vol.. LVII, pp. 169-222.

Lacaille, A. D. (1954) "The Stone Age in Scotland" Oxford University Press, pp. I-XXII and 345.

Lacaille, A. D. and Grimes, W. F. (1955) "The Prehistory of Caldey", Arch. Cambrensis, vol. CIV, pp. 85-165.

Lacaille, A. D. and Grimes, W. F. (1961) "The Prehistory of Caldey Part 2". Arch. Cambrensis, vol. CX, pp. 30-70.

Layard, N. F. (1927) "A Late Palaeolithic Settlement in the Colne Valley, Essex". Ant. J. Vol. VII, pp. 498-514.

Leach, A. L. (1913) "Stone Implements from Soil Drifts and Chipping Floors etc., in South Pembroke", Arch. Cambrensis, vol. LXVIII, pp. 391-432.

Leach, A. L. (1916) "Nanna's Cave, Isle of Caldey". Arch. Cambrensis, vol. LXXI, pp. 155-180.

Leach, A. L. (1917) "Supplementary Notes on Nanna's Cave, Isle of Caldey", Arch. Cambrensis, vol. LXXII, pp. 71-79.

Leach, A. L. (1918) "Flint Working Sites on the Submerged Land (Submerged Forest) bordering the Pembrokeshire Coast", Proc. Geol. Assoc., Vol. XXIX, pt. 2, pp. 46-64.

Leach, A. L. (1933) "Stone Implements from the Nab Head, St. Brides, Pembrokeshire". Arch. Cambrensis, vol. LXXXVIII, pp. 229-236.

Livens, R. G. (1972) "The Irish Sea Element in the Welsh Mesolithic Cultures" in F. Lynch and B. Burgess, Eds. "Prehistoric Man in Wales and in the West". Adams & Dart, Bath, pp. 19-29.

Liversage, G. D. (1961) "Larnian Flints on the Leinster Coast", J. Royal Soc. Antiq. Ireland, vol. 91, pp. 109-116.

Lowe, V. P. W. (1966) "Observations on the dispersal of Red Deer on Rhum", in P. A. Jewell and C. Loizos, (eds.). "Play, Exploration and Territory in mammals", Symp. Zool. Soc. Lond. (1966) No. 18, pp. 211-228.

Lynch, F. (1970) "Prehistoric Anglesey". The Anglesey Antiquarian Society, pp. I-XVI and 302.

Mellars, P. A. (1975)"Ungulate Populations, Economic Patterns, and the Mesolithic Landscape". in J. G. Evans, et al., eds. "The Effect on man on the Landscape: the Highland Zone", C.B.A. Research Report no. 11, pp. 49-56.

Mellars, P. A. (1976) "Settlement Patterns and Industrial Variability in the British Mesolithic" in G. de G. Sieveking, et al. (1976) "Problems in Economic & Social Archaeology", Duckworth vide esp. pp. 375-399.

Miles, H. (1972A) "Excavations at Rhuddlan 1968-71: Interim Report", J. Flints. Hist. Soc. Vol. XXV, pp. 1-8.

Miles, H. (1972B) "Rhuddlan". Current Archaeology, vol. III, May 1972. No. 32. pp. 245-248.

Moore, J. W. (1950) "Mesolithic Sites in the Neighbourhood of Flixton, North-east Yorkshire", P. P. S. Vol. XVI, pp. 101-108.

Moore, P. D. (1972A) "The Influence of post-Weichselian climatic fluctuations upon Forest composition and Development in Mid-Wales", in J. A. Taylor (Ed. 1972) "Research Papers in Forest Meteorology: An Aberystwyth Symposium", pp. 20-30.

Moore, P. D. (1972B) "Studies in the Vegetational History of Mid-Wales. III: Early Flanrian pollen data from West Cardiganshire", New Phytologist, vol. LXXI, pp. 947-959.

Moore, P. D. (1977) "Vegetational History", in D. W. Bowen, ed. "Studies in the Welsh Quaternary", Cambria IV, pp. 74-83.

Moore, P. D. and Chater, E. H. (1969A). "The Changing Vegetation of West-central Wales in the light of Human History". J. Ecol. vol. LVII, pp. 361-379.

Moore, P. D. and Chater, E. H. (1969B) "Studies in the Vegetation History of Mid Wales. I. The Post-glacial period in Cardiganshire", New Phytologist, vol. LXVIII, pp. 183-196.

Musson, C. R. (1975) in Lynch, F. (Ed.) "Brenig Valley Excavations 1974", Denbighshire Historical Transactions, vol. XXIV, pp. 1-25.

Nédervelde, Br. James Van: and Davies, M (1978) "Caldey Island Caves 1977", pp. 8, (Privately Printed.)

Noe Nygaard, N. (1975) "Two Shoulder Blades with Healed Lesions from Star Carr", P. P. S. vol. XLI, pp. 10-16.

Norman, C. (1975) "Four Mesolithic Assemblages from West Somerset", Somerset Archaeology and Natural History, vol. CXIX, pp. 26-37.

Palmer, S. (1878) "On the Antiquities found in the Peat of Newbury", Trans. Newbury and District Field Club, vol. II, pp. 123-134.

Palmer, S. (1969) "A Mesolithic Site at Portland Bill 1966", Proc. Dorset Nat. Hist. and Arch. Soc., vol. XC (1968), pp. 183-206.

Palmer, S. (1976) "The Mesolithic Habitation site at Culverwell, Portland, Dorset: Interim Note". P.P.S. Vol. XLII, pp. 324-327.

Peake, H. (1931) "The Archaeology of Berkshire", Methuen & Co., London, pp. IX-I and 260.

Pearsall, W. H. (1950) "Mountains and Moorlands", Collins New Naturalists Series pp. 1-XV and 312.

Pennington, W. (1965) "Pollen Analysis at a microlithic site at Drigg", Trans. Cumberland and Westmoreland Antiq. and Arch. Soc., vol. LXV, N.S. pp. 82-85.

Pye, R. (1959) "Report on Prehistoric Finds in North West Herefordshire", Trans. Woolhope Nat. Field Club Vol. XXXVI, no. 1 (1958), pp. 80-83.

Rees, J. (1973) "Flint Finds from the Nab Head", "Archaeology in Wales" C.B.A. Group 2, No. 13, Item 31.

Robinson, R. S. G. (1949) "The Prehistoric occupation of Cefn Hill near Craswell", Trans. Woolhope Nat. Field Club, vol. XXXII (1946-48), pp. 32-37.

Rogers, E. H. (1946) "The Raised beach, Submerged Forest and Kitchen Midden of Westward Ho! and the submerged Stone Row of Yelland". Proc. Devon. Arch. Explor. Soc., vol. III, pp. 109-135.

Rutter, J. G. (1948) "Prehistoric Gower: The Early Archaeology of West Glamorgan". Welsh Guides, Swansea, pp. 1-90.

Savory, H. N. (1961) "Mesolithic site on Craig-y-Llyn (Glam)". Bull. Board Celtic Studies, vol. XIX, pt. II, pp. 163-165.

Savory, H. N. (1961-8) "Recent Archaeological Excavation and Discovery in Glamorgan". Morgannwg, vols. V (1961), to XII (1968).

Seddon, B. (1962) "Late-glacial Deposits at Llyn Dwthwch and Nant Ffrancon, Caernarvons", Phil. Trans. Royal Soc. Series B, vol. 244, pp. 449-482.

Shotton, F. W., Williams, R. E. G. and Johnson, A. S. (1974) "Birmingham University Radiocarbon Dates VIII", Radiocarbon, vol. XVI, no. 3, pp. 285-303.

Smith, G. (1926) "Prehistoric Remains at Bryn Newydd, Prestatyn", Proc. Llandudno, Colwyn Bay and District Field Club, vol. XIII, pp. 62-67.

Smith, I. F. (1963) "Transversely Sharpened Axe from Cornwall". P.P.S. Vol. XXIX, p. 429.

Smith, I. F. and Simpson, D.D. A. (1966) "Excavation of a Round Barrow on Overton Hill, North Wiltshire, England". P.P.S. vol. XXXII, pp.122-155.

Switsur, V. R. and Jacobi, R. M. (1975) "Radiocarbon dates for the Pennine Mesolithic". Nature, vol. 256, pp. 32-34.

Taylor, H. (1928) "Second Report on the Excavations at King Arthur's Cave", Proc. Univ. Bristol. Spel. Soc., vol. III, no. 2 (1927), pp. 59-83.

Taylor, J. A. (1973) "Chronometers and Chronicles: a study of Palaeo-environments in West Central Wales". Progress in Geography No. 5, pp. 250-334.

Taylor, J. A. (1976) "Upland Climates " in T. J. Chandler, and S. Gregory, "The Climate of the British Isles", Longmans: vide pp. 264-287. 287.

Thomas, R. and Dudlyke, E. R. (1925) "A Flint Chipping Floor at Aberystwyth", J.R.A.I., vol. LV, pp. 73-89.

Wainwright, G. J. (1959) "The Excavation of a Mesolithic Site at Freshwater West, Pembrokeshire." Bull. Board Celtic Studies, vol. XVIII, Pt. II, pp. 196-205.

Wainwright, G. J. (1960) "Three microlithic industries from South-west England and their affinities", P.P.S. vol. XXVI, pp. 193-201.

Wainwright, G. J. (1963) "A Reinterpretation of the Microlithic Industries of Wales", P.P.S. vol. XXIX, pp. 99-132.

Wainwright, G. J. (1967) "Coygan Camp: A Prehistoric, Roman-British and Dark Age Settlement in Carmarthenshire". The Cambrian Archaeological Association, pp. 1-XII and 213.

Wainwright, G. J. (1970) "Excavations at the Nab Head Mesolithic Site, Pembrokeshire 1970" . Bull. Board of Celtic Studies, vol. XXIV, Pt.1, p. 95.

Webley, D. P. (1976) "How the West was won: Prehistoric land-use in the Southern Marches" in G. C. Boon and J. M. Lewis, Eds. Op cit. supra. pp. 19-35.

Weissner, P. (1974) "A functional Estimator of Population from Floor Area", American Antiquity, vol. 39, no. 2, pp. 343-350.

Wheeler, A. (1978) "Why were there no fish remains at Star Carr ?". Journal of Archaeological Science, vol. V, no. 1, pp. 85-89.

Wheeler, R. E. M. (1925) "Prehistoric and Roman Wales", Oxford.

White, R. B. (1975) "Trwyn Du" in "Archaeology in Wales" C.B.A. Group 2, no. 15, p. 25.

Woodhead, N. and Hodgson, L. M. (1935) "A Preliminary study of some Snowdonian Peat", New Phytologist, vol. XXXIV, pp. 263-282.

Woodman, P. C. (1974a) "Settlement Patterns in the Irish Mesolithic". Ulster J. Arch., vol. XXXVI-XXXVII, pp. 1-16.

Woodman, P. C. (1974b) "The chronological position of the latest phases of the Larnian". Proc. Royal Irish Academy, vol. 74c, pp. 237-258.

Wymer, J. J. (1958) "Excavations at Thatcham, Berks. 1958", Trans. Newbury District Field Club, vol. X, no. 4, pp. 31-48.

Wymer, J. J. (1962) "Excavations at the Maglemosian sites, at Thatcham, Berkshire, England". P.P.S. vol. XXVII, pp. 329-361.

Wymer, J. J. (1977) "Gazetteer of Mesolithic Sites in England and Wales". C.B.A. Research Report No. 20.

CHAPTER FIVE

THE NEOLITHIC IN WALES

H. N. Savory

Abstract

Neolithic studies in Wales, as elsewhere, have greatly benefited from the development of radiocarbon dating and palaeobotany in recent years. The former has set human cultures in a truer perspective and the latter has partially reconstructed their environment. But it is still difficult to define precisely the cultural content in Wales of the theoretical phases of the Neolithic—Early (c. 4500-3750 B.C.), Middle (c. 3750-3250 B.C.) and Late (c. 3250-2000 B.C.). It is now, at any rate, clear that the earliest tree-clearance and settlement by agricultural communities using Neolithic artefacts did not occur appreciably later in Wales than in England or Ireland, and must have begun before the end of the 5th millennium B.C. What is still uncertain on available evidence is the extent to which clearance was due to cereal agriculture rather than to pastoralism. It is thought, however, that some upland tracts in Wales, which were originally tree-covered, acquired their covering of blanket-peat mainly as a result of extensive clearance by Neolithic and Early Bronze Age Man, followed by climatic deterioration from the late 2nd millenium B.C. onwards.

The stone artefacts of Neolithic people in Wales relate largely to hunting, tree-clearance, wood-working and pastoralism, together with weaving and leather-working. Chance finds of them, and of Neolithic sites generally, are much more abundant than Mesolithic ones, and show that the interior uplands of Wales were by no means uninhabited, even outside areas where traces of megalithic tombs now remain, but the population was evidently not so dense, nor the settlements so large as in some parts of lowland Britain. These artefacts, like the associated pottery and architecture, domestic and sepulchral, relate to continental types and suggest the presence of at least a leaven of immigrants. However, whilst the oblong or trapezoidal house-plans relate ultimately to those of central Europe and eastern France, the megalithic architecture relates to that of Atlantic Europe, which arose from a practice there of mass communal inhumation, ultimately of Mediterranean and Near Eastern origin. Neolithic culture in Britain thus represents a synthesis which was probably achieved mainly in western France with the help of local Early Neolithic, and surviving Mesolithic, groups. It was here, at any rate, that the oblong house, derived from central Europe, seems to have inspired the chambered long cairn which in the Middle Neolithic became characteristic of the main (eastern) group of megalithic communal tombs in Wales, while growing regionalism led to the more purely 'Atlantic' passage graves establishing themselves in a western group, orientated towards Ireland, which lasted far into the 3rd millennium B.C.

The Late Neolithic of Wales is characterized by the development of new regional ceramic styles, the decay of the megalithic tomb-building tradition in the eastern parts of the area and the spread of the practice of cremation and 'Henge' monuments northwards through the Marches into North Wales, probably in connection with the flourishing of the Graig Lwyd axe industry. West Wales, however, is now linked, by its 'Henge' types and ceramic styles, with southern Ireland rather than with Wessex. There is still insufficient evidence to determine the contribution of surviving Mesolithic communities to the Neolithic of Wales, but it is clear that there was a long overlap in the late 3rd millennium and early 2nd millennium between cultural traits conventionally assigned to the Neolithic on the one hand and the Early Bronze Age on the other.

Introduction

There can be no doubt as to the transformation which Neolithic studies in Britain have undergone during the past two decades. The greatest contribution, certainly, is the provision, through radiocarbon dating, of a revised and extended chronological framework which, though still problematic in detail, at least shows that the Neolithic, in Wales, as elsewhere in Britain, must have begun before the 5th millennium B.C. was ended and must have lasted at least two thousand years—five or six times as long as Piggott was willing to believe in his classic study of 1954 (Piggott, 1954, 379-81). This has given the prehistorian the deeper perspective which has proved essential to the understanding of the relationship of various cultural groups which previously seemed to jostle one another on the same narrow horizon. Hardly less important, however, has been the contribution, over a somewhat longer time, of the palaeobotanists who have reconstructed the natural environment which the first agriculturalists encountered in Wales and shown how Man and nature interacted. Over the same period, too, the geologists have been quietly building up, through petrological study of stone implements, a picture of Neolithic industry and trade, and have begun to show us, more recently, that some Neolithic pottery may have to be seen as part of that picture. The list could, of course, be lengthened.

The prehistorian, however, has not lost his indispensable function amidst all these activities. Only he can say whose were the mysterious hands which first began in earnest to clear the natural woodland of Wales, or erected the extraordinary structures of stone, earth and timber, the remains of which we still see; by what means they accomplished these things; to what extent they belonged to newcomers in the land; and whence any such newcomers and their culture came. He depends for his answers to these questions on the study of a steadily growing body of material from chance finds of objects and of sites,

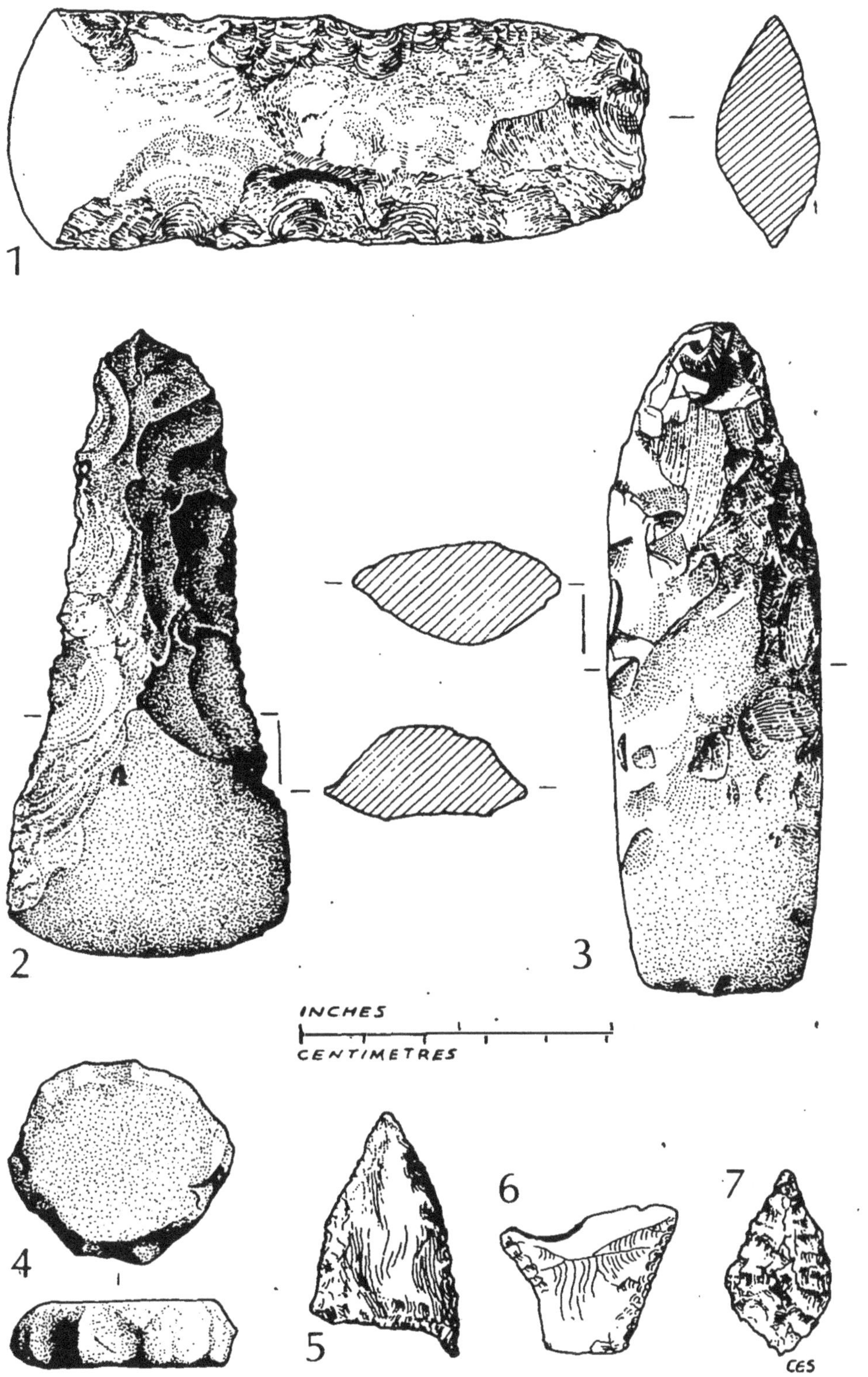

Fig. 5.1 A selection of Welsh Neolithic tools. 1. Axe-head (flint); 2. adze-head (flint); 3. chisel-head (flint); 4. sandstone lias; 5-7. flint arrowheads: lop-sided, petit-tranchet derivative and leaf-shaped. 7 from Ty-isaf (Talgarth) (after Grimes); 2 from Mynydd Merthyr, Aberdare; 3 from Caerau, Ely (S. Glam.); 5 from Newborough, Anglesey (Gwynedd); 4 from Penywyrlod, Talgarth; 6 from Bryn Celli Ddu, Anglesey, (Gwynedd) (after Lynch).

and upon field surveys or excavations of the latter. The deepened chronological perspectives and relatively precise absolute dates for particular sites he now has from radiocarbon dating have certainly helped him in his analysis of cultural groups, but have not removed all the problems which arise from the essentially fragmentary and one-sided nature of the archaeological evidence which normally survives from communities as remote from us in time, and in the nature of their lives and thought, as the Neolithic. As a result, though the new perspectives have encouraged prehistorians to differentiate Early, Middle and Late Neolithic phases, it is difficult to define precisely the contents of these phases, partly because particular types of object or structure which are thought to define different cultural groups still seem to overlap in time, partly because such types do not necessarily appear everywhere contemporaniously. This means, for example, that in spite of the lengthened chronology, we still cannot see very clearly the relationship, in the different phases, between those hunting and gathering communities of Mesolithic tradition, whose survival during at least part of the Neolithic is sometimes assumed, and those 'Secondary Neolithic' communities which are generally supposed to have evolved from them. Clearly there is still a considerable overlap in most parts of Britain between cultural traits which are classed as 'Late Neolithic' and those which are classed as 'Early Bronze Age'.

The result of the persisting uncertainties has been that some prehistorians have proposed, in despair, the total abandonment of the old 'Three Age' terminology in favour of conceptual frameworks related to supposed phases in the development of the economic and social life of prehistoric communities. Such systems, however, are premature because we simply do not know enough, as yet, about the development of those life-styles in particular environments. ApSimon, (1976, 41-9) has dealt with these difficulties and proposed that the Early-Middle-Late Neolithic terminology should be retained (with the addition of 'earliest' and 'final') as a convenient way of referring to chronological brackets with equivalent values in radiocarbon and calendar years (Earliest-Early would be in fact c. 4500-3750 B.C., Middle c. 3750-3250 B.C. and Late-Final c. 3250-2000 B.C.). I agree that the 'Three-Age' terminology must be retained, but the radiocarbon dates so far available, especially in Wales, are far too few and unevenly distributed among Neolithic phenomena for one to fit these very precisely within the brackets in particular areas. I propose, therefore, in this brief review to deal first with the general characteristics of Neolithic culture in Wales, then with the megalithic phenomena which centre on the Middle Neolithic phase, and finally, the developments of the later Neolithic.

The Foundations of Neolithic Culture in Wales

The 19th century museum archaeologists who seized upon the polished stone axe, adze- and chisel-heads so well represented in their collections as the type fossils by which an early period of human culture would be defined, were not, in fact, making a bad choice. For these implements were, after all, the most obvious of the new types of tool which appear in the New Stone Age and are related to three fundamental changes in human activity, i.e. the switch from hunting and gathering to the clearance of forest to provide grazing for domesticated animals; the preparation of the ground for domesticated plants where conditions were favourable; and the construction of the more solidly built timber-framed dwellings which tree-clearance made possible and

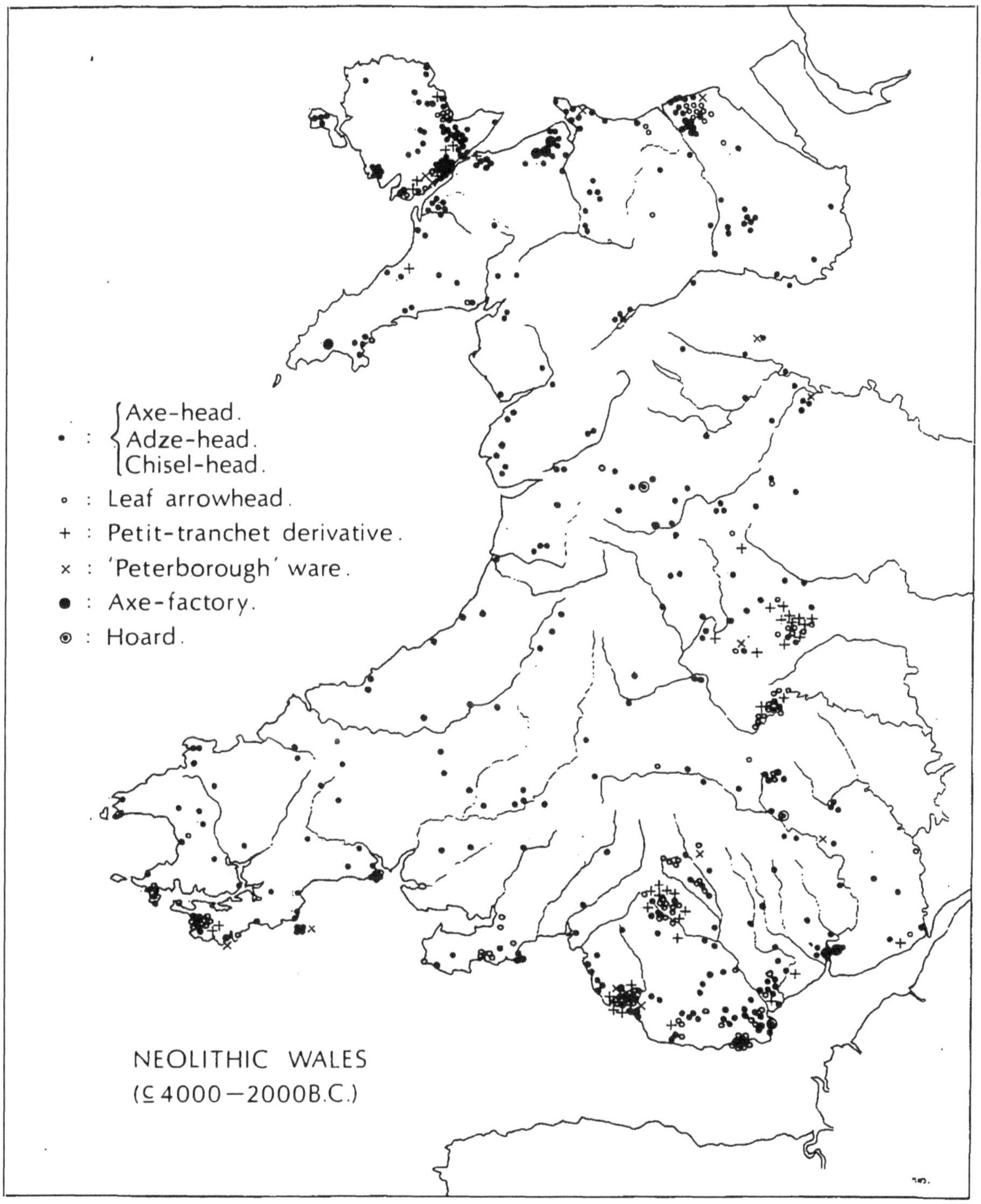

Fig. 5.2 Recorded Welsh finds of Neolithic artifacts

a more settled mode of life desirable. The great variations in size and shape of these new tools (Fig. 5.1) do indeed suggest a variety of uses; finds, moreover, are sometimes associated, in tombs, settlements, and surface scatters of flints revealed by modern ploughing, with a range of smaller tools made of flint or chert. These, themselves, constitute a new 'industry', distinct from Mesolithic industries which went before, and are largely based, in Wales, upon better sources of supply made available, so it would seem, by the growth of trade with distant flint mines, first in south-eastern England and later in East Anglia, and an industrial pattern which was soon extended to suitable lithic materials in Wales itself. Finally, we now have a new artefact, pottery, which is often associated with Neolithic stone implements in burials and settlements and because of its infinite possibilities of typological variation has often been used, in the past, to define cultural groups within the Neolithic.

All the museum curator's characteristic Neolithic artefacts have been found in Wales, and an attempt has been made here to plot the distribution of the most distinctive of them (Fig. 5.2). This distribution presumably reflects, to a degree, the incidence of settlement, although allowance must be made for the greater chance of discovery in certain areas of intense modern disturbance of the soil e.g. around Cardiff, and the effects of Neolithic industry and trade e.g. around the 'axe-factories' near Penmaen-mawr (Gwynedd). The distribution, allowing for this, suggests a concentration near the coasts and along main river valleys, with scanty occupation of the interior uplands, except for the main Glamorgan uplands, where Forestry Commission ploughing has been relatively well watched in recent years. Until recently, most prehistorians would have interpreted this distribution in terms of 'Highland Zone' retardation, diagnozing a delayed and partial penetration of the country by Neolithic settlers spreading from Southern England. Now, however, the few radiocarbon dates we have, which relate to an early Neolithic stage in Wales, prove not to be greatly out of line with the earliest dates of the same kind from southern England. We have 3290 bc ± 150 (NPL-223) from an open settlement preceding the first 'Henge' at Llandegai (Gwynedd) (Houlder, 1976, 58) and 3050 bc ± 95 (NPL-132) from an occupation layer at Coygan (Dyfed) (Wainwright, 1967, 14) —dates which imply activity, in calendar terms, about the beginning of the 4th millennium B.C. That this activity was not confined to the southern and northern sea-boards of Wales is suggested by a date obtained from the minor bones of various members of the community which erected the great chambered long cairn at Penywyrlod (Talgarth) in the Black Mountains: 3020 bc ± 80 (HAR-674) which is very close to the classic date for the elm decline (pp. above). Such dates are, of course, subject to the reservations which are commonly made about single radiocarbon dates from a given archaeological deposit, but the general impression they give is surely confirmed by the comparable range of dates now available not only from Britain but from Ireland. The very numerous radiocarbon dates from Ballynagilly (Tyrone) suggest an initial Neolithic occupation soon after the middle of the 5th millennium B.C. followed by evidence for large-scale exploitation of the surrounding countryside on the horizon suggested by our early Welsh dates (Smith, 1974, 100-3).

A roughly simultaneous onset of the Neolithic through most parts of Britain can, of course, be understood now through the work of climatologists and palaeobotanists, who have brought home to us how misleading the crude application of Highland Zone concepts to the Western side of Britain can be in the

Plate 5.IA Axe with birchwood handle preserved, from Aberavon (Mid-Glam.)

B Long cairn with main chamber partly exposed, from Penywyrlod, Talgarth (Powys)

Atlantic and Sub-boreal climatic phases with which we are concerned. During these climatic phases the temperature in Wales would have been somewhat higher than it is now (Webley, 1976, 19-22) and it is better to speak of an Atlantic Zone where coastal tracts at this time would have had a great abundance of fowl, fish and shellfish (Evans, 1975) and some at least of the interior would have had a fertile soil of the brown earth type and shelter from the wind and excessive wet by a canopy of trees (Taylor, 1975). It was the activities of Man in removing this canopy—on a small-scale, perhaps, already at the end of the Mesolithic and on a much larger scale from the early 4th millennium B.C. onwards, which would have promoted the subsequent formation of heath and blanket peat on the uplands, especially in phases of heavier rainfall (Moore, 1973). We have also to remember that much of the submerged land surface along the northern, southern and western coasts (Taylor, 1973) of Wales would, in Neolithic times, still have been above water, tree-covered and valuable to those who depended upon woodland animals, wild or semi-domesticated. The early axe-head from Aberavon (Plate 5.IA), with its rare, partially-preserved handle of birch-wood, the survival of which we owe to preservation in a coastal peat-deposit, is a reminder of these conditions and perhaps indicative of the type of tree available in this habitat. In view of Webley's maps and my own (Fig. 5.2), one can no doubt reasonably suppose that the initial Neolithic settlement of Wales began in the coastal tracts, especially if one assumes that it was initiated by colonists from outside Wales, many of whom would have arrived by sea-routes from the West Country or beyond.

That the beginnings of Neolithic culture in Wales were in fact due to colonists, rather than the independent and isolated response of its Mesolithic inhabitants to an evolutionary process stimulated by their environment, is suggested by a number of features of their material equipment which surely indicate a close relationship with other groups in southern Britain and even on the Continent. Among these are the distinctive forms of the pottery and the dwellings; the latter in particular, with their rectangular, ridge-roofed and sometimes aisled structure are already present at Llandegai (Houlder, 1976, 58) as at Ballynagilly and Peterborough (Smith, 1974, 104) and present a contrast to the flimsy round huts of their Mesolithic predecessors. We must, however, note a probable difference in the density of settlement between parts of lowland England for example, and Wales, reflected not only in the numbers of chance finds but the fact that no Neolithic settlements with traces of more than one or two houses have so far come to light in Wales, still less any causewayed camps like those of southern England, or a true hill-fort, with numerous houses, like the one on Carn Brea in Cornwall (Selkirk, 1974). It may well be that the relatively unfavourable conditions affecting growth of vegetation and animals, particularly on the uplands of Wales as tree-clearance by man progressed upslope, limited the size of immigrant communities and brought about a shift in the balance of their economics from agriculture to pastoralism. Whereas it has been generally assumed that much of the forest clearance undoubtedly carried out by Neolithic Man, in Ireland as in lowland Britain, was for purposes of cultivation, and palaeobotanists in these areas have produced plenty of evidence of cereal crops, Webley (1976, 26) has deprecated the assumption of widespread Neolithic agriculture on the Welsh uplands in spite of the evidence for it at Llanelwedd in the Wye Valley near Builth Wells which Crampton (1967) produced (Spurgeon, 1966). Cereal

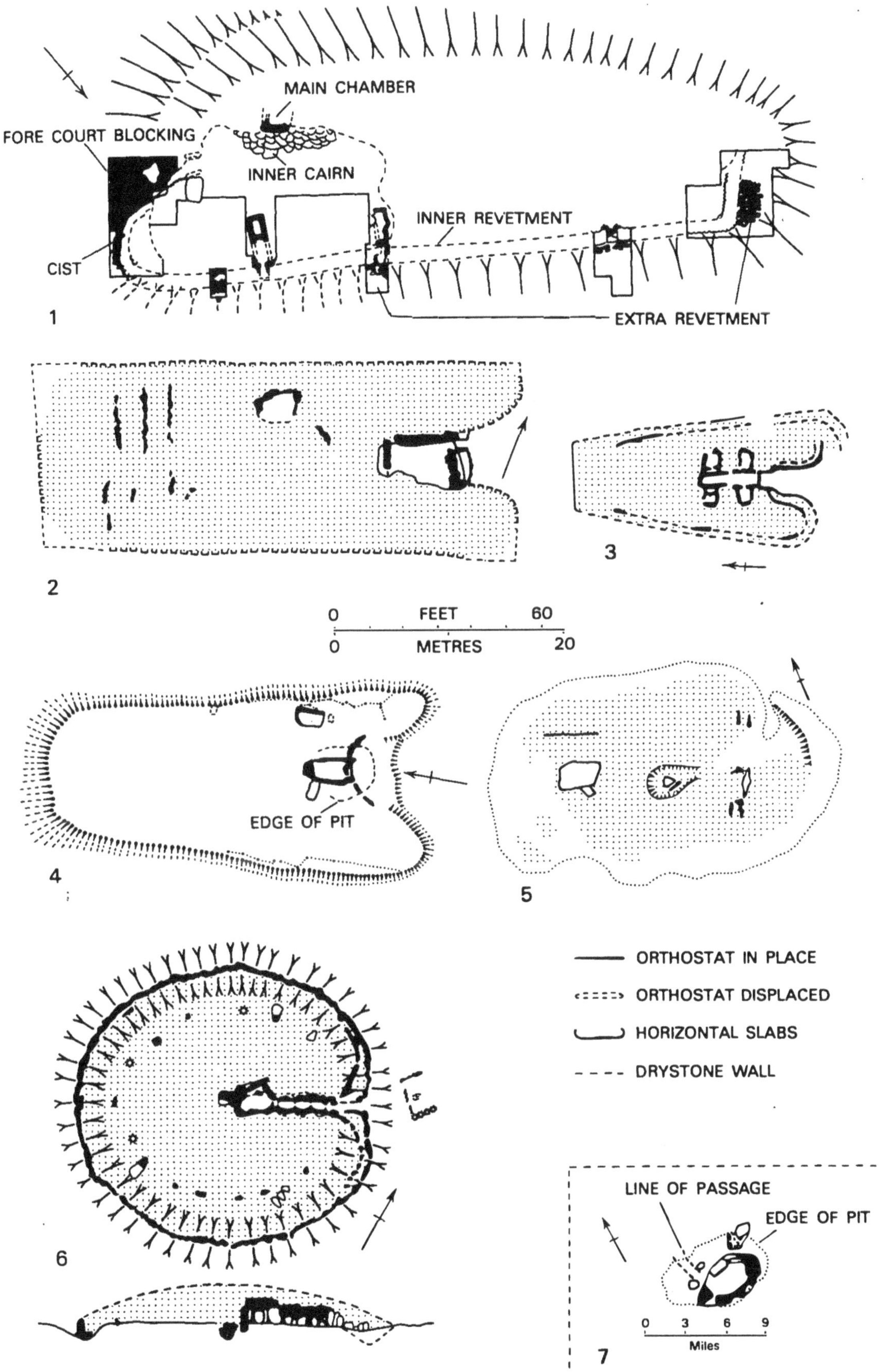

Fig. 5.3 Representative plans of Welsh megalithic tombs. 1. Penywyrlod, Talgarth (Powys); 2. Tinkinswood (S. Glam.) (after Ward); 3. Parc-le-Breos (W. Glam.) (after Atkinson); 4. Pentre Ifan (Dyfed) (after Grimes); 5. Carnaddau Hengwm, N. (Gwynedd) (after Lynch); 6. Bryn-celli-ddu (Gwynedd) (after Hemp); 7. Longhouse (Dyfed) (after Lynch).

agriculture certainly seems not to have begun in the neighbourhood of Carneddau Hengwm in Ardudwy until long after the Neolithic (Moore, 1973).

Who the Neolithic newcomers to Wales were is a question which is answered, up to a point, by a study of their skeletal remains, which have been recovered in relative abundance from ossuaries in megalithic tombs and caves. Like similar remains from other parts of the British Isles they seem in general to represent a robust variant of the great Mediterranean group which was widespread in Western Europe (Cave, 1938, Coon, 1939. Brothwell, 1973). This does not take us very far, and when we examine the archaeological evidence for their background further afield in Europe, we find that different traits in the earlier Neolithic culture of Britain have different origins on the Continent. While the rectangular, and even trapezoidal plans of the houses (Fig. 5.3) and the causewayed camps of the lowlands, have fore-runners in Neolithic groups in the Low Countries, the Rhineland and north-east France, which were evidently strongly influenced by the 'Danubian' houses of central Europe, the associated pottery in Britain, as in Ireland, relates best to the earliest Neolithic wares of Atlantic Europe, even though much of it shows the effects of local specialization. Moreover, the method of hafting stone axeheads, as at Aberavon (Plate 5.IA), without the antler sleeve characteristic of west-central Europe, appears to be that practised in Iberia and Brittany (Del Val Caturla, 1948; Savory 1971). Above, all, however, it is the practice of mass communal burial, as opposed to family or single burial, which links Early and Middle Neolithic Britain with the Atlantic sea-boards of Europe. This is something more fundamental than the typology of tomb architecture, which has so obsessed megalithic specialists, and it is now becoming increasingly clear, as radiocarbon dates accumulate, that it is a trait which was distributed originally along a line partly of maritime and partly of overland contact through the Mediterranean from the Levant (not the Aegean) to Iberia, and thence along the Atlantic sea-boards of Europe, only later spreading eastwards into the Paris Basin and across the north European plain (Savory, 1977).

What came into the British Isles, then, in the Early Neolithic, was a synthesis of cultural elements derived partly from the interior of Europe and partly from its south-west margin. One must, therefore, await the steady progress of later prehistory in northern and western France, which should in due course tell us how the traditional oblong and trapezoidal house of the living in central Europe was converted into the House of the Dead, as represented by the chambered and 'unchambered' long barrow and cairns of France and the British Isles. But the synthesis varied, of course, among different groups in France and Britain alike, and their movements followed different routes. One can thus see that there were two main groups of early colonizers in Wales, each with their own distinctive ceramics and megalithic architecture —an eastern one which had connections with Wessex and west-central France, and a westerly one which had links with Ireland (below, p. 220).

Just as the synthesis of cultural elements, which I have outlined in Atlantic coastal France, probably represents, in large measure, the acculturation of local Early Neolithic and even Mesolithic people through contact with Danubian agriculturalists immediately to their east in the 5th millennium B.C., as well as with the earliest communal-burying groups of south-west Europe, so it may well be that our own British synthesis involved the more or less gradual incorporation of local Mesolithic groups and adaptation of the newcomers to a different environment—one which in Wales, especially, might involve a greater emphasis on pastoralism.

The question of Mesolithic survival in Wales is an important one, because it has been held by a number of specialists (Piggott, 1954, 275-87; Case, 1961, 231-20; Houlder, 1976, 57, 61) that various elements in the material culture of the later Neolithic of the British Isles represent the contribution of descendants of Mesolithic hunters and gatherers who had absorbed some Neolithic arts but preserved some of their cultural identity. It is suggested, moreover, particularly by Piggott and Houlder, that this contribution might have included axe-factories and trade in their products at least initially, since the migratory habits of such people would lend themselves to this process of development. If one accepts Webley's interpretation of the Neolithic economy in Wales (1976, 20-30), it is easy to visualize the Mesolithic seasonal movement from winter quarters at Merthyr Mawr to summer hunting and fishing grounds on the Glamorgan Uplands, with perhaps, a little burning of woodlands to improve the feed of deer or Bos primigenius, evolving into the transhumance of Neolithic pastoralists, with more extensive clearance for their domesticated cattle, about the time of the 'Elm Decline' early in the 4th millennium B.C. One could even imagine that the search for beach-flint substitutes, which the late Mesolithic communities at Glamorgan, for example, evidently conducted during their seasonal sojourn near the out-crop of Carboniferous Limestone and chert in the interior of South Wales could have pointed the way to early exploitation of similar local substitutes in north Pembrokeshire and the Mynydd Rhiw in the Lleyn, in the 4th millennium B.C. ' Unfortunately, however, no direct radiocarbon dates have so far been secured for the late Mesolithic narrow-blade industries of Wales, and at present the best evidence for their survival down to the late 4th millennium B.C. seems to be the occurrence of characteristic microliths at the base of a later Neolithic occupation layer at Ogmore (Glamorgan) which has yielded two such dates (p. 228 below). On the other hand, it must be pointed out that Webley's reconstruction would imply that long-distance transhumance was already being practised in the early 4th millennium by the 'pure' Neolithic community which built the Severn-Cotswold tombs, and the flint industries of the earlier and later Neolithic communities are alike sharply contrasted with those of their Mesolithic predecessors and are often perfectly distinct from them stratigraphically. It must be remembered too, that the Mesolithic type of transhumance in South Wales could not have coexisted for long with the development of the coastal tract by Neolithic farmers, and it is probably best to envisage an Early Neolithic phase in which Mesolithic hunters were absorbed into the local farming communties.

The Communal Tomb in Wales

In Wales, as in other parts of Britain and Western Errope, the practice of mass communal burial (most strikingly in megalithic chambers), as distinct from single or family burial, is the feature of Neolithic culture which has attracted most attention. The weakness of any attempt to estimate, from the recorded contents of known tombs, the total population in the areas where they occur, or to argue thereafter that they were reserved for an 'elite' (Atkinson,

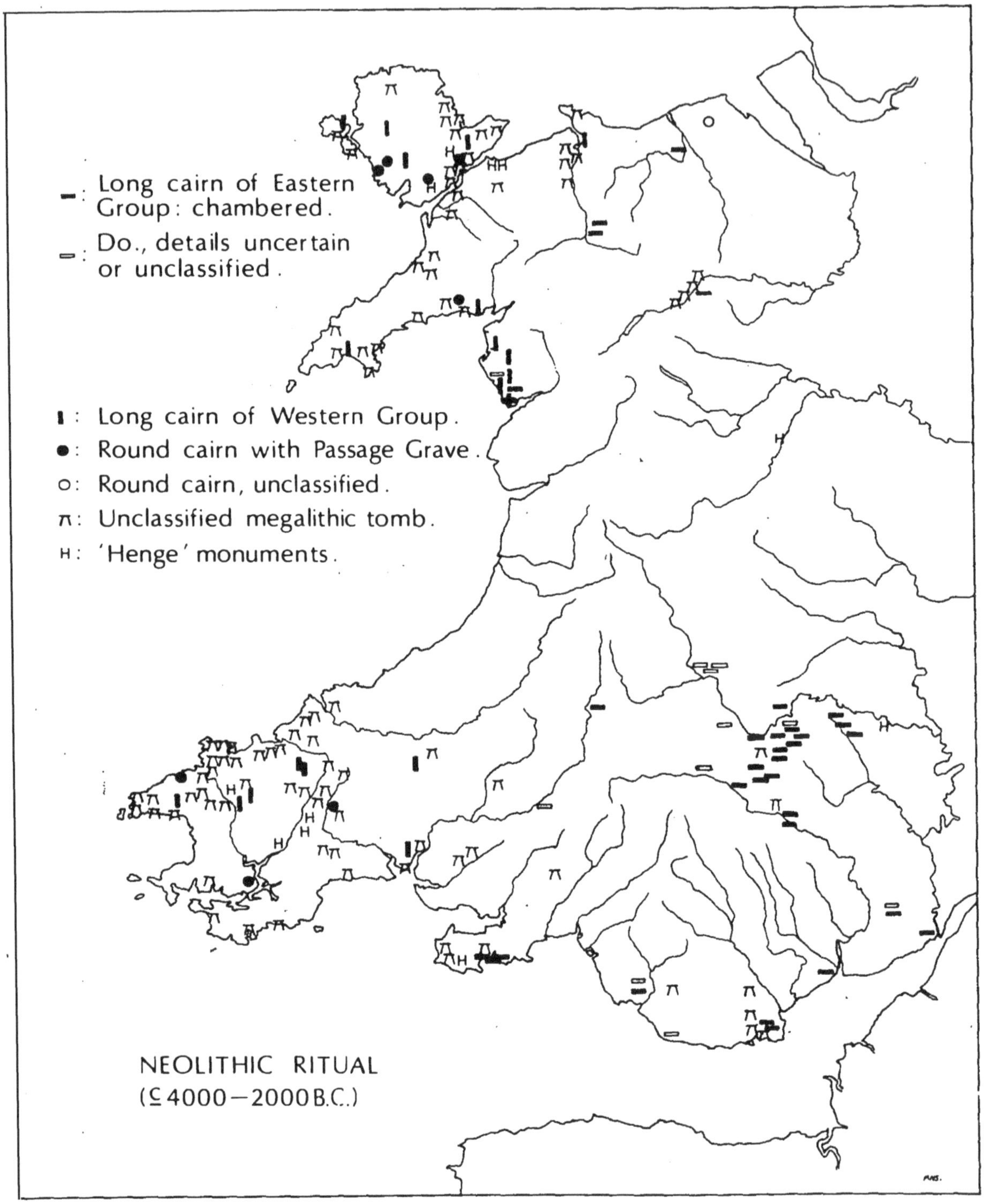

Fig. 5.4 Distribution of megalithic monuments in Wales.

1968), is that we do not know how many have been completely destroyed, or have not been recognized for what they are, nor do we know how representative they are of Neolithic ossuaries in general: many of these, in areas without caves, may not have been capable of leaving any more trace than the burials of other long periods of British prehistory. In any event, it is emerging that they were not much used in most parts of Wales during the later Neolithic, or anywhere at the very beginning of the Neolithic. Their use in Wales seems to have centred on a middle phase when regionalism had already developed in the Welsh Neolithic. The distribution of the varied types of megalithic tomb reflects this regionalism (Figs. 5.3 and 5.4). The main group is best represented in south-east Wales, and is clearly an extension of the 'Severn-Cotswold' group which is so well developed between the upper Thames and the Severn. Outlying members of this eastern group are known from North Wales and Carmarthenshire and it is probably the earliest, being related to primary Neolithic settlement of South-West England and the more fertile parts of Wales. A western, coastal group is typologically much more diversified and seems to represent, in large measure, a secondary settlement of the area from across the Irish Channel, as some of the associated objects suggest, and one lasting much longer than the Severn-Cotswold group.

The Eastern Group. The basic features of the Severn-Cotswold group are wedge-shaped cairns, ceremonial forecourts varying in orientation from NNW through E to SSE, chambers which are usually roughly rectangular with entrance passages of varying length and a tendency to transeptal arrangements in more elaborate cases, and great virtuosity in the use of drystone walling for revetment walls (often double) and entrance passages. Inhumation is invariable, but the skeletal material found in the chambers is almost always fragmentary, representing reburial of remains originally deposited either in other chambers or somewhere outside the cairn. The single large chamber at Tinkinswood (Glam.) (Fig. 5.3) contained portions of at least 50 individuals while several small chambers at Ty-isaf (Breck.) yielded fragments of over thirty. Until recently end-chambered cairns with functional forecourts, like Tinkinswood or transepted sites like Parc-le-Breos (Gower) (Fig. 5.3), with similar entry, were thought to be primary, partly because of their location near the Bristol Channel, partly because of obvious continental parallels, but now the radiocarbon date from Penywyrlod (Talgarth) (3020 ± 80 b.c.) (HAR - 674) proves to be earlier than the only date from an end-chambered, transeptal tomb-Wayland's Smithy (Berks.) (2820 ± 140 b.c.) (I-2328) (Atkinson 1965). Penywyrlod, as it happens, was, until recent damage, an exceptionally large and well preserved example of a supposedly late and locally developed type of cairn, which is characteristic of the Black Mountains and the Cotswolds, with multiple chambers that are entered from the side of the cairn and not through the forecourt (Pl. 5I, Fig. 5.3) (Savory 1973a). The two Severn-Cotswold sub-groups are now probably best regarded as parallel series of purely local significance, with common roots in western France, where it is now clear, from a large number of radiocarbon dates, that long cairns with laterally-entered chambers were built as early as the latter part of the 5th millennium B.C. (l'Helgouach, 1965 and 1973). It is also clear that divergent regional groups existed in northern Brittany, lower Normandy, near the Loire mouth, and in Poitou-Charente, the last two of which appear to have contributed most to the Severn-Cotswold tradition—the first with a tendency to transeptal chambers and the second with rectangular chambers which were sometimes entered from a terminal forecourt, sometimes laterally. Double revetment walls are a feature of well-preserved cairns, round or long, in the Atlantic region of France, and there is no need, (pace Corcoran, 1969, 84-6 and Lynch, 1976, 71) to regard them as evidence of two-period construction at British sites where no actual indications of such construction have been observed by their excavators.

The scanty grave-goods from Severn-Cotswold tombs—pottery related mainly to the 'Abingdon' tradition of southern England, e.g. leaf arrowheads, polished flint axe-heads (Fig. 5.1)—are not at variance with a general background in the pure Neolithic of western France about the beginning of the 4th millennium B.C., earlier than the appearance in that area of more elaborately shaped and decorated pottery. In particular, mysterious artefacts found on some Breconshire sites and in the derivative tomb at Pant-y-Saer in Anglesey —sandstone discs of graduated sizes—have parallels from Breton megaliths and appear to be common on Neolithic settlement sites in south-west France, notably at St. Michel-du-Touche, near Toulouse (Meroc, 1967; Phillips, 1975, 90-2). Their development would be parallel to that of the eastern long barrows of lowland England which lack stone chambers for local geological reasons and yield traces of wooden chambers to careful excavators (Ashbee, 1970, 33-54). The dates obtained for a laterally-chambered tomb at Ascott-under-Wychwood (Oxon.)—2943 bc ± 70 and 2785 bc ± 70 (BM 491b-492) (Selkirk, 1971) are only a little later than Penywyrlod (Talgarth) and suggest that true Severn-Cotswold tombs belong mainly to the early and middle parts of the 4th millennium B.C. Tombs in North Wales and Ardudwy which relate to them (below p. 220) may be somewhat later but here we encounter a problem in searching for intermediate stages in mid-Wales, which now lacks surviving Neolithic communal tombs of any kind, although chance finds of artefacts and one or two vestiges of settlement sites show that this region was not uninhabited (Figs. 5.2, 5.4). Perhaps the lack of suitable stone for megalithic construction may have encouraged ossuaries of a different type, like those recently found in the ditches of a cause-wayed camp of Hambledon Hill (Dorset). Since a cause-wayed camp and ensuing fort, both Neolithic, have recently been explored on Crickley Hill near Cheltenham, in Severn-Cotswold territory (Selkirk, 1975), similar sites may still await discovery in the upper valley of the Severn. The abundant finds from the thorough excavation in 1977-8 by W. J. Britnell for the Clwyd-Powys Archaeological Trust of a damaged tomb of this group at Gwernvale (Breck.), will, when fully published (Cambrian Monographs) throw much light on the cultural sequence chronology and economy of the Severn-Cotswold people (Britnell, 1978).

The Western Group. The westward zone of chambered cairns is far more complex and includes many monuments which are now too badly wrecked for easy classification. They have been much discussed recently by a number of specialists (in particular, Grimes, 1965; Lynch 1969 and 1976). It is, clear, at any rate, that we are dealing, in this area, with two main groups—one of long cairns and the other of round cairns—which probably lasted longer than the Severn-Cotswold group, late enough for some of them to receive cremation burials of later Neolithic character. The long cairns, when sufficiently well preserved, prove not to have the distinctive features of the eastern group; one of the best known of them, Pentre-ifan (Pemb.) (Fig. 5.3), has a concave forecourt lined with an orthostatic facade which rises in height towards the tilted capstone of the rectangular chamber, the front of which is blocked by a tall orthostat. This'portal' arrangement relates to a whole series of 'Portal Dolmens', generally without well-preserved cairns, that are scattered along Irish Channel coasts from Cornwall to Antrim. Pentre-ifan, however, with a few other tombs in west and north Wales, also has features, particularly of forecourt arrangement, which relate to the Court Cairns of northern Ireland. Powell and F. Lynch have argued that Portal Dolmens may be the earliest tombs in Wales and Ireland and were inspired by theoretical wooden prototypes in southern England derived from similar structures in the long cairns of N. Germany and Poland (Powell 1969, 264-71, and Lynch 1976, 63-71). This view has not won much acceptance in Ireland, where Portal Dolmens are generally held to derive from Court cairns (Herity, 1977, 85-92), and the idea that Kujavish type long cairns in Poland could have inspired the earliest communal tombs in the west (Daniel, 1970) seems to me to be fallacious. These eastern long cairns, though undoubtedly inspired, like western ones, by long houses of Danubian tradition, were originally built to cover single burials (Jazdzewski, 1973, 63f) and only later had communal burial chambers inserted in them as a result of influences coming from Atlantic Europe(Savory, 1977).

In the absence of radiocarbon dates for Portal Dolmens, Powell and Lynch have based their argument for an early date partly upon the supposed typological priority of pottery found in the earlier of the two Portal Dolmens at Dyffryn Ardudwy (Powell, 1973) over pottery from the earliest settlement at Llandegai (Caern.) and tombs of the eastern group. This western pottery belongs, however, to a lengthy Irish sequence, which is distinct from the Abingdon series found in the eastern group, and chronological conclusions drawn from the evolution of rim forms in two parallel regional groups are highly precarious. Ceramics, as Peacock has shown (1969, 145-9), had already begun to be industrialized in the Neolithic, with the products of particular manufacturing centres travelling considerable distances. Lynch, however, has also argued (1976, 69-71) that the southern of the two long cairns at Carneddau Hengwm (Ardudwy) originally consisted of a Portal Dolmen in a small cairn, which was afterwards lengthened to contain a lateral chamber inspired by Severn-Cotswold influences. Without excavation, however, it is surely hazardous to guess whether the Portal or the Severn-Cotswold elements in this cairn were primary. Certain it is, especially after study of Gresham's detailed plans and descriptions (1967, 11-15), that both the cairns have many features of the eastern group and that the great size of the southern cairn, with the various dry-stone walls, which might be interpreted from surface indications either as the remains of entrance passages of of cross-walls, suggests that it is substantially a Severn-Cotswold construction. It could be argued, indeed, that Portal Dolmens represent the further development of tendencies—"dummy" entrances and heightening of the portal—which already appear in some chambers of the eastern group. Above all, however, the siting of Portal Dolmens in less desirable settlement areas near Cardigan Bay, avoiding the good land of South Pembrokeshire, Arvon and Anglesey, makes it unlikely that they are particularly early.

In North Wales, the influence of the eastern group is also discernible not only in late-seeming Anglesey tombs like Pant-y-Saer and Bryn-yr-Hen-Bobl

(Lynch, 1969, 161-6) but in the remarkable wedge-shaped long cairn, with a deep dummy forecourt and lateral chamber, at Capel Garmon (Denb.) (Lynch 1969, 143f). Here the chamber has transepts—a good Severn-Cotswold tradition—but these have the rounded plan which suggests contact with a different tradition, that of the 'Boyne' Passage Graves. This tradition with its round covering cairns is, in fact, well represented in north-west and south-west Wales (Fig. 5.4), but most of its examples are more or less wrecked and unexcavated, so that attention has been concentrated on the two fine excavated sites in Anglesey—Bryn-celli-ddu (Fig. 5.3) and Barclodiad-y-gawres. These undoubtedly represent a late stage of the great Passage Grave tradition of Atlantic Europe of which the primary focus in the 5th and 4th millennia B.C. was in Iberia, although it soon spread to western France, including Brittany (Savory, 1975 and 1977; l'Helgouach, 1973), and later flourished exceedingly well in east-central Ireland. An abundance of radiocarbon dates from the renowned sites at New Grange and Knowth (O'Kelly, 1973; Eogan, 1974) show that the main tombs here were built in the latter part of the 4th millennium B.C. Their tendency to contain large numbers of cremation burials suggests, that their use continued into the later Neolithic (below, p. 225). It is not surprising, therefore, that the repertoire of carvings with which some of these tombs are decorated, should include much that is not typical of the earlier art either in Iberia or Brittany, although in fact the carvings on the Anglesey monuments have features which relate to the Iberian tradition (Savory, 1973). None the less, Bryn-celli-ddu is now known to have been built over a superseded Henge monument which must have been an early example of a specifically later Neolithic type (C. O'Kelly, 1969) (p. 225 below). It does not follow, however, that all 'Boyne' Passage Graves in the British Isles belong to this late horizon, considering how early some of the ancestral tombs in Brittany area. There are, as Lynch points out (1976, 75-7), indications that some relatively small and simple passage graves, with polygonal chambers—like Broadsands (Devon) Longhouse (Pembs.) (Fig. 5.3) and the earliest of a sequence of tombs at Trefignath, Holyhead (Ang.) recently excavated by C. Smith for the Ancient Monuments Branch (Wales) (publication pending) may be not much later in date than the earliest of tombs of the eastern group. These would represent the earliest megalithic colonisation of the Irish Channel area, made by Atlantic sea-routes, before the introduction of cremation to that area by the builders of the early Henges at Landegai and Bryn-celli-ddu.

The Later Neolithic

General. We have seen how, for the 4th millennium B.C., a study of the forms and grave goods of communal tombs in various parts of Wales enabled us to trace, in faint outline, the shaping of two provinces separated by the major north-south watershed—a large one orientated towards England and a smaller one in the west based on Atlantic and Irish Channel contacts. In the 3rd millennium it is no longer possible to continue this approach throughout Wales because, in the area of Severn-Cotswold influence, the tombs seem generally to have gone out of use until the Beaker phase at the very end of the Neolithic. It is rather from surface finds, settlements and cave deposits in the main area that we gain material for comparison with finds in the later tombs of north-west Wales, and can establish that some of the cultural changes which affected England also affected most parts of Wales. In this way we know, for example, that much of Wales shared in the development of a later Neolithic flint industry in which, notably, the leaf-shaped arrowheads of the earlier phase were largely replaced by transverse and pointed arrowheads of 'petit-tranchet derivative' type (Fig. 5.1). These forms seem at present to be particularly common as surface finds in Radnorshire and on the uplands of Glamorgan, (Fig. 5.2) but this may reflect to some degree the recent incidence of keen collectors in those areas. Another common feature is the spread to most parts of Wales of the 'Peterborough' style of pottery (sensu lato) which Smith (1974, 111-3) no doubt rightly, treats as the outcome of local evolution from British Early Neolithic forms but which involves a marked deterioration in fabric and a tendency to heavy, complicated rim forms and elaborate impressed decoration of various kinds (Fig. 5.6). Such ware is now known not only from Bryn yr-Hen-Bobl in Anglesey, where it may be secondary to the tomb, but from Gop cave in Flintshire, Ysgwennant in Denbighshire, newly discovered settlement sites in Radnorshire, at Abergavenny, on Cefn Cilsanws near Merthyr Tydfil and at Ogmore near Bridgend, as well as from caves in Caldey and near St. Govan's Head in south Pembrokeshire. It is important, however, to notice that this Welsh 'Peterborough' has a distinctly regional character, without much of the cord-impressed decoration or incised trellis and herringbone decoration found in eastern England. Perhaps significantly, it has not so far appeared in the Cardigan Bay area, and its place is taken at Dyffryn Ardudwy by coarse bucket-and barrel-shaped urns related to Irish wares like 'Lough Gur II' (Powell, 1973, 39-46)—a tradition which continues into the Bronze Age on both sides of the Irish Channel. On the other hand, it has not so far been found on megalithic sites in the eastern group, only at later sites in the western group.

The Axe Industry. The dates of the factory on Mynydd Rhiw (Caern.) (Group XXI) (Houlder, 1961), which sent its products as far as South Wales, and the centres of production, in North Pembrokeshire, of implements of petrological groups VIII (distributed to North as well as South Wales), XIII (rare in the Neolithic) and XXIII (mainly occurring in mid and south Wales), have not yet been fixed. There is no doubt, however, that the axe production which was centred on Graig Lwyd, near Penmaen-mawr (Group VII), flourished mainly in the Middle and Late Neolithic, as is suggested by the association with a petit-tranchet derivative arrowhead at Llandegai (Houlder, 1976, 61) and radiocarbon dates at Shapwick Heath (Som.) (Q.430) and Upware (Cambs.) (Q-31, 499, 580-1) (RCAM, Caern, 1956, xli-lviii). This last group of 'factories' undoubtedly had a considerable output which dominated the North Wales market and was widely distributed, not only to South Wales but to the rest of Britain, particularly Wessex. Whatever the precise mechanism of this distribution, Brigg's theory of local manufacture from erratics (1976, 268-71) can hardly account for the widespread distribution of the products of leading British factories like Langdale, Graig Lwyd and West Cornwall, whatever may be one's suspicions about the occasional implement of groups XIII ar XXIII. The fact that products of different factories often occur at the same site (not only at Henges, like Llandegai, which might attract visitors for social as well as economic reasons, but also at ord inary settlement sites like Ogmore or Merthyr Mawr or near English flint mines) simply means that the market and the movements of individuals were free, and the hoards, which begin to appear at this time, foreshadowing the bronze hoards of later times, emphasise this point. The find at St. Cyres, Penarth, in which two very large polished axe-

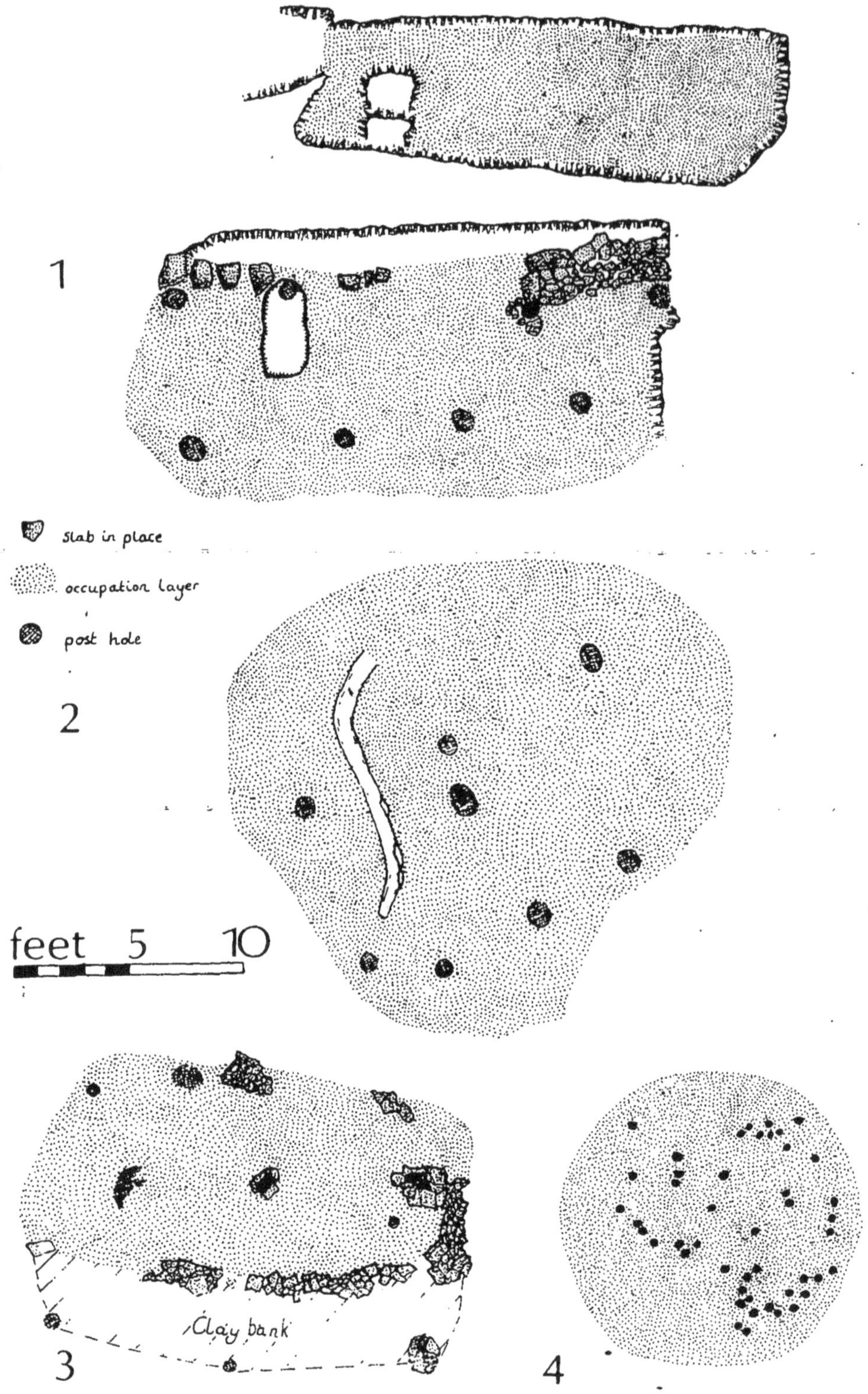

Fig. 5.5 Selected Neolithic house-plans. 1-2. Clegyr-boia, St. Davids (Dyfed), (after Mrs. Audrey Williams; 3. Mt. Pleasant, Nottage (Mid-Glam.) (after Savory); 4. Cefn Cisanws, Vaynor (Mid-Glam.), (after Webley).

heads, one of flint and the other of Pembrokeshire quartz-dolerite (XXIII b), are associated, is a good example (Savory, 1957).

Ceremonial Sites. It is clear that many communal tombs in the western group, especially those of 'Boyne' affinity, with cremated burials, were in use in the later Neolithic. It seems likely that a number of related tombs in Carmarthenshire and the Gower are also late, but I cannot follow Corcoran in seeing true 'Portal Dolmens' in south-east Wales (1969, 275f, 288f). The Dee Valley group of megalithic tombs may also be late, but, in the heartland of the Severn-Cotswold group, nothing has so far been found to suggest use of long cairns for burial in the 3rd millennium, or of the practice of cremation, although Peterborough ware and petit-tranchet derivatives occur in the area. On the other hand, the two extraordinary ritual long mounds, which Spurgeon explored at Llanelwedd near Builth (Spurgeon, 1965-9), embody a number of Severn-Cotswold features in their structure, but no burials. However, the D-shaped pit under the centre of the first mound seems to anticipate a tradition which frequently appears on Early Bronze Age burial sites in Wales, and may therefore suggest a relatively late date. It seems, however, that cremation did not become universal in the later Neolithic, because various ossuaries discovered in caves in the limestone belts in north and south Wales seem to belong to this horizon, as the pottery from Gop Cave (Flints.) and Rhos-ddigne (Denb.) for example, suggests. The associated material at the Tooth Cave, Ilston (Gower) (Webley, 1967) indeed seems to belong to a transitional Late Neolithic —Early Bronze Age phase. This skeletal material appears to be of normal local Neolithic type, not foreshadowing the 'Planoccipitale Steilkopt' type associated with Beaker pots (Gerhardt, 1974).

Whatever happened in the interior of Wales during the later Neolithic— and it must be admitted that we do not know very much about it as yet—'Henge' monuments with associated cremation burials (whatever may have been the other functions of these site) existed in north-west Wales at a date (late 4th millennium) which, on available radiocarbon evidence, puts them in the van of progress, especially in relation to central Wessex, where cause-wayed camps were slow to be replaced by monuments like Stonehenge I. Accordingly, Burl (1976, 24-30) seems inclined to agree with Houlder (1976, 59f) that the Menai Sites—Llandegai I and Bryn-celli-ddu—may have played a key role in spreading the 'Henge' idea to Wessex, perhaps in connection with the dissemination of Group VII axes. Once established, the 'Henge' tradition lived on to the end of the Neolithic at Llandegai II and Castell Bryngwyn (Anglesey) (Wainwright, 1962). At present this group of 'Henges' appears rather isolated, but some links with Wessex are beginning to appear in the west Midlands and near Welshpool in the Upper Severn Valley, where unpublished air-photographs suggest the former presence of related sites. Burl's idea, however, that the 'Henge'—cremation complex may have started in Ireland, tempting though the relative proximity of the Boyne to the Menai Straits may make it seem, encounters the difficulties that there seems to have been hardly any exchange of group VII and Tievebulliagh axe-heads between these centres, and the origins of Neolithic cremation seem to lie in the timber ossuaries of lowland Britain and northern France rather than Ireland. Moreover, Houlder points out that the imported axe-heads at Llandegai I came from West Wales and Cumbria, areas which also have 'Henges' and stone circles. Clarification of this problem

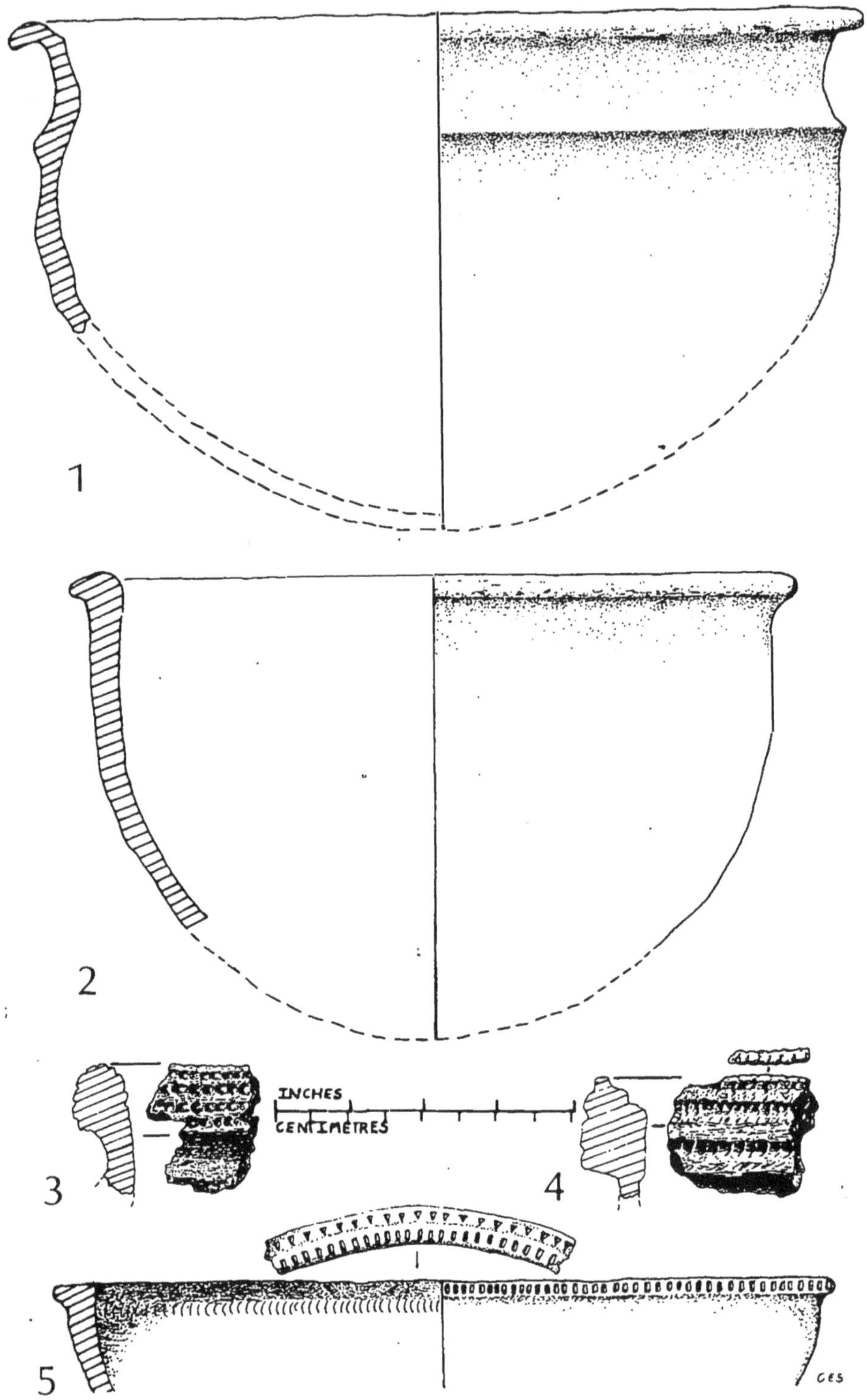

Fig. 5.6 Selected Neolithic pottery. 1. Clegyr-boia, St. Davids (Dyfed), (after Mrs. Audrey Williams); 2. Ty-isaf, Talgarth (Powys) (after Grimes); 3-4. Bryn yr Hen Bobl, Anglesey (Gwynedd) (after Hemp); Mt. Pleasant, Nottage (Mid-Glam.) (after Savory).

may come with the exploration of the 'Henge'-like sites on the Boyne, which has, as yet, barely begun (Daniel 1964, 87-90 and Sweetman, 1976).

Elsewhere in Wales stone circles are puny, and seem to represent the tail-end of the 'Henge' tradition in the Early Bronze Age. Only in west Wales are there a few small 'Henges' (Grimes, 1963; 141-3; Savory, 1963, and Burl, 1969, 15f and 1976, 255-7) which have features (orthostats lining the entrance and the inner slope of the bank) which definitely relate them to an Irish type, best represented by the Grange circle ('Lios'), Limerick (O'Riordain, 1951). All these sites, however, appear to belong to the end of the Neolithic and the opening of the Early Bronze Age—to a horizon, in fact, when southern Ireland and south Wales shared a certain community of ceramic tradition (p. below). What appears to be another 'Henge' site has recently been detected on an air-photograph taken over Llanddewi in west Gower (RCAM (Wales), 1976, No. 47) —an area liable to be linked with Pembrokeshire. The horizon to which these monuments belong is very probably that of the blue-stone circles at Stonehenge and possibly that of the blue-stone-axe-heads found in Wessex—but I cannot review here the arguments for and against the possibility of human agents for the first part of the journey involved. At the moment, the archaeological evidence relates rather to local connections between south Wales and southern Ireland. The extreme rarity in Wales of the 'Grooved Ware' so plentiful on 'Henge' sites in England towards the end of the 3rd millennium speaks against a close relationship between 'Henge' communities in south Wales and Wessex at the time, though the ubiquitous users of 'Beaker' pottery were soon to use the Bristol Channel coasts as a route to southern Ireland.

Settlements. Apart from the 'Henge' concentration in North-West Wales, which may indicate seats of political and economic, as well as religious power, the evidence we have for settlements in Wales in the later Neolithic comes mainly from a series of huts in South Wales, the earliest of which are probably the two encountered, about 20 m. apart, in trenches cut by Mrs. Audrey Williams to explore a small Iron Age hill-fort at Clegyr-boia, St. Davids (Williams, 1952). The main structure here (Fig. 5.5) was oblong, some 7 m long and 3 m wide, with two rows of posts along one of the long sides. The other hut had a rounded outline, about 7 m. in diameter, and a central roof support. The pottery associated with both these huts, like that found on megalithic sites of the western group, represents the continuation of the superior 'Western Neolithic' tradition of Ireland, with occasional rim decoration which relates to a southern Irish (Lough Gur) late Neolithic style (Fig. 5.6); other huts may have escaped excavation on this site. The other south Wales huts, as far as we know, were isolated: a fine one, with the foundations of dry-stone walling preserved under an Early Bronze Age round barrow at Mount Pleasant, Nottage (Glam.) (Savory, 1952) (Fig. 5.5), was oblong, 6 m. by 3 m. with a ridge-roof, and door near the western end of the north wall; another, recessed into a slope on Cefn Glas near the head of the Rhondda Fach, also had traces of dry stone walling and a ridge roof and could be dated 2160 b.c. $\pm$70 (HAR-744) by charcoal incorporated in the make up of its floor-level. The most interesting feature of this unpublished site is its trapezoid plan—the end of a long Neolithic tradition, encountered, as already mentioned, in many of the Danubian long-houses in the Low Countries and northern France. Finally, there is the small, oval, flimsily-built hut on Cefn Cilsanws, Merthyr Tydfil

(Webley, 1958) (Fig. 5.5,4). Since this yielded a few scraps of the regional type of 'Peterborough' ware (which occurred abundantly in the unpublished occupation layer explored by Webley in the sand-dunes at Ogmore, Glam., in association with microliths, a polished flint perforated mace of 'Thames' type and some cord-impressed potsheads of final Neolithic type—but no trace of well-built huts), it is tempting to see here evidence of the descendants of Mesolithic food-gatherers still moving with their herds from winter quarters near the sea to summer camps on the uplands, contented throughout with lightly-built huts for shelter. But the two radiocarbon dates from near the base and near the top of the layer at Ogmore, respectively 2709 bc ± 52 (BM 1112) and 2370 bc ± 80 (HAR 1140), suggest a long occupation for this site, and it cannot be asserted that microliths remained for long in the repertoire of the people who used Peterborough pottery. The main flint industry at Ogmore was later Neolithic in character, as was that at Cefn Glas, where, however, no pottery occurred. The association of two types of hut at Clegyr-Boia warns us that the same community might use different structures for different purposes, and this lesson is confirmed by contrasting structures in similar upland positions on Cefn Cilsanws and Cefn Glas. All these communities in the 3rd millennium B.C. were surely fully Neolithic in their mode of life, with some adaptation to local circumstances, but the constrasting pottery at Mount Pleasant—"Western Neolithic" with decorated rims which relate, like Clegyr-boia to southern Irish forms (Lough Gur and Dalkey Island), —together with forms of secondary Neolithic ware, including cord-impressed (Fig. 5.6)—suggests that at some stage new groups moved into Glamorgan, probably from West Wales,and mingled with the locally established communities.

The Transition to the Early Bronze Age

The new, extended radiocarbon chronology has not exorcised the old difficulty of fixing the boundary between the Neolithic and the Bronze Age, because it is clear that certain cultural traits which are conventionally classed as Early Bronze Age were present, in Wales as in other parts of Britain, towards the end of the 3rd millennium alongside ones conventionally classed as Neolithic. This overlap is particularly noticeable along the southern sea-board of Wales, where traces of the movement which brought the users of the earliest Beaker pottery from Northern France and Wessex to southern Ireland have recently come to light—pottery on Caldey, a continental type of archer's wrist-guard from Llantrithyd (Glamorgan)—to reinforce the early Beaker sherds found at Tinkinswood. No doubt more evidence of this kind lies in the submerged land-surface off the present coast; but centuries would have passed before the interior of Wales knew a culture in which the old and new elements had been merged in a true Early Bronze Age. There could not be a sharp break, because the Neolithic itself represents the beginning of a continuous development, leading up to the barbaric agrarian culture which the Romans converted to their own uses in the 1st century A.D.

BIBLIOGRAPHY

ApSimon, A. (1976) "A view of the Early prehistory of Wales" in Boon and Lewis (eds.), Welsh Antiquity, National Museum of Wales, Cardiff. 37-53.

Ashbee, P. (1970) The Earthen Long Barrow in Britain, London.

Atkinson, R. J. C. (1965) "Wayland's Smithy", Antiquity, (Cambridge) 39, 126-33.

Atkinson, R. J. C. (1968) "Old Mortality: some aspects of Burial and Population in Neolithic England", in Studies in Ancient Europe (Essays presented to Stuart Piggott), (Leicester University Press) 83-93.

Briggs, C. S. (1976) "Notes on the Distribution of some Raw Materials in Later Prehistoric Britain", in Burgess and Miket (eds.), Settlement and Economy in the Third and Second Millennia B.C. (=BAR33), (Oxford), 267-84.

Britnell, W. J. (1978) "Gwernvale Chambered Tomb", in Archaeology in Wales, 17, no. 14.

Brothwell, D. R. (1973) "The Human Biology of the Neolithic population of Britain", in Schwedetzky (ed.), Die Anfänge des Neolithikums vom Orient bis Nordeurope, Vol. VIIIa, 280-99.

Burl, H. A. W. (1969) "Henges: Internal Features and Regional Groups". Arch. Journ. (London), 126, 1-28.

Burl, H. A. W. (1976) The Stone Circles of the British Isles (Yale University Press, New Haven and London).

Case, H. (1961) "Irish Neolithic Pottery: Distribution and Sequence., Proc. Prehist. Soc. (Cambridge), 27, 174-233.

Cave, A. J. E. (1938) "Report on the Neolithic Skeletons", in Keiller and Piggott, "Lanhill Long Barrow", Proc. Prehist. Soc. (Cambridge). 4, 131-50.

Coon, C. S. (1939) The Races of Europe. (Macmillan, New York).

Corcoran, J. X. W. P. (1969) "The Cotswold—Severn Group", in T. G. Powell (ed.), Megalithic Enquiries in the West of Britain (Liverpool University Press), 13-104.

Crampton, C. B. (1967) "Analysis of Pottery and Soil buried beneath a Pillow Mound at Llanelwedd, S. Radnorshire", BBCS, 22 (iv), 273f.

Daniel, G. E. (1964) (with O'Riordain, S. P.) New Grange and the Bend of the Boyne, Thames and Hudson, (London).

Daniel, G. E. (1964) (with O'Riordain, S. P.) "Megalithic Answers", Antiquity (Cambridge), 44, 264-9).

Evans, E. E. (1975) "Highland Landscapes : Habitat and Heritage" in "The Effect of Man on the Landscape of the Highland Zone". Evans, Limbrey and Clare (eds.) (CBA, London, Research Report No. 11), 1-5.

Gerhardt, K. (1974) "Anthropotypologie der Glockenbercherleute in Ihren Ausschwärmelandschaften", in Glockenbecher Symposium, Oberried, 1974, 147-64, and 482.

Gresham, C. A. (1967) (with Bowen, E. G.) A History of Merioneth, Vol. 1, Dolgellau.

Grimes, W. F. (1963) "The Stone Circles and Related Monuments of Wales", in Foster and Alcock (eds.), Culture and Environment - Routledge and Kegan Paul (London), 93-152.

Grimes, W. F. (1965) "Neolithic Wales" in Foster and Daniel (eds.), Prehistoric and Early Wales, Routledge and Kegan Paul (London), 35-69.

l'Helgouach, J. (1965) Les Sépultures Mégalithiques en Armorique, Rennes.

l'Helgouach, J. (1973) "Les Mégalithes de l'Ouest de France, Evolution et Chronologie", Daniel and Kjaerum, (eds.), Megalithic Graves and Ritual (Copenhagen), 203-19.

Houlder, C. H. (1961) "The Excavation of a Neolithic Stone Implement Factory on Mŷnydd Rhiw in Caernarvonshire", Proc. Prehist. Soc. 27, 108-43.

Houlder, C. H. (1976) "Stone Axes and Henge Monuments", in Boon and Lewis (eds.), Welsh Antiquity, National Museum of Wales, Cardiff, 55-62.

Lynch, F. (1969) "Megalithic Tombs in North Wales" in T. G. E. Powell (ed.), Megalithic Enquiries in the West of Britain, Liverpool University Press, 108-48.

Lynch, F. (1976) "Towards a Chronology of Megalithic Tombs in Wales", in Boon and Lewis (eds.), Welsh Antiquity, National Museum of Wales, Cardiff, 63-79.

Meroc, L. (1967) "St. Michel-du-Touche", Gallia Préhistoire (Paris), 10, 393-7.

Moore, P. D. (1973) "The Influence of Prehistoric Cultures upon the initiation and spread of blanket bog in upland Wales", in Nature, 241, No. 5388, 350-3.

O'Kelly, C. (1969) "Bryn Celli Ddu, Anglesey. A Reinterpretation". Arch. Camb. (Cardiff), 118, 17-48.

O'Roirdain, S. P. (1951) "Lough Gur Excavations. The Great Stone Circle (B) in Grange Townland". Proc. Roy. Irish Acad. 54C, 37-74.

Peacock, D. P. S. (1969) "Neolithic Pottery Production in Cornwall". Antiquity, (Cambridge), 43, 145-9.

Phillips, P. (1975) Early Farmers of West Mediterranean Europe, Hutchinson (London), 1975.

Piggott, S. (1954) Neolithic Cultures of the British Isles (Cambridge).

Powell, T. G. E. (1969) "Some Points and Problems", in Powell (ed.). Megalithic Enquiries in the West of Britain (Liverpool University Press), 108-48.

Powell, T. G. E. (1973) "Excavation of the Megalithic Chambered Cairn, at Dyffryn Ardudwy, Merioneth, Wales", Archaeologia (London), 104, 1-49.

Savory, H. N. (1952) "The Excavation of a Neolithic Dwelling and Bronze Age Cairn at Mount Pleasant Farm, Nottage, (Glamorgan)". Trans. Cardiff Nat. Soc., 81, 75-89.

Savory, H. N. (1963) "Excavation at a Third Round Barrow at Pen-dre, Letterston, (Pemb.) 1961". Bull. Board Celtic Stud. (Cardiff) 20 (iii), 309-25.

Savory, H. N. (1971) "A Neolithic Stone Axe and Wooden Handle from Port Talbot". Ant. Journ. (London), 51, 296f.

Savory, H. N. (1973a) "Penywyrlod : a New Welsh Long Cairn". Antiquity (Cambridge), 47, 187-92.

Savory, H. N. (1973b) "Serpentiforms in Megalithic Art : a link between Wales and the Iberian North-West". Cuadernos de Estudios Gallegos (Santiago de Compostela), 84, 80-9.

Savory, H. N. (1975) "The Role of the Upper Duero and Ebro Basins in Megalithic Diffusion" Bol. del Seminario de Estudios de Arte y Arqueologia (Valladolid) 40-41, 159-74.

Savory, H. N. (1977) "The Role of Iberian Communal Tombs in Mediterranean and Atlantic Prehistory", in Markotic (ed.), Ancient Europe and The Mediterranean (Studies presented in honour of Hugh O'N. Hencken). (Aris and Phillips, Warminster).

Selkirk, A. (1971) "Ascott-under-Wychwood". Current Archaeology, 24, 7-10.

Selkirk, A. (1975) "Crickley Hill". Current Archaeology, 50, 83.

Smith, I. (1974) "The Neolithic" in Renfrew (ed.) British Prehistory—A New Outline (Duckworth, London), 100-36.

Spurgeon, C. J. (1965-9) "Llanelwedd Mound". Archaeology in Wales, 5,13; 6,11; 7,10; 8,14; 9,16.

Sweetman, J. (1976) "An Earthen Enclosure at Monknewtown, Slane, Co. Neath". Proc. Rev. Irish. Acad. (Dublin), 76C, 3, 25-72.

Taylor, J. A. (1973) Chronometers and Chronicles: a Study of the Palaeo-environments of West Central Wales, Progress in Geography, Vol. 5, International Reviews of Current Research, 248-334.

Taylor, J. A. (1975) "The Role of Climate Factors in Environmental and Cultural changes in Prehistoric Times", in Evans, Limbrey and Clare (eds.), The Effect of Man on the Landscape : Highland Zone (CBA, London, Research Report 11), 6-19.

Val Caturla, Del (1948) "Cueva de los Blanquizares de Lebor, Totana, Murcia". Cuadernos de Historia Primitiva, 3, 5-36, pls. IX-X.

Wainwright, G. J. (1962) "The Excavation of an Earthwork at Castell Bryn-gwyn, Llanidan Parish, (Anglesey)". Arch. Camb. (Cardiff), 111, 25-56.

Wainwright, G. J. (1967) Coygan Camp : A Prehistoric Romano-British and Dark Age Settlement in Carmarthenshire. Cambrian Arch. Association (Cardiff).

Webley, D. P. (1958) "A Cairn Cemetery and Secondary Neolithic Dwelling on Cefn Cilsanaws Vaynor, (Breck.)". Bull. Board Celtic Stud. (Cardiff), 18 (iv), 79-88.

Webley, D. P. (1967) "Tooth Cave, Ilston, Gower. An Early Bronze Age Occupation". Bull. Board Celtic Stud. (Cardiff), 22(iii), 277-83.

Webley, D. P. (1976) "How the west was won : prehistoric land-use in the Southern Marches", in Boon and Lewis (eds.), Welsh Antiquity, National Museum of Wales, Cardiff, 19-35.

Williams, A. (1952) "Clegyr Boia, St. David's (Pembrokeshire). Excavation in 1943". Arch. Camb. (Cardiff), 102, 20-47.

CHAPTER SIX

BRONZE AGE MONUMENTS IN WALES

Frances Lynch

Abstract

Bronze Age burial and ritual activities reveal both the continuities from the Neolithic and the new trends of the Bronze Age. Individuality and an interest in wealth and new resources, together with an expansion into upland areas represent the new; but in the mature Early Bronze Age multiple burials, protracted funeral processes and close links between ceremony and burial (best seen in the Ring Cairn group of monuments) reveal the strength of older traditions in Wales. Various classes of burial and ceremonial monuments in Wales are illustrated.

The Bronze Age is traditionally considered to have been a period very different from the preceding Neolithic in its social structure and ethos. The peaceful Neolithic communities without weapons or defences, dominated by collectivism, hard work and religion are contrasted with the flashy, aggressive Bronze Age chieftains with their concern for individual wealth and status and their restless pursuit of fine herds and excitement. Such a view cannot be entirely wrong, but its universality and its permanence may be called into question when one looks more closely at some of the evidence from burial and ceremonial sites which are normally thought to reflect these changes most clearly.

This is not to say that no changes can be observed between the Neolithic and the Bronze Age because it is certainly true that emphases are being shifted and directions slightly altered once we come within the sphere of the Beaker-users. The burials of these people, whether they were all newcomers or indigenes converted to a new view of life, are normally described as single, but might be better called 'individual' for what actually distinguishes them from the tradition of anomymity, that was established in the earlier Neolithic, is the fact that the dead person may be recognised; the bones are not mixed and prized possessions may be retained for use in an after-life in which we may assume that the dead person intends to keep his own role and personality. The concept of private property is also reflected in these graves and, while it is unlikely that the idea of private wealth is new, it has not previously been formally identified. The concentration of power and wealth into fewer hands and the use of this wealth to obtain exotic trinkets from distant parts may be

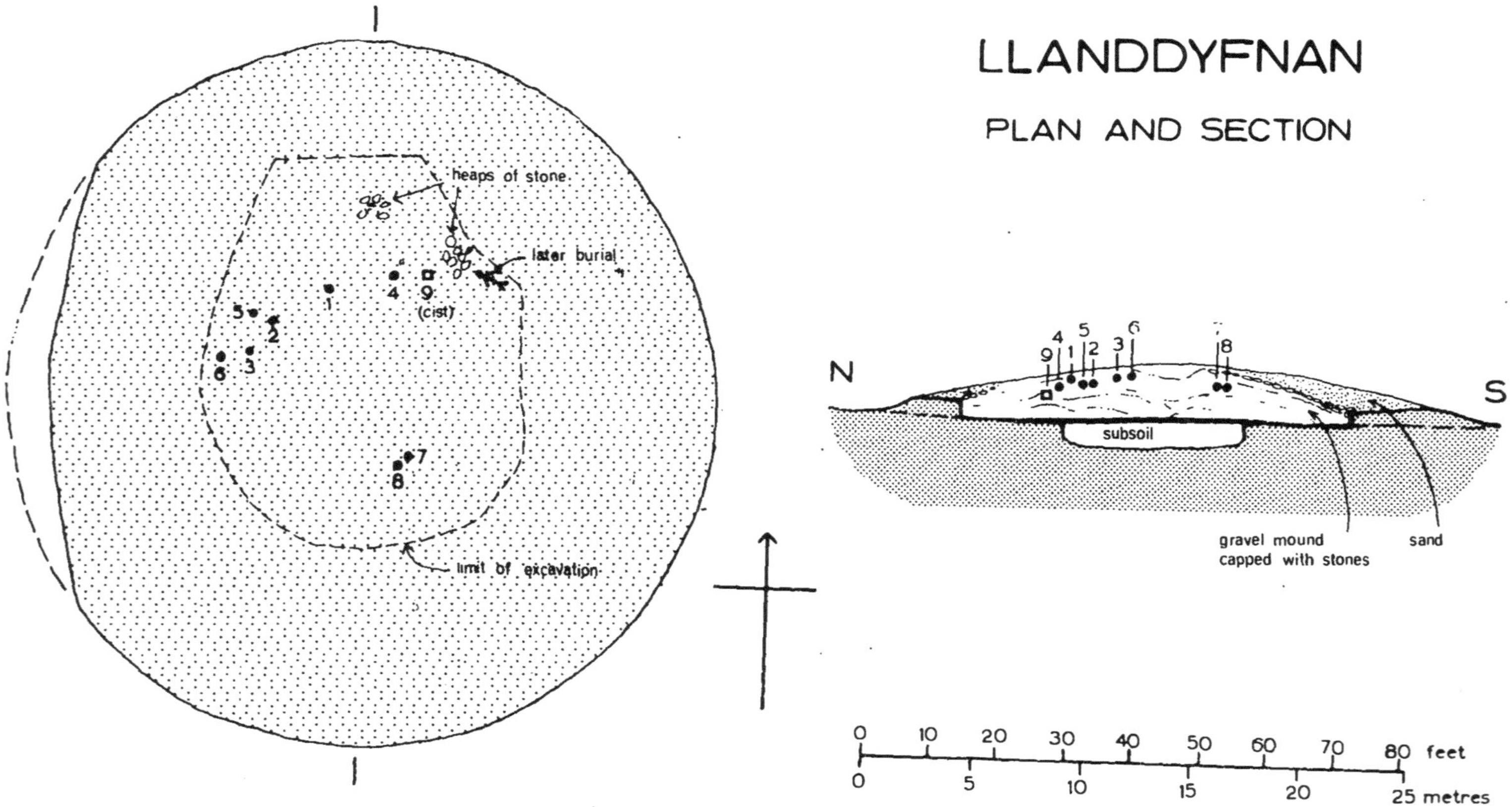

Fig. 6.1 Llanddyfan: plan and section

to some extent the outcome of the new reliance on less universally available resources and the resultant trading empires, and also perhaps of the colonisation of new upland areas encouraging the development, in some parts, of a pastoral economy, more volatile and mobile than before. In these new lands new monuments are built which reflect a growing distinction between ceremonial and burial, but a distinction which is by no means complete.

In Wales, it is difficult to judge the sharpness of the initial changes since our knowledge of the Late Neolithic is hazy and dateable sites are few and far between. Some Henge Monuments exist in north Wales and in the Vale of Glamorgan (Houlder, 1968 and RCAM 1976) but the fact that the Bryn Celli Ddu Passage Grave is built on top of one of these monuments (Lynch 1970, 55-65) suggests that older traditions persisted here longer than elsewhere. Consequently, it is difficult to judge the exact chronological relationships between the Long-Necked Beaker burials (Griffiths, 1957), the later megalithic tombs and the cremation burials in urns or food vessels (Lynch, 1975a) which predominate in the mature Early Bronze Age. But looking at this period of round barrows and stone circles on the moors, superficially so different from the long cairns on the coasts and valleys, many essential continuities may be recognised, albeit in a new form.

In the megalithic tombs the predominant ritual is multiple burial, whether inhumation or cremation; the bones are always anonymous and often actually mixed together—the group or family absorbing the individual. This system of burial no doubt reflected the community's attitude during life and it is this attitude, it is suggested, which is eroded during the Bronze Age, giving rise to the practice of individual burial. The urn burial is undoubtedly a system of individual burial but in Wales and in several other parts of Britain and Ireland it is combined with a continuing consciousness of the importance of the group or family.

In southern England it is usual to find that the round barrow has been designed to cover a single burial, whether or not others have been subsequently dug into the mound once it was built. This is not universally the case but it is common, whereas in Wales it is very uncommon. Barrows and cairns are normally found to cover four or more cremations and, where excavation evidence allows, it is possible to demonstrate that these multiple burials have been made as a group during a single funeral ceremony (Fig. 6.1). At sites with a large number of burials, two or more groups may be recognised, implying more than one funeral ceremony, but the importance of the group ideal over the individual is maintained (Lynch, 1971). Since these multiple Bronze Age burials tend to predominate in areas such as Anglesey where we know that the traditions enshrined in the megalithic tombs persisted into the Late Neolithic (Lynch, 1970), it is reasonable to suggest that this system of burial reflects the survival of older social structures in the area. The comparative rarity of weapons in Wales at this time may be another indication of a less sharply stratified society.

Another point which is raised by these group burials in the Early Bronze Age concerns the variety of pottery which may be used by what is assumed to be a single community. All the various styles of urn and food vessel may be found together in one mound e.g. Treiorwerth, Anglesey where an urn of

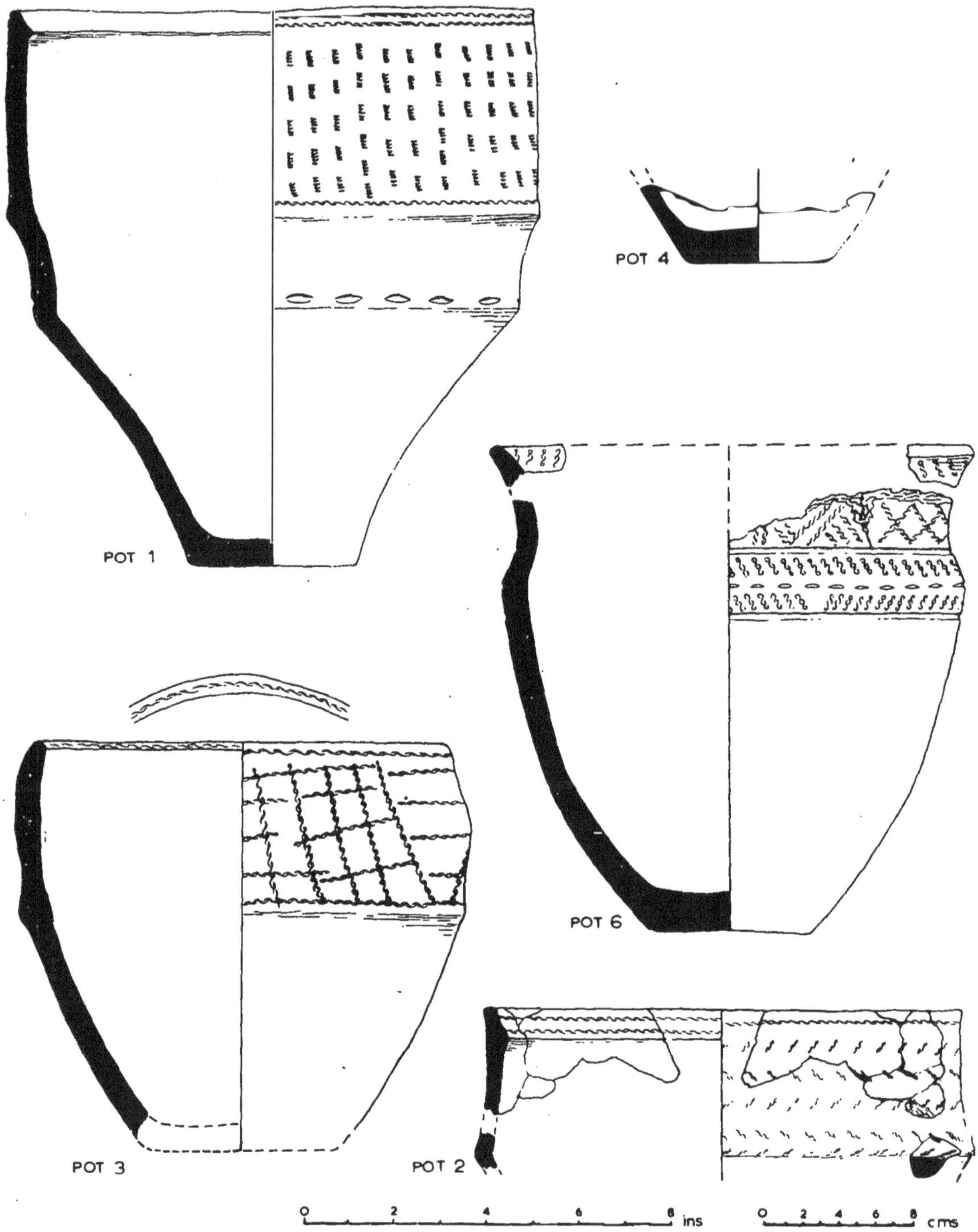

Pottery from Treiorwerth, Anglesey. Pots 1, 3 and 6 are primary.

Fig. 6.2

Longworth's Secondary Series can be shown to be contemporary with a Cordoned Urn and a Food Vessel (Lynch, 1972 and Fig. 6.2). This is not the place to go into the implications of this mixture of styles beyond suggesting that in Wales there can be little, if any, social difference between the users of the different types of pots, nor much chronological significance in their variation.

The relatively small size of the round barrow and the dominance of the idea of single burial has led to the assumption that the Bronze Age burial monument was a structure built fairly rapidly immediately after the funeral of the person it was intended to commemorate. By contrast the various types of Neolithic monument have always been recognised as the result of a protracted series of rituals of which the construction of the mound was but the final, or almost final, stage. Again, more careful and complete excavation of round barrows and cairns in recent years has shown that in this aspect, too, the Bronze Age monuments are closer to their Neolithic predecessessors than was previously thought. Even the individual grave may be dug long before it is finally filled in (Thomas, 1967; Smith and Simpson, 1966) and, with monuments where several bodies are found scattered on the old ground surface within a demarcated area (Greenwell, 1877), one may suspect that a long time may have elapsed before they were covered over. The presence of stake circles beneath certain barrows cannot always be explained in purely practical terms of revetment and must imply some rather elaborate ceremonies taking place on the spot before the mound was raised. Recent excavations in the Brenig Valley, Denbighshire (Lynch et al., 1974) have shown that these stake circles may have various functions within the same cemetary and that some at least indicate the dedication of an area for burial for a decade or more before the mound was built.

In Wales during the Neolithic period the megalithic tomb in its various forms seems to have been the only 'public' monument, the burial place being the sole expression of community consciousness, fulfilling both social and religious roles. The adaptation of these monuments in the course of their long history (Corcoran, 1972) indicates that they had an importance over and above their use as containers for bones and we must suppose that they were the focus of all the major ceremonies expressive of the religious feelings of the time. In the later Neolithic, the identity between burial and religious ceremonial may have been weakened by the adoption of the Henge Monument as the centre for ceremonies not so intimately connected with death. In the Early Bronze Age the tendency to separate burial and other ceremonial may have grown with the proliferation of various forms of circular monument which may have played a more specialised role in the religious practice of the day. But the association of these open ring monuments in cemetery groups with burial cairns, the obvious inter-connection of their architectural features and the occasional conversion from ceremonial site to burial monument (Savory, 1961), show clearly that the dead still held an important place in the contemporary religious framework.

Wales is particularly fortunate in possessing a very varied and well preserved series of these circular monuments, most of which involve an open central space or an architecturally dominant circle of upright stones (Fig. 6.3). These monuments have been very variously described in the past; their broad relationships have been recognised but the more precise classifications which

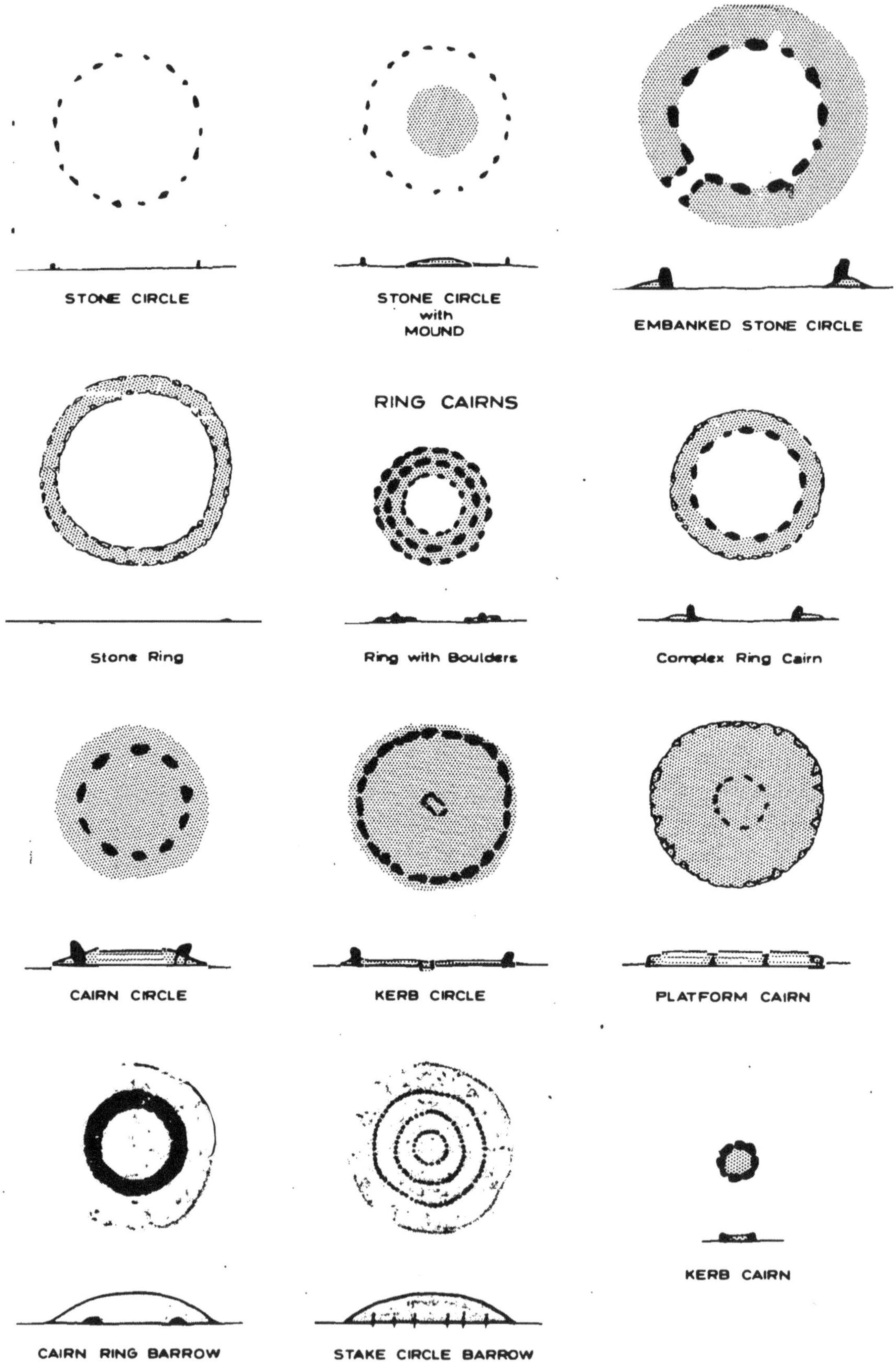

Fig. 6.3 Types of Bronze Age monuments in Wales: definition diagrams

are possible have been blurred. Space does not allow a detailed description of each type to be given, but Fig. 6.3 should enable the salient features to be recognised and distinguished (Lynch, 1972). The Cairn Circle, Kerb Cairn and Platform Cairn are solid monuments, and burial cists may be recognised at the centre of some damaged examples—so it may be reckoned that these lie on the spectrum close to the normal burial cairn. But their usually impressive siting (Lynch, 1975b) and the information from the few that have been excavated (Lynch, 1974) suggest that they were reserved for burials which were in some way exceptional. The Ring Cairns, either low banks of stone without architectural elaboration, or more complex varieties involving spaced uprights around the inner edge, are the most common and perhaps the most ambiguous. Several have been excavated with varying results; some have contained token burials or unusual burials while others have contained normal cremations but, more importantly, also large quantities of charcoal, buried with care and ceremony in such a way as to suggest that this was the primary purpose of the monument (Griffiths, 1960; Lynch, 1972; Lynch, 1973). The Embanked Stone Circles are normally large monuments with impressive stones. Those that have been excavated (Griffiths, 1960; Savory, 1961) indicate a closer connection with burial than is to be found in the Open Stone Circles which consist only of spaced uprights—in Wales usually extremely small—without any enclosing bank. However, even these monuments, where excavation has yet to provide any clue to their use, and the siting and design of which have led to suggestions of an astronomical purpose, maintain a connection with the dead since several contain a small mound/barrow in the centre.

The majority of the Open Stone Circles are sited on open moorland above 1,000 ft. (305 m) and many of them lie on natural cross-routes (Grimes, 1963), suggesting that they might have been meeting places for groups from quite a wide area (Fig. 6.4). Many of them are inconspicuous, but the presence of tall outliers, visible from a distance, lends support to this view of their role in society, a role which, in this part of the country, may formerly have belonged directly to the burial monuments.

The fact that many of the burial cairns and circles of the Bronze Age are to be found on what is now moorland, and was already at that time open land with incipient podzols suggests both an increase in population over the Neolithic and a shift in agricultural practices. The wandering life of the herdsmen on these hills would naturally lead to a concern for the sky and the guiding stars and for sources of water. Gradually, these new interests may have increased in importance, eclipsing the role of the burial monuments of which fewer and fewer were built as the period progressed (Burgess, 1976). The retreat from the central moorlands, which is suggested by the distribution of later hilltop settlements, may have been caused by the degradation of the soils due both to over-exploitation and/or a deterioration of the climate. In either case these environmental troubles may have hardened the trend towards the propitiation of water spirits which may lie behind the votive offerings of the Late Bronze Age and Iron Age.

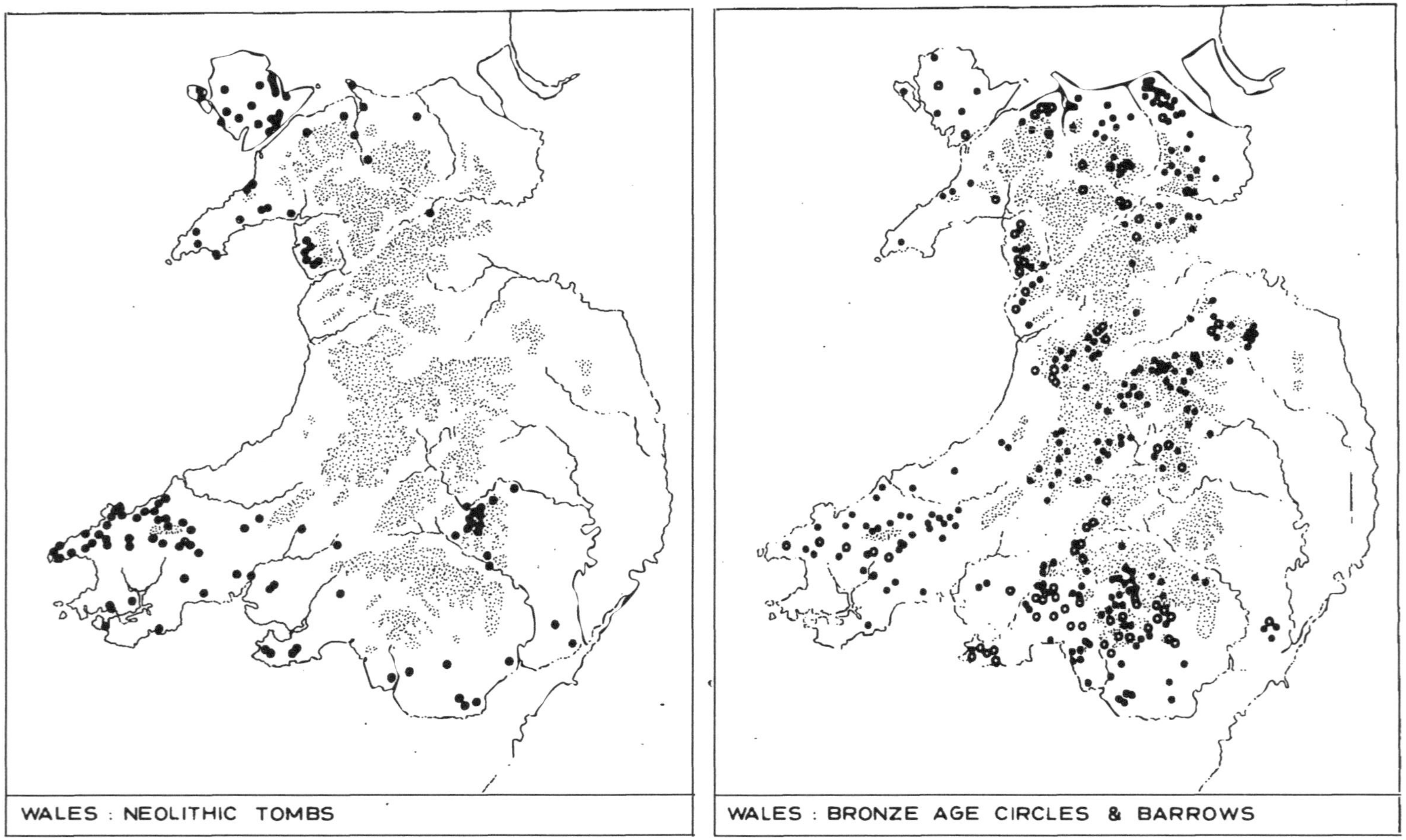

Fig. 6.4 Wales: Neolithic tombs and Bronze Age circles and barrows

BIBLIOGRAPHY

Burgess, C. B. (1976) 'Burials with metalwork of the later Bronze Age in Wales and beyond' in Boon, G. C. and Lewis, J. M. eds. Welsh Antiquity Cardiff, 1976. 81-104.

Corcoran, J. X. W. P. (1972) 'Multiperiod Construction and the Origins of the Chambered Long Cairn in Western Britain and Ireland' in Lynch, F. M. and Burgess, C. B. (eds.) Prehistoric Man and Wales and the West Bath, 1972, 31-63.

Greenwell, W. (1877) British Barrows, Oxford 1877.

Griffiths, W. E. (1957) 'The Typology and Origins of Beakers in Wales' Proc. Prehist. Soc. XXIII, 1957, 57-90.

Griffiths, W. E. (1960) 'Excavation of Stone Circles near Penmaenmawr, N. Wales' Proc. Prehist. Soc. XXVI, 1960, 305-39.

Grimes, W. F. (1963) 'Stone Circles and Related Monuments of Wales' in Foster, I. Ll. and Alcock, L. (eds.), Culture and Environment London, 1963, 93-152.

Houlder, C. H. (1968) 'The Henge Monuments at Llandegai' Antiquity, XLII, 1968, 215-21.

Lynch, F. M. (1970) Prehistoric Anglesey, Llangefni, 1970.

Lynch, F. M. (1971) 'Report on the Re-excavation of two Bronze Age Cairns in Anglesey: Bedd Branwen and Treiorwerth' Arch. Cambrensis, CXX, 1971, 11-83.

Lynch, F. M. (1972) 'Ring Cairns and Related Monuments in Wales' Scottish Archaeol. Forum 4, 1972, 61-80.

Lynch, F. M. (1975a) 'Brenig Valley Excavations 1974' Trans. Denbighs. Hist. Soc. 24, 1975, 13-37.

Lynch, F. M. (1975b) 'The Impact of Landscape on Prehistoric Man' in Evans, J. G., Limbrey, S. and Cleere, H. (eds.), The Effect of Man on the Landscape: Highland Zone, London 1975, 124-7.

Lynch, F. M. et al (1974) 'Brenig Valley Excavations 1973: Interim Report' Trans Denbighs. Hist. Soc. 23, 1974, 9-64.

R.C.A.M. (1976) 'An Inventory of the Ancient Monuments in Glamorgan, Vol. I, Part I, H.M.S.O. Cardiff 1976.

Savory, H. N. (1961) 'Excavations at a third Round Barrow at Pendre, Letterston, (Pembs.)' Bull. Board of Celtic Stud. XX, 1961, 309-25.

Smith, I. F. and Simpson, D. D. A. (1966) 'Excavation of a Round Barrow on Overton Hill, North Wiltshire' Proc. Prehist. Soc. XXXII, 1966, 122-55.

Thomas, N. (1967) 'A double Beaker Burial on Bredon Hill, Worcs.' Trans. Birmingham Arch. Soc. 82, 1967, 58-76.

CHAPTER SEVEN

THE BRONZE AGE IN WALES

Colin Burgess

ABSTRACT

A radical review of the period is suggested in the light of radiocarbon dates and recent large-scale excavations. Higher population figures and more sophisticated social systems must be envisaged. Clearance for agriculture, already well advanced in the Neolithic, escalated in the Early Bronze Age, and a much more open and organised landscape is likely over the country as a whole. The 'Bronze Age' is an increasingly outmoded concept. Continuity must be stressed at its beginning and at its end, and the major break in traditions occurs in its middle. Migration and population movement must be regarded as minor, not major agencies of change in the period. Societies were confined to their territories, and innovations were spread by culture contact at territorial boundaries.

A chronological framework for the Welsh Bronze Age is proposed. The period now includes much that is traditionally 'Neolithic' at one end and 'Iron Age' at the other. It begins with the introduction of copper and Beakers circa 3000/2500 B.C., but the Early Beaker period saw 'Neolithic' societies little affected in fundamentals, even though by circa 2000 BC they were fully bronze-using. Many 'Neolithic' traditions were severed only at the beginning of the Late Beaker period, circa 2000 B.C., and it was only at this point that such classic Early Bronze Age indicators as food vessels and cinerary urns emerged. Lynch discusses the complex EBA burial/ritual situation (see Chapter six).

These EBA traditions everywhere ended circa 1500 B.C. in social upheaval. Little is known of the ensuing MBA in Wales, apart from the metalwork. The metallurgy of the EBA and MBA is discussed. Growing numbers of burials, settlements and field systems of these periods are known elsewhere, and need to be sought for in Wales.

A major upheaval affected the whole of the Old World circa 1200 B.C. In southern Britain every aspect of life was affected, and climatic deterioration and even French immigrants are possible. There were major changes of land use, including abandonment of fields on upland and lowland alike. In Wales the changes of this Penard period are best seen in the building of hill-forts. These first defences were not maintained in the Isleham period, circa 10th-8th centuries B.C. though many hill-tops continued to be occupied. A further period of unrest, coincides with the rise of Hallstatt C power in central Europe, circa 700 B.C. A new era of hill-fort building and rebuilding began, and further climatic deterioration is likely. The metalworking traditions of the later BA are discussed. The regional industries of the Ewart Park phase, circa 9th-8th centuries B.C., anticipate the tribal dispositions of Roman times. Iron came with Hallstatt 'C' influence from circa 700 B.C., affecting especially South Wales, where the Llynfawr deposit shows precocious local iron-working in the 7th century B.C. By circa 600 B.C. iron had largely replaced bronze.

In an account of the Bronze Age in Wales which appeared more than a quarter of a century ago the difficulty of dividing the Neolithic from the Bronze Age was already appreciated (Grimes, 1951, 47). Nevertheless the arrival of the Beaker Folk and of metal was thought to institute a major break in cultural traditions, while at the other end of the period, the appearance of iron and of Hallstatt settlers seemed an equally significant dividing point. These were the traditional parameters of the Bronze Age, and they remain so today.

The 'Bronze Age' now appears a much more arbitrary slice of prehistory, for the introduction of copper and bronze did not lead to or coincide with commensurate social, economic and political changes. Copper and iron were not at first appreciably superior to the stone and bronze they replaced, nor did they immediately transform the lives of ordinary people. Furthermore, the nature and impact of Beaker and Hallstatt arrivals in Britain has recently been questioned. The last decade has seen growing disenchantment with the concept of folk-movement as a major cause of prehistoric culture change (Clark, 1966), so much so that today scarcely any prehistoric migrations to Britain seem acceptable. Among the casualties have been both the Beaker and Hallstatt 'invasions', removing two of the most crucial props on which the entity of the Bronze Age has traditionally rested. The recent trend has been to stress the concept of continuity, both from Neolithic to Bronze Age and Bronze to Iron. A more significant break in prehistory now appears to have occurred in the middle of the traditional Bronze Age, at about 1200 B.C. The day is therefore fast approaching when our prehistory will be divided on a very different basis from that of the traditional 'Three Ages' system.

But for the moment the 'Bronze Age' stubbornly lingers on, and therefore in this paper is taken to mean that part of prehistory in which copper and bronze were in use for everyday artefacts. The introduction of copper can now be set at least as early as c. 2500 B.C. (c. 2000 b.c.), while iron had effectively replaced bronze in most regions by c. 600 B.C. This two thousand year Bronze Age is now seen to incorporate on the one hand much that has for long been regarded as 'Neolithic', including many chambered tombs, and 'Late Neolithic' wares, and on the other much traditional 'Iron

Age' evidence, including many hill-forts and much supposedly 'Iron A' pottery. Such a radical re-appraisal of the evidence has stemmed largely from three developments. Firstly, radiocarbon dating has been applied to Bronze Age samples on an appreciable scale only in the last dozen years. Secondly, the boom in large-scale excavation, necessitated by the crisis in rescue archaeology, has revolutionised our understanding of many categories of sites. Thirdly, there has been a general change in attitude to the problems of prehistory, in keeping with the fundamental changes in archaeological philosophy which have so characterised the 1960's and 1970's. As a result the Welsh Bronze Age must be re-evaluated in the light of radically revised estimates of population and cultural levels, and new concepts of social and economic systems, and of mechanisms of change and culture contact.

SOCIAL SYSTEMS AND POPULATION

Traditionally, prehistoric Europe has been seen as a thinly populated and under-developed landscape, in which volatile peasant populations were forever on the move. In particular, prehistory has been peopled with 'cultures', which appear and disappear in rather mysterious fashion, although more often than not invasion and migration are called to account.

A very different world must now be envisaged, in which the landscape was much more highly developed, population levels immensely greater, and society more settled, structured and organised than has usually been conceived. If any single discovery were needed to emphasize this concept and point out the bankruptcy of traditional approaches to our prehistoric past, then it is the uncovering in recent years of great Neolithic and Bronze Age field systems, often extending over hundreds of acres, in areas as far apart as western Ireland, Wessex and the East Midlands. As recently as 1972 an Early Bronze Age population of only 2000 was calculated for England and Wales (Atkinson, 1972), but this is now seen to be perhaps fifty, or even a hundred times too small (Green, 1974). Indeed, by the end of the Bronze Age populations may already have reached those of Domesday times (Stanford, 1972b).

On the other hand this growth of population should not be seen as a smooth and uninterrupted process. There is increasing evidence for periodic agricultural crises in prehistoric as in historic times, resulting in temporary population standstill or even decline. One such crisis may have developed in the centuries before c. 2500 b.c., in the conventional Middle Neolithic, and may well have contributed, to the sweeping social and economic changes observed from c. 2500 b.c. (Whittle, 1978). Another may have set in after c. 1300 b.c. (Bradley, 1978) and played a part in equally far-reaching changes in society and the landscape. These culminated in the dramatic upheavals of the Penard period between c. 1200-1000 B.C.

It seems likely that from an early stage of the Neolithic some parts of Britain were divided up into territories with defined frontiers. Renfrew (1973) has argued convincingly that by the Middle Neolithic Wessex was divided into five such territories or chiefdoms, each covering several hundred square miles and incorporating centres of varying importance to serve as foci for populations scattered in farmsteads and hamlets. By the Early Bronze Age the whole of Britain may have been carved up in such a way, and a similar

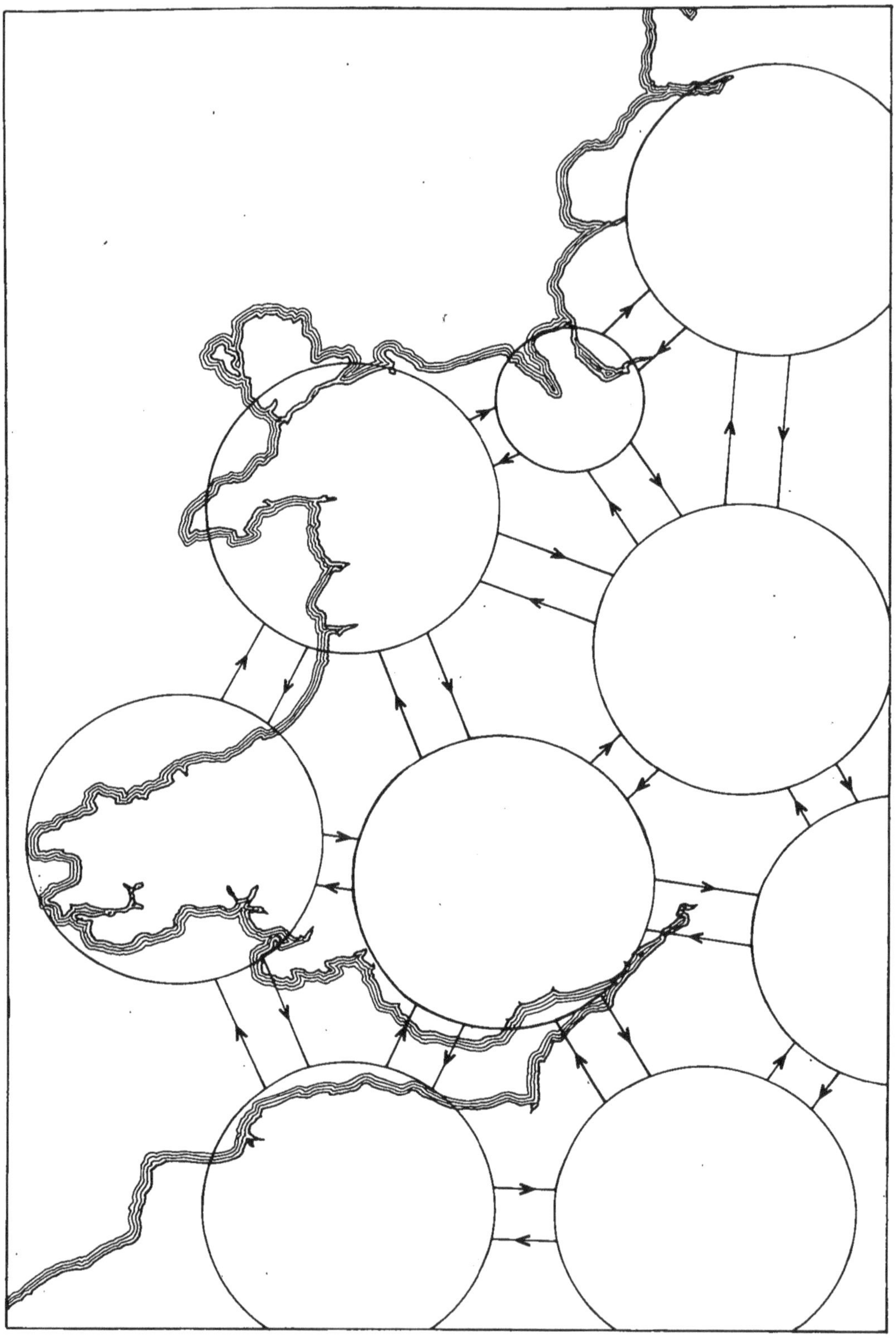

Fig. 7.1 Constant culture contact in Wales and the West

territorial structure would make sense of the evidence from Ireland and the the Continent. It provides a most convincing framework against which to view the main characteristics of the Bronze Age.

The plausibility of such a model is emphasized by the scale of Neolithic and Bronze Age monuments, and the variability of burial rites. These imply not only that the land had been divided into territories, but that these were peopled by highly organised, structured societies, dominated by capable ruling elites. Fixed territories mean more static societies, in the sense that movement may occur freely within their boundaries but not normally across them. On the other hand each territory will have been in a state of constant culture contact with adjoining territories, through contacts along its borders both formal and informal. The territories can be imagined as a cellular structure covering the whole of Britain. As each cell (or territory) develops in sympathy with its neighbours, an innovation in any individual cell repercusses on adjacent cells, and ripples through the structure until its force is dissipated. Those cells it touches will together form a tradition block. As impulses can be generated at any point in the cellular structure, some tradition blocks will coincide exactly or closely, while others will overlap only partly or not at all. Cells, i.e. territories, which consistently share a number of fundamental traditions will form a community of tradition. In many ways this will approximate to the traditional 'culture', although not to all those claimed as such, nor to the interaction sphere of American archaeologists, (Streuver and Houart, 1972).

Such a model can be applied equally at the European, British and Welsh levels (Fig. 7.1). Thus, in the central European Bronze Age, instead of sucessive Únětice, Tumulus and Urnfield 'cultures', or successive Únětice, Tumulus and Urnfield stages of an expanding and contracting central European culture (Gimbutas, 1965), a central European community of tradition of fluctuating extent should be envisaged. The heart-land, where there was the greatest and most enduring coincidence of tradition blocks, was south Germany/Czechoslovakia, but eventually, at the Urnfield stage, territories throughout the region between the Marne and the Carpathians and the Vistula and the Adriatic were drawn into this community of tradition.

With Beakers we are clearly dealing with a tradition block covering much of western Europe, but not a community of tradition (Burgess and Shennan, 1976). Equally the south English Deverel-Rimbury 'culture' is only a local branch of an international ceramic tradition block. This was characterised by handled biconical or carinated vessels, decorated with incised multiple zig-zags and hatched triangles (i.e. Globular Urns), contrasting with the much coarser, more variably sized bucket and barrel-shaped vessels with applied cordon and finger-tip ornament (i.e. Bucket and Barrel Urns). This ceramic tradition was common to much of the area from Britain to central Europe and from the Low Countries to the Rhone, territories which otherwise belonged to a number of different communities of tradition.

Such a model also makes sense of the varied burial customs and funerary wares of Early Bronze Age Britain. Traditionally separate Beaker, Food Vessel and Urn cultures have been seen to spread in turn over the country, leaving unanswered questions of origins and endings. Instead, we must think

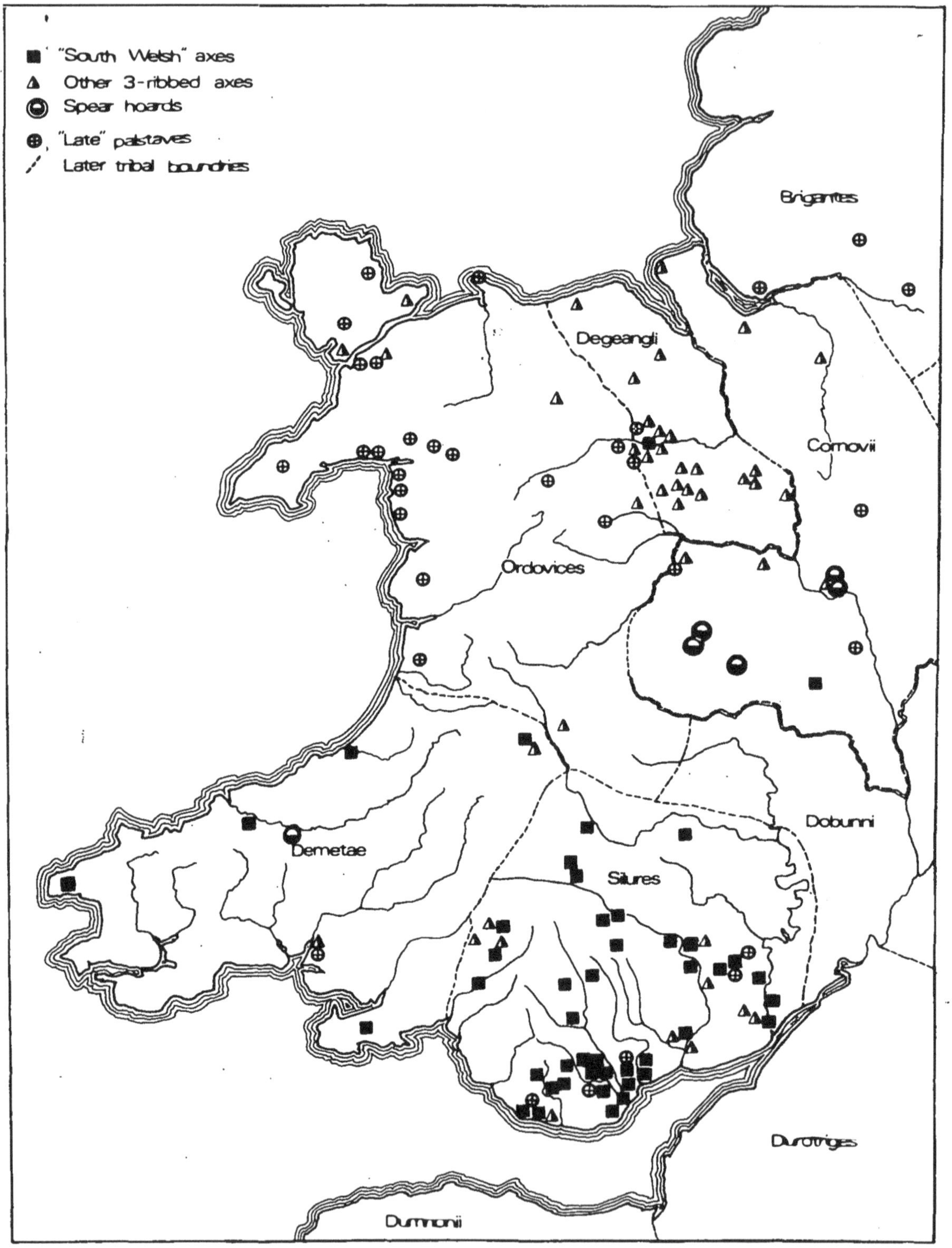

Fig. 7.2 Late Bronze Age metal work and later tribal dispositions in Wales.

of three distinct sepulchral traditions, the later stages of the Beaker tradition overlapping with the early development of the Food Vessel and Urn traditions, which spread from different territorial starting points. The Beaker tradition was the first to succumb, then food vessels, for it was urn cremation that proved to be the dominant and expanding force in the centuries towards the mid Second Millennium B.C. In all of this it was not the people within the territories who changed but their ideas and motivations, which adapted successively to each innovation.

As yet no attempt has been made to isolate prehistoric territories in Wales on the Wessex model. The region as yet has neither the causewayed camps nor the great enclosures of Durrington Walls type, which in Wessex have been interpreted as centres in the earlier and later Neolithic respectively. Whether these were territorial centres, as recently conceived for Wessex, or situated at territorial boundaries to serve as meeting places for different peoples, is not yet clear, but in either case a similar range of activities could have been involved. There appear to be no Welsh sites which could have served such purposes before the later Neolithic, when the henge and cursus complex at Llandegai, Caenarvonshire (Houlder, 1968, 1976) might thus be interpreted. On the other hand chambered tombs may have served similar duties as ceremonial centres and meeting places on a more local level. These were obviously much more than mere burial vaults. Chambered tomb groups, like long barrows in Wessex, may provide clues to Neolithic territories. Their mainly lowland distribution reminds us that exploitation of upland areas was slow and limited in the Neolithic. After the arrival of Beakers and metallurgy the gathering pace of upland clearance can be deduced from the distribution of round cairns and ring monuments (Lynch, 1972). These concentrate as much in highland settings as valleys and coasts, forming notable upland complexes such as those at Brenig and Penmaenmawr in north Wales, and Trecastle Mountain in the south. Territories which in the Neolithic had kept mainly to low ground may well have expanded into adjoining uplands in the Early Bronze Age thus becoming geographically more diverse.

Some of the best evidence for territorial divisions comes in the Late Bronze Age, and from metalwork not field monuments (Fig. 7.2). Twenty years ago Savory (1958, 49) pointed out the similarity between certain metalwork distributions and the tribal dispositions known from Roman times. The period in question is the Ewart Park phase, the 9th and 8th centuries B.C., and the most striking example is the concentration of south Welsh socketed axes in south-east Wales, corresponding very closely with the area occupied by the historical Silures. But in addition the north-western area, later inhabited by the Ordovices, is remarkable for its numbers of 'late' palstaves compared with socketed axes, while the north-east, later the territory of the Deceangli, is distinguished by a mixture of socketed axe types, some of them brought from other parts of Britain. The Marches, from the Mersey to the upper Severn, were inhabited by the Cornovii in Roman times, and, in the Ewart Park phase, by the spear warriors of the Broadward complex (Burgess, Coombs and Davies, 1972). Finally, the lands of the Demetae in the south-west stand out in the Ewart Park phase if only because they have produced little metalwork of any kind. Clearly, these were much larger territories than might be postulated for the Early Bronze Age, but equally Wales at the

THREE AGE PERIOD	PERIOD	C14 DATE bc	CALENDAR DATES BC	METALWORK NATIONAL	METALWORK WALES	BURIAL	BURIAL SITES	RITUAL	RITUAL SITES	POTTERY	SETTLEMENTS ETC
LATE NEOLITHIC	MELDON BRIDGE	2500	3200	?		LATE CHAMBERED TOMBS	BARCLODIAD Y GAWRES	HENGES CURSUS STONE RINGS STONE & TIMBER ROWS	LLANDEGAI A	LOCAL PETERBOROUGH	
						CREMATION CEMETERIES	LLANDEGAI A			PLAIN ROUND-BOTTOMED BOWLS	
		2150	2750	KNOCKNAGUE	MOEL ARTHUR	CAVE INHUMATIONS	GOP CAVE				
LATE NEOLITHIC, BEAKER/COPPER AGE/EARLY BRONZE AGE	MOUNT PLEASANT			FRANKFORD	TONFANNAU	LATE CHAMBERED TOMBS	BRYN YR HEN BOBL	HENGES CURSUS STONE RINGS STONE & TIMBER ROWS		EARLY BEAKERS	BRYN YR HEN BOBL
				MIGDALE-KILLAHA		CREMATIONS INHUMATIONS WITH BEAKERS	LLANDEGAI B			LOCAL PETERBOROUGH	NEWBOROUGH WARREN
					PONT CARADOG		SUTTON 268 I		LLANDEGAI B CASTELL BRYN GWYN CARNEDDAU HENGWM BRENIG 44, 51		
EARLY BRONZE AGE		1700	2000	COLLEONARD-WILLERBY WOLD		MAINLY CREMATIONS	SUTTON 268 II			LATE BEAKERS	
	OVERTON				DEVELOPED FLAT AXES	INHUMATIONS WITH LATE BEAKERS	PENTRAETH LLANDYFNAN BREACH FARM	RING CAIRNS PLATFORM CAIRNS		CINERARY URNS & FOOD VESSELS	BRENIG 51
		1500	1700								
	BEDD BRANWEN			ARRETON-INCH ISLAND	EBNAL	CREMATIONS	COLWINSTON CAE MICKNEY SUTTON 268 III RHOSCROWTHER BRYN CRUG LLANFACHRETH BEDD BRANWEN			CINERARY URNS	
				ACTON PARK I	MOELFRE UCHAF						
		1200	1400	ACTON PARK II	ACTON PARK		ELBA WORKS				
MIDDLE BRONZE AGE: MBA 1				TAUNTON-BISHOPSLAND						COARSE BUCKET & BARREL SHAPED VESSELS	
MBA 2	KNIGHTON				CEMMAES-DEANSFIELD	CREMATIONS - LOCAL EQUIVALENT OF DEVEREL-RIMBURY					
		1050	1200				BROMFIELD	WATER RELIGION ?			
				PENARD I	PENARD						FIRST HILL-FORTS
MBA 3	PENARD				MOLD BRACE	INHUMATION ?					DINORBEN, FFRIDD FALDWYN
				PENARD II	FFYNHONNAU	MOSTLY CASUAL?				'POST-DEVEREL RIMBURY TRADITION'	BREIDDIN
LATE BRONZE AGE: LBA 1		850	1000	WILBURTON-WALLINGTON	GWITHIAN AXES GUILSFIELD						HILL-TOP SETTLEMENTS, PALISADES, OPEN, NO RAMPARTS.
	ISLEHAM		900	EWART PARK	LLANTWIT-BROADWARD-ORME'S HEAD	CASUAL INHUMATION ?	OGO, LLANYMYNYCH HILL		BROADWARD ?		MOEL-Y-GAER
LBA 2		550	700	LLYNFAWR	LLYNFAWR CARDIFF TINTERN				LLYNFAWR ?		RENEWED HILL-FORT BUILDING: AT OLD SITES (FFRIDD FALDWYN, DINORBEN), & NEW SITES (MOEL-Y-GAER, MOEL HIRADDUG)
LBA 3	LLYNFAWR	450	600								

Fig. 7.3 Chronological Scheme for the Bronze Age in Wales.

time of the Roman conquest contained more tribes than the few whose names have reached us.

2. A CHRONOLOGICAL REVIEW OF THE WELSH BRONZE AGE

Any scheme for dividing the Bronze Age has to contend with the vagaries of the evidence. In the earlier periods the evidence suggests a preoccupation with burial and ritual, as can be seen in the numbers, and often grandiose scale, of burial and ritual monuments. This concern faded after 1500 B.C., and by the 12th-11th centuries had disappeared. From this point onwards practically nothing is known about burials, and the evidence is dominated increasingly by metalwork on the one hand, and by settlements on the other, especially hill-forts and hill-top occupation sites. In the following review (Fig. 7.3), the treatment of the earlier periods follows on from an outline already sketched out elsewhere (Burgess, 1976a). Calendar dates derived from radiocarbon dates are generally based on the historical calibration proposed by McKerrell (1975).

The Meldon Bridge period: c. 2500-2150 b.c./c. 3250-2750 B.C.

This was a true Late Neolithic period in the sense of being pre-metal. Case (1977) has hinted that All-Over-Corded Beakers may have emerged in this period on the strength of the sherds from the Giants' Hills long barrow, and noted that vessels decorated with cord impressed lines were in use in many parts of the country. Rather than take this as evidence for AOC Beakers originating in Britain, a more practical solution would be to interpret the Giants' Hills sherds not as AOC Beakers, but as some local corded equivalent to the pre-AOC Beakers (e.g. PFB Beakers) of the Low Countries.

This period differed markedly in its characteristics from earlier Neolithic phases, enough to suggest that a major break in social and ritual traditions can be set around 2500 b.c. Whether these changes were triggered off by fresh incomings of ideas or people from the Continent, or were connected with the agricultural problems and population decline postulated by Whittle (1978) in the centuries preceding c. 2500 b.c., is not yet clear. Any such crisis must have been followed in the Meldon Bridge period by stabilisation then recovery, for the scale of its monuments speak of a populous, not depopulated land. The emergence of henges, cursus monuments and stone rings shows a sweeping change in ritual attitudes, especially set alongside the abandonment of some old-established burial and ritual practices. In southern England no more causewayed enclosures were built (although they remained in use), and long barrows became a thing of the past. Increasing use of round burial mounds can be demonstrated in some regions, both in megalithic and non-megalithic areas. Activity continued at chambered tombs of all types, but new construction tended to favour round mounds. In particular, this was the era of the passage graves, from Ireland to the Orkneys (Herity, 1974; Renfrew, 1976). In eastern and southern England the growing popularity of round barrows is seen best in the Duggleby tradition of east Yorkshire (Mortimer, 1905, 23-42; see also Manby, 1974, 92-101). A similar predilection for round barrows and cairns may be postulated in other regions, but the whole question is made difficult by a lack of relevant radiocarbon dates and datable grave goods. Cremation cemeteries were another novelty of this period, frequently associated with henges and other enclosures. The great timber-walled enclosure at Meldon Bridge, Peebles, typifies these developments, with its mixture of domestic, ritual and sepulchral features, including cremation burials. Several radiocarbon dates range from c. 2736-2150 b.c. The Meldon Bridge pottery is in a local Peterborough style, and epitomises the blossoming of decorated pottery in this period (Burgess 1976c). Peterborough and Grooved wares now dominated the ceramics of Britain in this period but present a dilemma for the prehistorian. For they survived, like so much else, to play an equally significant role in the subsequent Mount Pleasant period. As a result much of the content of these two periods cannot be assigned securely to one or other of them unless associated with Beaker sherds of radiocarbon dates.

This is as much a problem in Wales as in other regions. For example, the inhumation burials found in caves such as Gop Cave and Rhos Ddigre in north-east Wales (Daniel, 1950, 46) have associations which are nominally 'Neolithic', but what this means in absolute terms is not clear. The later development of Welsh chambered tombs presents similar problems. Continuing activity at the various long tombs may be postulated, but there is no evidence that any were built as late as this. On the other hand, some chambered tombs covered by round mounds were late constructions, especially on Anglesey (Lynch, 1976). Closer dating of individual examples is difficult, but Barclodiad y Gawres is likely to belong to the Meldon Bridge period on the strength of its affinities with Irish cruciform passage graves.

Early development of henges and cremation cemeteries in Wales is suggested by the dates of c. 2790 b.c., c. 2530 b.c., and c. 2470 b.c. for the Class I Llandegai A henge in Caernarvonshire, which yielded a number of cremation burials. Whether there was any interest in round mound burial, to match the Yorkshire Duggleby tradition, is uncertain. In the absence of grave goods such sites have always been ascribed to Beaker influence, but strictly they are of unknown date. The northern cave burials, and the inhumations inserted late in some chambered tombs, suggest inhumation traditions existed by the Meldon Bridge or Mount Pleasant periods, but when and where they originated cannot yet be established.

The Mount Pleasant Period: c. 2150-1700 b.c./c. 2750-2000 B.C.

This period saw the arrival of copper metallurgy, combed and All-Over-Cord Beakers, events which have long been closely connected, but which now demand more complex explanations. The Continental background of most British Beakers still seems to lie in the lower and middle Rhine basin (Case, 1977; Clarke, 1970; Lanting and van der Waals, 1972), and the British Beaker tradition undoubtedly played a part in the development of Irish Beakers, particularly the later forms. But the once favoured, though recently unpopular, idea of an Atlantic source for Irish metallurgy and at least some early Irish combed Beakers now deserves reconsideration. Bound up with this is the possibility of dual origins for the early metalworking which sprang up in many parts of Britain, including Wales. This would involve on the one side influence from the massive Irish industries and, on the other, central European influence, transmitted via the Rhine at the same time as, and possibly in company with, the Beaker tradition.

Whatever their origins, metal and the Beaker tradition were no more than a gloss on existing institutions, which survived largely unchanged from the Meldon Bridge period. The most eye-catching burial novelty of the period, individual inhumation accompanied by a Beaker, in a flat grave or under a round mound, has traditionally been ascribed entirely to Beaker influence, but this view must now be treated with caution. The deposition of the Beaker with the dead was novel, but certainly in Yorkshire, and probably in southern England, both individual and multiple inhumation in flat graves and under round barrows were already well-established traditions when combed Beakers arrived (Burgess and Shennan, 1976, 316-17).

Lanting and van der Waals (1972) have demonstrated the length of time it took for combed Beakers to catch on outside Wessex. In other regions All-Over-Corded Beakers dominated the first stages of development (Lanting and van der Waals' steps 1 and 2), and these were placed only occasionally with the dead. Thus a general survival of established burial customs must be assumed, and these lasted even when combed Beakers, and Beaker-accompanied inhumation, finally took hold, with the development of step 3 vessels, c. 2000/1900 b.c. In the latter part of the period there was a spectacular upsurge in interest both in Beakers (step 4, mostly with short necks) and Beaker-accompanied burial.

Meanwhile in western and northern areas activity continued at chambered tombs of all types. Passage graves and other round mound tombs were still being built as far apart as Ireland (Mound of the Hostages, Tara, Co. Meath) and Sutherland (Embo) (Smith, 1974). Some at least of the numerous Irish wedge tombs may have been built in this period, while continuing construction in Wales is indicated by the two notable tombs of Bryn yr Hen Bobl and Bryn Celli Ddu on Anglesey. The first was built on occupation material which included much Peterborough ware of a local style and a Beaker sherd, while the latter was built on top of an already existing henge (Lynch, 1970). This has often been taken as evidence of the antipathy of chambered tomb and henge traditions, but equally might involve a fusion of the two.

Development of henges and cursūs continued apace, and at Llandegai a second henge was built, of the double entrance class II form. The possible class II henge at Castell Bryn Gwyn, Anglesey, should date to the end of this period, for in addition to Peterborough sherds of the Fengate style and Beaker sherds, it also produced a bronze awl (Wainwright, 1962; Lynch, 1970, 64-7). This cautionary message is emphasized by another example of bronze in a 'Neolithic' context, at the Mount Pleasant enclosure in Dorset, where the main ditch primary silts yielded Grooved Ware and a radiocarbon date of c. 1778 b.c., from just below a developed bronze flat axe (Burleigh et al. 1972).

The development of great centres of Durrington Walls-Mount Pleasant type has not yet been matched outside southern England but other types of site may have served a similar function. For example, the intense 'domestic' activity around the great passage graves at Knowth and Newgrange in Co. Meath (O'Kelly, 1973; Eogan, 1976) can be explained in these terms, and in Wales a variety of monuments may have fulfilled this role. Many of the stone circles, avenues, rows and monoliths of Wales and other upland regions were built in the Mount Pleasant period, and at least some of the other ring monuments towards its end. As Lynch records (see chapter six), the nodal siting of many of these complexes argues powerfully for their central role within the community.

Over Britain as a whole, this period saw inhumation burial reach its apogee. Many established burial rites persisted, but new elements sprang from the growing influence of the Beaker tradition. Continued use of chambered tombs was only part of the story, and probably catered just for certain privileged elements in society. Beakers were now frequently used in the activity at chambered tombs, for example at Ty Newydd on Anglesey (Lynch, 1969) and Tinkinswood, Glamorgan (Savory, 1955, pl. v). The use of inhumation and cremation at tombs of different types illustrates the prevailing diversity. Thus on as small an island as Anglesey cremation was practised at the passage graves, but inhumation occurred at sites such as Bryn yr Hen Bobl and Lligwy (Lynch, 1970).

Nevertheless, inhumation was probably a minority rite in Wales to judge from the comparatively small number of Beaker inhumations from the region and the almost total absence of food vessel inhumations. It is therefore curious that cremation burials should be so difficult to find compared with the well-known cave and chambered tomb inhumations. Cremation cemeteries can be postulated but remain elusive, though the Llandegai B henge yielded cremations in a wooden box (radiocarbon date c. 1790 b.c.; Houlder, 1976, 59). Also relevant is an old discovery of cremations in 'urns' at Eddisbury, Cheshire, for the only one which survives is a Grooved Ware vessel (Smith, 1974, 119).

The Beaker tradition never achieved the success in Wales it enjoyed in southern and eastern Britain. The slightness of its early impact can be measured by the sparsity of All-Over-Corded finds, confined largely to Anglesey, and especially the coastal settlement at Newborough Warren. The paucity of early combed Beakers (steps 1 and 2) is common to all regions outside Wessex, but the general burgeoning of interest in combed Beakers indicated by the greatly increased numbers of step 3 and 4 vessels, saw little response in Wales. It is noteworthy that most of the early (step 1-4) Beakers from Wales have come from the south, in contrast to the northern concentration of All-Over-Corded sherds. The contrast between the AOC contexts, in settlements and chambered tombs, and the individual inhumation graves of the southern combed Beakers reflects the national situation. The finds from barrow Sutton 268', Glamorgan, (Fox, 1943) epitomise the southern trends, and will serve in more general terms to illustrate the changing pattern of Early Bronze Age burial traditions. The first phase here is typical of early Beaker burials: an inhumation in a rock-cut pit, accompanied by a step 2 Beaker and barbed and tanged arrowheads, was covered by a small, ditch-girt cairn.

Irish influence undoubtedly played an important part in the establishment of Welsh copper-working, which, although tiny by Irish standards, quickly took on a character of its own. Actual Irish imports are few, but include the important hoard of Lough Ravel-type axes from Moel Arthur, Flintshire (Forde-Johnston, 1964). Metal analysis suggests that there may originally have been more Irish imports, but that some were melted down to provide metal for local products. Undoubtedly, local ores were also used. The principal workshops were in the south-west and the north-east, producing mainly broad-butt flat axes, and especially small examples of the straight-sided, thin-butted Growtown form. The only other copper implements known from Wales are a few halberds. No doubt this limited Welsh production was influenced by the remarkable popularity of halberds in Ireland, but in fact most of the Welsh examples are of British, and specifically non-Irish, types.

Bronze was in sporadic use early enough to be deposited in graves with step 3 Beakers at Dorchester, Oxon, and Sewell, Beds. (Clarke, 1970, 574, n. 57). By c. 1900/1800 b.c. the first bronze industries were in operation, termed Migdale in Britain and Killaha in Ireland (Burgess, 1974). Bronze working was soon established in Wales, but the production range remained limited. English and Scottish workshops now had a greatly expanded repertoire, including a variety of trinkets derived from central European Únětician metal-working in its Reinecke Al stage, but little of this filtered through to Wales apart from the new narrow-butt, Migdale-type flat axes. Stone moulds for these from Bettws-y-Coed and the Walleybourne, Shropshire (Thomas, 1972), point to production in the north and the Marches, but the main distribution of axes is in the south. The fact that only one axe of the contemporary Irish Killaha type has been found in Wales illustrates the weakness of Irish influence.

Before the end of this period even more advanced bronze-working ('Stage V', 'Colleonard', Burgess, 1974) had begun, to judge from the Castell Bryn Gwyn awl and the Mount Pleasant axe. The latter is a more advanced version of the Migdale axe, with transverse face bevel and 'rain pattern', a form which in Wales is again found mainly in the south. But these developments belonged in the main to the subsequent Overton period.

The Overton Period: c. 1700-1450b.c. /c. 20th-18th centuries B.C.

The important changes of this period are epitomised by a Wiltshire burial mound, West Overton G6b (Smith and Simpson, 1966). The period appears to have begun with a major social and spiritual upheaval, involving the severing of old traditions and the development of new ones. Symptomatic is the blocking up of chambered tombs throughout Britain, many of which had been a centre of activity for a thousand years or more. At the same time long-established 'Late Neolithic' ceramic traditions petered out, notably Grooved and Peterborough wares, though they provided the basis for the new wares which characterise this period. The appearance of a wide range of specialised funerary wares, 'food vessels' and 'cinerary urns', is an indication of dramatically changed attitudes to burial. For centuries only Beakers had been deposited with the dead, and then only the privileged few, but now a great range of food vessels as well as Beakers were placed in graves. Similarly cremation burials for centuries had been placed in holes, bags and boxes, very rarely in pots, but now cinerary urns, of a wide variety of forms, became commonplace in some sectors of society. The implied diminution of the status of Beakers is confirmed by the associations of many late Beakers. In place of the eye-catching associations with warrior equipment, late Beakers are as likely to be buried with craftsmen's tools.

We see these changes at West Overton G6b. The primary and most important burial, at the bottom of a central grave, is still with a Beaker, but accompanied by a set of leatherworker's tools. The numerous burials that were interred subsequently, probably over a space of months, show the much greater mixture of inhumation and cremation which characterises this phase. Two of the cremations were in Collared Urns.

This barrow also produced beads appropriate to the so-called 'Wessex Culture'. A major upheaval in Wessex society is indicated by the character of the Bush Barrow series of graves, which betray the rise of a new and ostentatious aristocracy (Burgess, 1974). Although Beakers were still in widespread use, the new élite eschewed them. Some indication of the effectiveness of the new order may be provided by events at Stonehenge, where at the beginning of the period the bluestone monument was abandoned in the middle of construction, and replaced by the much more gradiose sarsen structure. Nevertheless these developments at Stonehenge, and the contemporary alterations at Mount Pleasant, show that these enclosure sites and the whole complex of henges, circles, rows, avenues and cursūs, still played a central role in society.

In Wales the events at Sutton 268' were symptomatic of the new order. Three new burials were inserted into the Beaker cairn, an inhumation and two cremations. One of these was accompanied by one of the new Pygmy Cup accessory vessels and a bronze riveted knife, while the other was in a Collared Urn. Whereas combed Beakers had hitherto been found mainly in south Wales, the late, necked Beakers (steps 5-7) are as common in north Wales as the south. The Beaker tradition was now more popular in Wales than before, but still weak compared with other regions. Inhumation burial must have spread in Wales as part of the Beaker tradition, for it is very rare outside Beaker graves. Beaker designs show that, as might be expected, south Welsh Beakers reflect influence from Wessex and south-west England, while northern examples owed most to Yorkshire styles.

Food vessels (Savory, 1957) are rare in Wales and idiosyncratic, and are almost invariably found with cremations. Not unexpectedly they come mostly from the north which would have been most open to influences filtering in from Yorkshire. A few examples with false-relief ornament from western districts betray Irish influence, but Wales, in strong contrast to western Scotland, has yet to produce a single convincing example of the Irish Bowl which so dominated the Irish food vessel tradition.

The poor showing of Beakers, food vessels and inhumation in Wales was due to the popularity of cremation, which was now an expanding force throughout southern and western Britain. Undoubtedly, it was already widespread in Wales and on the increase, but without radiocarbon dates it is difficult to distinguish in the case of individual burials those which belong to this period, and which to the succeeding Bedd Branwen period. Herein lies the value of the Brenig sequence of dates (see Lynch, Chapter Six), which show that some Collared Urn cremations belonged to the Overton period and were centuries earlier than others.

Some of the best evidence for continuing interest in ring and linear monuments comes from Llandegai, where late Beakers were found in pits inside Henge B (Houlder, 1968). At Brenig the ring cairn 44 and platform cairn 51 were built and saw their initial use in this period on the strength of radiocarbon dates between c. 1680-1500 b.c. (Lynch, 1975). Unfortunately there has been little modern excavation of such sites in Wales, and the Brenig results show that finds do not necessarily relate to construction and initial use. Thus the late Beakers found in the Carneddau Hengwm circles, Merioneth (Grimes, 1963, 119-20), show these sites were in use in the Overton period, while the cinerary urns and radiocarbon dates from the Penmaenmawr circles (Griffiths, 1960) indicate activity in the Bedd Branwen period, but in neither case does this rule out earlier building and use.

We have seen that new advances in bronze-working, Stage V, emerged towards the close of the Meldon Bridge period. Two developed flat axe traditions crystallised in Wales in the ensuing Overton period, both sporting novelties such as a transverse face bevel and slightly flanged face edges (betraying the improved knee-shaft hafting method), and both frequently decorated, especially with 'rain pattern'. In the south the new axes clearly derived from the old Migdale axes, but in the north and west the Irish form is common, characterised by an ultra-narrow butt and straight, sometimes almost parallel sides. Some were no doubt Irish imports, but others were copied locally to judge from their metal content.

Welsh metalworking remained limited in its range and modest in its output. Halberds had vanished, but a range of riveted knives and daggers were produced which were lacking in standardisation as befitted the fragmented and individual Welsh industry. Other tools included single-pointed awls, simple tanged chisels, and small, narrow flat 'axes' like those occurring in Bush Barrow graves of the early Wessex Culture (Burgess, 1974). The last two types occur together in grave groups at Breach Farm, Glamorgan (Grimes, 1938), and Llanddyfnan, Anglesey (Lynch, 1970, 140-43), but both of these are likely to have been deposited in the ensuing Bedd Branwen period, when for a time metalworking continued essentially unchanged.

The Bedd Branwen Period: c. 1450-1250 b.c. /1700-1400 B.C.

This period was characterised by the total domination of cremation burial. At its beginning only eastern areas, from Yorkshire to the Moray Firth, where food vessel inhumation had been most firmly rooted, still retained a strong inhumation tradition, but even these bowed to cremation by c. 1400 b.c. Throughout Britain and Ireland inhumation became an eccentric rite. The Anglesey cemetery mound of Bedd Branwen (Lynch, 1971) is typical of the period, yielding cremations dated from c. 1403-1274 b.c.

The diversity of cremation sites and customs is bewildering. Some burials were interred individually under new mounds, as at Brenig 40, while some existing mounds were modified and new burials added, as at Sutton 268. Cremation cemeteries have been found under mounds, as at Colwinston, Glamorgan (Fox and Fox, 1941), incorporated in mounds, as at Llanddyfnan (Lynch, 1970, 136-142), and inserted into earlier mounds as at Rhoscrowther, Pembrokeshire (Fox, 1926). Cemetery mounds frequently show two or more phases of burial, as at Bedd Branwen and Treiorwerth, Anglesey (Lynch, 1971). North Wales lay within the north-western province characterised by flat as well as mound cemeteries, epitomised by sites such as Cae Mickney, Anglesey (Lynch, 1970, 160-63). Cremations were also interred within stone circles and other ring monuments, as at Penmaenmawr (Griffiths, 1960) but were not always primary to such sites.

The cinerary urns of Wales are predominantly Collared Urns, with some food vessel urns ('enlarged food vessels'), confirming that local traditions were firmly rooted in Britain. Irish influence was slight in comparison, limited to a few flamboyant 'Encrusted Urns' from western sites such as Penllwyn, Cards., Preseli, Pembs., and Plas Penrhyn, Anglesey (Fox, 1959, 35-37, pl. 'B), perhaps the few Cordoned Urns from northern and southern sites such as Colwinston and Llanddyfnan, and Necked Urns from north Welsh sites such as Llangwm, Denbs (Grimes, 1951, fig. 28).

Knowledge of burial customs in this period is complicated by uncertainties about the beginnings of Deverel-Rimbury burial. In southern England it now seems certain that Deverel-Rimbury pottery was in widespread domestic use by the Bedd Branwen period, and already in funerary use among those of lower status (Barrett, 1976a). But, although its Grooved Ware origins are now assured, it is still not clear when it first emerged. The coarse bucket and barrel-shaped vessels associated with cremations in many parts of Britain now look increasingly like local equivalents of the Deverel-Rimbury phenomenon. Best known are the groups found in the East Midlands (May, 1976, 77-89), Yorkshire (McInnes, 1968), and western Scotland (Morrison, 1968), and at least the first and last of these had clearly emerged by the Bedd Branwen period. In all probability most regions will prove to have similar complexes. 'Deverel-Rimbury' cemeteries have been reported from the Marches, notably at Bromfield, Salop (Stanford, 1972a), and Deverel-Rimbury pottery has been found at Rhuddlan, Flints (Miles, 1972), so these can be added to old 'Deverel-Rimbury' finds such as those recorded by Savory (1958). Some at least of this material belongs to later periods of the Bronze Age, but it is not clear how much of it there is, and how far back its beginnings can be set.

Cremations from circles and ring monuments, as at Penmaenmawr, can be interpreted either as proof of the continuing veneration of such sites, or, as the Brenig evidence may hint, that they were now being used in a new, or at least additional, role. Activity continued at ring sites and enclosures throughout the country, but it was not always the activity for which the sites had been designed. Henge building had now stopped, and only the most modest ring sites were constructed as late as this, and now much more explicitly for burial purposes. The great era of public monuments was clearly over.

By c. 1250 bc, perhaps rather earlier in the north and west, and later in the south, especially Wessex, the concern for burial and ritual, which for so long had been expressed in great monuments, had reached a low ebb. The construction and use of stone rings and rows cannot be traced beyond this point, and by implication, venerable traditions of celestial observation disappeared with them, together with the remarkable store of skills, in field geometry, engineering and mensuration, which were implicit in their construction. Practical difficulties may have played a part in this process. If the climate had already begun to deteriorate, and knowledge remains very sketchy in this area, then encroaching blanket peat and cloudy skies may already have made the sites increasingly difficult to use (Burgess, 1974). But a more complex interaction of factors is more likely to have been responsible. The country may now have been entering one of those periodic times of agricultural crisis which demanded considerable adjustments in land-use and in settlement and social systems (Bradley, 1978). Deteriorating climate may increasingly have influenced the situation but is unlikely to have caused it. Social upheaval may have been considerable. After 1250 b.c. the whole paraphernalia of privileged burial, epitomised by the great variety of cinerary urns, came to an end, and perhaps a sudden end, to judge from the ubiquity and completeness of the phenomenon. Only in the Wessex uplands did the old order linger on in the guise of the Aldbourne-Edmondsham group of the later Wessex Culture (Burgess, 1974). But Wessex increasingly was becoming a backwater, and the Deverel-Rimbury groups who previously may have constituted the less ostentatious majority came more and more to the fore. A similar social turnabout may be postulated in Wales as in other regions, but the evidence is not yet as clear as for Wessex. If circular and linear monuments and all they stand for had been a preserve of the privileged using the labour of the underprivileged, then it is not surprising that the dramatic changes revealed in the sepulchral record coincide with the end of megalithic traditions.

Stage VI metalworking was still practised in the early part of the Bedd Branwen period, as can be seen from the bronzes included in the exceptionally rich Breach Farm and Llanddyfnan graves. From c. 1600 B.C. influences emanating from Continental workshops at the Reinecke A2/B1 (or A3) stage led to the important metallurgical advances represented by the Arreton tradition in southern and eastern England (Britton, 1963; Burgess and Cowen, 1972; Gerloff, 1975). This introduced new methods, notably hollow-casting, new alloys, and many new and superior products, including flanged axes, riveted daggers with complex grooved and ribbed blades (especially the Camerton, Snowshill and Hammersmith types), tanged spearheads, socketed, peg-hole spearheads, and dress pins based on Continental forms. These innovations rapidly spread via the Marches and north Wales to Ireland, but seem to have left the rest of Wales and much of Britain north of a Tees-Morecambe Bay line, relatively untouched. Older metalworking traditions may well have persisted in these areas.

In Ireland the Inch Island tradition emerged as the local response to these novelties, still using stone moulds, unlike the Arreton workshops, and adapting the new fashions to local taste. Thus, Irish flanged axes are subtly different from the Arreton examples (Burgess and Cowen, 1972), and Irish socketed spearheads generally have socket loops ('end-looped' type, Burgess, 1962a). Wales, between Arreton and Inch Island, may have played a key role in the metallurgical developments of the period, having elements appropriate to both traditions. The crucial Ebnal hoard, Salop (Burgess and Cowen, 1972), combines Arreton axes and daggers with an end-looped spearhead, a unique shouldered tool and a trunnion tool. End-looped spearheads have been regarded as an Irish invention on grounds of sheer numbers, but there is no reason why Welsh craftsmen should not have been the first to add the loops to the loopless Arreton form, passing the idea on to Ireland. Certainly, the examples found in Wales were produced in the region and not imported to judge from the Bodwrdin mould and local peculiarities of form and metal content. Similar doubts surround the origin of trunnion tools and some other products of this period, notably Class Ib razors. The existence of a Welsh equivalent of Arreton and Inch Island metalworking, which we can term the Ebnal industry,

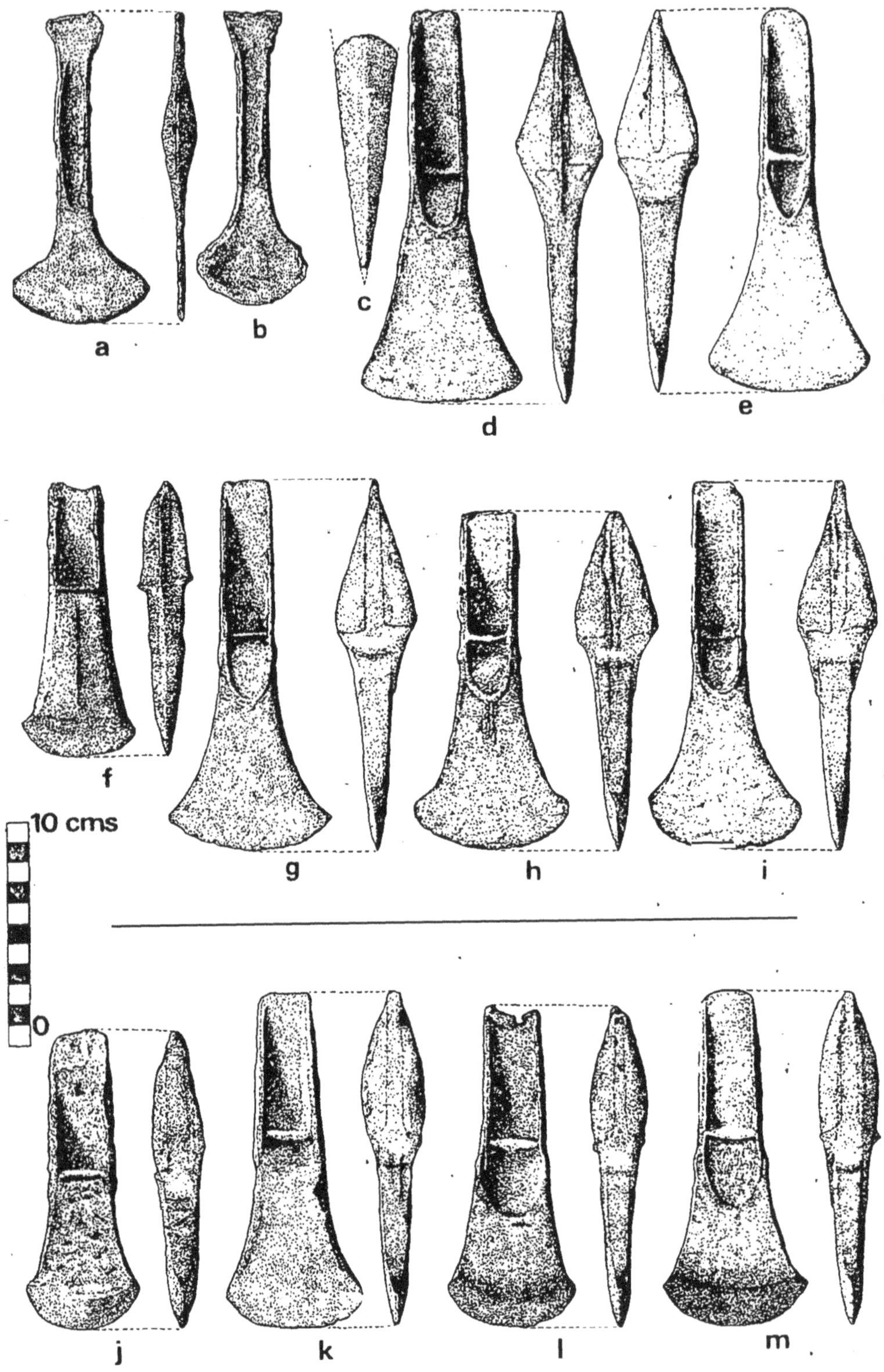

Fig. 7.4 Metalwork of the Acton Park phase. a-i, Acton Park hoard, Denbs.; j. Tenby, Pembs.; k. Charelau, Ang.; l. Llanderfel, Merion.; m. Llandwrog, Caerns. (b, c, after Barnwell, Archaeol. Cambrensis, 4s. VI, 1875) (after Stanford, 1972a).

is beyond doubt, however, as can be deduced from idiosyncracies of form and alloy. Of all the Arreton-type material from Wales the axes in the Menai Bridge hoard (Lynch, 1970) look most like English examples, and it therefore comes as no surprise to find that they alone are made of metal similar to standard Arreton metal.

If the Ebnal tradition penetrated Wales only peripherally, it is perhaps because it was overtaken by a precocious local development of even more advanced metalworking, characterised by palstaves, looped spearheads and dirks and rapiers, and traditionally regarded as 'Middle Bronze Age'. It is now clear that Welsh workshops took a leading role in the development of these new traditions in Britain and Ireland. The diagnostic products are widespread especially in eastern and southern England and Ireland, but few areas can show such a concentration as that in north Wales. Typical are developed flanged axes of various forms, including haft-flanged axes and bar-stop axes, two palstave forms, shield-pattern (Group I) and early midribbed (Group II), socket-looped and side-looped spearheads, and dirks and rapiers of Group I (ribbed and grooved blades) and Group II (lozenge-sectioned blades). Regional differences crystallised from the outset. Wales and lowland England followed the same course, as was usually the case in the Bronze Age, and concentrated on palstave development, whereas northern England and Scotland preferred haft-flanged axes. Ireland maintained a balance, developing both a distinctive palstave tradition (Burgess, 1974) and haft-flanged axes. Irish craftsmen also evolved the socket-looped spearhead with kite-shaped blade from the end-looped type, and persisted with it throughout the Middle Bronze Age. In contrast the side-looped spearhead with leaf-shaped blade became the standard British form. Group I and II dirks and rapiers were produced in Britain and Ireland, and regional forms can be identified.

The evidence, particularly the lack of hoards, does not allow us to establish beyond doubt where individual types were first developed, but distribution patterns suggest eastern England, north Wales and Ireland were leaders in the new traditions. North Wales is fortunate in having at least a few hoards of this period, after one of which, from Denbighshire (Fig. 7.4), we can term this new Welsh metallurgy the Acton Park tradition. Finds are concentrated heavily in the northern half of the region, and point to some remarkable connections with the Continent. The ultimate background of many of these developments lies in Early Tumulus Europe, but more immediately in the Ilsmoor group of northern Europe and the Tréboul group of north-west France (Burgess, 1974; Butler, 1963). The emergence of the new tradition was contemporary with Hachmann's Horizon III in northern Europe, and with Tumulus B1 in central Europe, for this is the period of the earlier Ilsmoor hoards in which we find shield-pattern and midribbed palstaves just like those which initiate the Acton Park series (Burgess, 1976b). Palstaves of local north European forms are absent in these hoards, and appear only in the later Ilsmoor hoards of Hachmann Horizon IV, contemporary with Tumulus B2, by which time the associated western palstaves are of much more advanced shield-pattern forms. This implies that the traditional derivation of western palstaves from northern palstaves should be reversed. The word 'western' must be used instead of 'British', because similar palstaves to those produced in Britain were also typical of north-west France.

In absolute terms this means that the earliest stages of Acton Park metal-working cannot have begun much later than c. 1500 B.C. and possibly rather before, i.e. while Arreton and Inch Island workshops were still operating in some areas, and while the ritual and sepulchral traditions of the Bedd Branwen period were still in evidence. The absence of Acton Park metalwork in graves is simply explained, for in most regions it had long been an eccentricity to put metal objects in graves. The notable exception was Wessex, but here a good case can be made out for a lingering survival of Arreton traditions and resistance to the new metalwork fashions.

Thus, in the final stages of the Bedd Branwen period, the early Acton Park output will have progressively challenged Arreton-Ebnal-Inch Island products and survivals from still older traditions. The novelties in question were primitive palstaves, near-palstaves and bar-stop axes, such as occur in early Ilsmoor hoards. They include broad-bodied shield pattern palstaves like those from Dinas Dinlle , Caerns. and Llandderfel, Merioneth (Fig. 7. 2),), midribbed palstaves of both the Chepstow and Liswerry forms (Burgess, 1964), and a hotch-potch of axes symptomatic of the prevailing atmosphere of experimentation, part bar-stop axe, part haft-flanged axe and not quite palstave, such as those from Broadwell Hayes, Tenby and Chwarelau, Anglesey. A variety of bar-stop axes and haft-flanged axes also have to be taken into account, mostly imports from, or influenced by, Irish workshops, such as the Caernarvon axe (Grimes, 1951, pl. V), or by north English sources. The first dirks and rapiers also reached the markets, mostly Group II, such as the dirk from Caerphilly (Grimes, 1951, Fig. 62.7), still with blade grooved in Arreton/Ebnal fashion. Group I is rare in Wales. The first socket- and side-looped spearheads will also have been available, the former imported from Ireland along with bar-stop and haft-flanged axes. This was certainly not a one-way traffic, for north-Welsh palstaves were carried across to Ireland, and there inspired the development of a distinctive Irish palstave series (Burgess, 1974).

While some of this early Acton Park production was in traditional metals, made, no doubt, both from local ones and re-used scrap, a remarkable new lead-bronze alloy was introduced and came increasingly to dominate Acton Park metallurgy in Wales. By the end of the Bedd Branwen period, when more sophisticated shield-pattern and midribbed palstaves were being produced, lead-bronze had become standard.

The Knighton Heath Period: c. 1250-1050 b.c./c. 1400-1200 B.C.

In southern Britain this period was dominated by Deverel-Rimbury ceramics and their regional equivalents. Of the old order only the Aldbourne-Edmondsham group of the Wessex Culture lingered on for a while in what had become an unremarkable region. The abandonment of the great centres of Wessex, and of the circular and linear monuments strewn throughout Britain and Ireland, is perhaps the best testimony of the profound social and spiritual upheaval that had taken place. If quantities of metal goods are any indication, then the foci of power and wealth had switched from traditional centres, such as Wessex and east Yorkshire, to new ones, notably north Wales and, increasingly, the Thames Valley and the Fens. The growing importance of the last two has been linked to the emergence of a water religion, a by-product of increased wetness, and possibly deteriorating climate (Burgess, 1974). Certainly from this time onwards increasing quantities of top quality metalwork found their way into rivers, pools, and other watery places.

The communities of southern and eastern England continued to inter their cremated dead in flat cemeteries, or dig them into existing mounds, and even raise small barrows. Some of the dead merited just holes or pits, but many were placed in globular, bucket-and barrel-shaped vessels of which the Deverel-Rimbury pots of Wessex are the best known. Radiocarbon dates are few, but several for the Knighton Heath cemetery, Dorset, range from c. 1205-1102 b.c. (Barrett, 1976a). It is likely that cremation persisted throughout Britain and Ireland, but as yet it is poorly documented in most regions. An equivalent to Deverel-Rimbury burial was certainly practised in the Marches at sites such as Sharpstones Hill and Bromfield in Shropshire (Stanford, 1972a), but sites and radiocarbon dates are still too few to permit analysis of the situation, and the picture in Wales is entirely unknown.

The pattern of settlement in the Deverel-Rimbury lands of southern England has become very familiar: ditch and bank and palisade perimeters enclosed settlements and farmsteads housing at most enlarged family groups in a few round timber houses. These have adjoining 'celtic' field systems and grazing in which mixed farming was practised. Extensive field systems were still being worked in eastern England, as at Fengate (Pryor, 1976), and the settlements of south-west England are also well known (Fox, 1964, 86-96). These varied in form and character according to local geography, but in upland areas they consisted of round stone buildings, frequently associated with stone-walled enclosures, yards, plots and fields. Recent work on the Dartmoor reaves (Fleming, Collis and Jones, 1973; Fleming, 1976) has revealed their role as land divisions in a tightly organised landscape. These south-western communities were characterised by Trevisker ware, and the presence of Trevisker pottery in south Wales, as in the Six Wells 271 barrow, Glamorgan (Fox, 1941), together with the well-known similarities between the burial mounds on both sides of the Bristol Channel, makes it all the more remarkable that nothing like the settlement pattern of the south-west has yet been found in Wales.

Practically nothing is known, therefore, about this period of Welsh prehistory apart from the metalwork. During its first stage, corresponding to Hachmann Horizon IV and later Ilsmoor hoards in north Europe, and Tumulus B2 in central Europe, the Acton Park tradition reached its apogee. Lead-bronze was now used throughout Wales. Dirks and rapiers of Group II remained the principal edged weapons, and side-looped spearheads were also in use. While it has always seemed logical to suppose these emerged in the Acton Park phase (Burgess, 1968a, 3; 1974, 201-2, 313, n. 227), some recent writers have disputed their appearance before the subsequent Taunton phase, when they first appear in British hoards (e.g. Ehrenberg, 1977, 8). However, several Welsh side-looped spearheads are made of a typical Acton Park lead-bronze, and the presence of a side-looped spearhead in a Tréboul-type hoard at Caen in Normandy (Edeine, 1961) confirms they were in use early enough to have been part of the Acton Park repertoire. The Caen hoard has other typical Acton Park forms, Group II dirks and a midribbed palstave of Liswerry type.

The typical Acton Park palstaves of north Wales are large, heavy and well-made, with high, often angular flanges. By comparison, south Welsh products tended to be smaller and lighter, with lower, more modest flanges. Also the south had relatively greater numbers of midribbed palstaves (Burgess, 1964). Haft-flanged axes seem to have disappeared as the Acton Park tradition grew in strength, for none have proved to be of Acton Park metal. Curious thin-bladed, flanged axe-shaped implements of uncertain purpose have been found in two Acton Park hoards, in the eponymous find (Fig. 7.4), and in the Moelfre Uchaf hoard, Denbs. (Davies, 1949, 434-6) associated with a bar-stop axe or Stegbeil.

The similarity of the shield-pattern palstaves in the Dutch Voorhout hoard to the north Welsh Acton Park form has long attracted comment (Butler, 1959), and the connection has now been confirmed by metal analysis. But whatever the explanation of this overseas connection, whether it indicates itinerant Welsh craftsmen, traders or products, it was not limited to north-west Europe. For the distinctive heavy Acton Park implement is also found in Brittany, including examples in the Tréboul hoard itself (Briard et al. 1956, 1959). Ilsmoor and Tréboul contexts provide plenty of parallels for other elements in the Acton Park tradition, for example the strange, thin-bladed implements, not to mention a variety of midribbed palstaves and bar-stop axes. Clearly, the nature of this triangular relationship will bear much more intensive scrutiny.

North Welsh influence in northern England and Ireland is indicated by scatters of heavy Acton Park shield palstaves, but it hardly touched lowland England, where shield palstaves are lighter, more graceful implements, with low, rounded flanges. Similar palstaves were standard in northern France, especially Normandy, indicating, in metalworking terms, a cross-Channel tradition block that lasted throughout the Bronze Age. It was this pan-Channel tradition that initiated the next changes in fashion. The western palstaves in the Frøjk-Ostenfeld hoards of Montelius II b-c, which followed the Ilsmoor group in northern Europe, are no longer shield-pattern and midribbed examples, but a new low-flanged form, sometimes with side-loop (Butler, 1963). This was a logical development from the shield-pattern and midribbed palstaves typical on both sides of the Channel, and formed the basic axe type of the new Taunton tradition in lowland England and the Baux-Saint-Croix/Mont Saint Aignan tradition in northern France. The weight of contacts between Britain and Europe had now passed from Wales to lowland England. The Taunton tradition in addition to its low-flanged palstaves had new dirk and rapier forms, Group III, with triple arris blade, and Group IV, with flattened mid-section, and introduced the leaf-shaped basal-looped spearhead as a thrusting spear. The side-looped form now became perhaps more wholly a throwing spear. The origin of these weapon developments cannot be ascertained. Ireland is a possibility, but lowland England had an atmosphere ripe for change, and the Continental contacts. Also from Europe came a range of ornaments which were extensively copied here and inspired the invention of distinctively British ornaments. The result was the 'Ornament Horizon' (Smith, 1959). This was largely a southern English phenomenon, but its ripples affected much of Britain and especially Ireland, where the ornaments were often copied in gold. With the ornaments came a range of tools, notably socketed chisels, punches and hammers, knobbed sickles and saws. Socketed axes of Taunton-Hademarschen form have also been regarded as one of these exotic tool types, but now look increasingly like a British response to the imported tool fashions (Burgess, 1976b, 71).

The background of these novelties must again be sought ultimately in central Europe, at the Tumulus B2/C stage, but more immediately in France, and in northern Europe in Montelius II b/c. The triangular relationship again complicates problems of origins. Traditional chronologies suggest the Taunton tradition emerged in lowland England during the 14th century B.C. The quantities of Acton Park metalwork in Wales suggest that in north Wales in particular the new fashions may have taken some time to catch on, but eventually the new Cemmaes-Deansfield tradition replaced Acton Park metalworking. It was certainly less dynamic than its predecessor, used lead-bronze more erratically, and lost much of its international standing. Wales in fact entered a period of marked conservatism, for the new bronze ornaments and tools remained practically unknown. Welsh workshops were still strongest in the north where the ores were to hand, and they had a limited repertoire consisting mainly of low-flanged palstaves and side-looped spearheads, the new Group III and IV dirks and rapiers, but only small numbers of the new basal-looped spearheads. To this phase, the later 14th and 13th centuries B.C., belong the few Welsh hoards of low-flanged palstaves notably those from Cemmaes, Montgomery, Harlech, Merioneth, and Deansfield, Caernarvon (Savory, 1958). This last find includes bronze moulds for casting the new palstaves. The Beddgelert rapier hoard (R.C.A.M. Caerns. II, liv), the largest rapier hoard ever found, also belongs to this phase. The outlandish socketed palstaves from north Wales (Lynch, 1970, 197) perhaps epitomise the failure of north Welsh workshops to keep pace with developments. The socket of the socketed axe is reproduced but combined with the triangular low-flanged palstave blade.

The Penard Period: c. 1050-850 b.c./c. 1200-1000 B.C.

This period marked a major watershed in later British prehistory. It coincided with the great disturbances which beset the east Mediterranean civilisations from the later 13th century B.C., and which culminated in the collapse of the Hittite and Mycenaean empire. It coincided too with the entry of the central European community of tradition into its Early Urnfield stage. The Early Urnfield warrior chiefs, the first armoured warriors north of the Alps, epitomise the disturbed conditions which affected much of the Old World. Barbarians were everywhere on the move, on land and by sea, but the written records also speak of natural calamities, of earthquakes and famine, and climatic change may have been a growing problem. It is not surprising that the ripples of these disturbances should reach out, permeate France, and then Britain and Ireland.

The western limits of the Early Urnfield province extended only to the Paris Basin. This means that the Early Urnfield material coming to light in increasing quantities in southern Britain can reflect only long distance raiding at most. More plausibly, it was received second-hand from those maritime French territories with which intimate connections had long been established. Two finds of bronzes from presumptive wrecks which have recently come to light off Dover (Coombs, 1975), and off Salcombe, Devon (YHN, 1977), both belong to the early part of the Penard period. They may reflect an increase in cross-Channel traffic, for they consist of north French metalwork, or north French versions of Early Urnfieid types, belonging to Bz D/Ha A1, Mostly, they are forms otherwise seldom or never found in Britain, such as median-winged axes, rod-tanged swords (Griffangelschwerter) and Rosnoën swords, suggesting shipments of French scrap or the belongings of French refugees.

These French/Early Urnfield imports are concentrated particularly in the Thames Valley, and scattered more sporadically through south-east and southern England, and south Wales. Ballintober swords, which represent the Irish-British response to rod-tanged swords, are distributed more widely through lowland England and Ireland.

Two south Welsh hoards epitomise the two phases into which the Penard period can be divided. The first, contemporary with Bz D-Ha A1 on the Continent, is characterised by the eponymous Penard hoard, Glamorgan (Crawford and Wheeler, 1920-21). It contains a Rosnoën blade, Ballintober swords, a barbed and tanged bronze arrowhead of familiar Early Urnfield type, a pegged leaf-shaped spearhead, indicating that this old method of shaft attachment was coming back into popularity, and a slender socketed axe derived from the Taunton-Hademarschen series. Bronze shields were also introduced in this phase, though they do not occur in hoards except with other shields. Of the Welsh examples, that from Harlech (Evans, 1881, Fig. 429) is closest to the Early Urnfield originals, such as have been found recently at two sites in the Thames Valley. The three other Welsh shields are unlikely to be much later than the Harlech specimen (Coles, 1962).

The second phase, contemporary with Ha A2, is represented by the Ffynhonnau hoard, Brecs. (Savory, 1958). The Continental imports in this find are pointed spearshaft ferrules and a tanged knife with re-curved blade, one of the most diagnostic of all Urnfield fashions. In contrast, a notched-butt dirk of Cutts type (Trump, 1962) was imported from Ireland (Burgess, 1968a, 7, 9, Fig. 5). These exotics were associated with 'transitional' palstaves, the first looped, narrow-blade type in Britain. The form had been developed by south English and north French craftsmen late in the Taunton phase, and subsequently became characteristic of the Penard period throughout Britain.

One of the most important Early Urnfield innovations of this second phase, surprisingly absent from Wales, was the leaf-shaped, flanged-hilt sword. The original imports, of the Erbenheim and Hemigkofen types (Cowen, 1951), are concentrated largely in the Thames Valley, but one has come from Co. Donegal (Burgess, 1974, 318, n. 274). The first local copies of these imports were unstandardised, but then the U-butt Hammersmith swords emerged as the first standardised British type (Burgess, 1968b, Fig. 1). All these early swords have a similar distribution, concentrated in the Thames Valley, with a thin scatter through the rest of lowland England and Ireland, even more sporadic finds in north Britain, but none as yet from Wales.

A hoard from Maentwrog, Merioneth (Hawkes and Smith, 1955), illustrates the indigenous weapon developments of the Penard period. Group IV dirks and rapiers continued, but now with new-look smaller butts, which are frequently notched. A novel thrusting spearhead was introduced, still basal-looped, but with triangular blade.

Some of the tools and implements of the Penard period were old-established forms which it would have been difficult to change, such as the trunnion tool found with transitional palstaves at Meole Brace, Shropshire (Chitty, 1925). Others were logical developments from well-established lines, such as 'transitional' palstaves, and many socketed tools, including hammers and chisels, and the slender 'axes' of Penard-Trawsfynydd type (Burgess, 1962b). The persistence of old 'Ornament Horizon' traditions in Wales is demonstrated by the Monkton Farm cave find, Pembs. (Laws, 1908), combining a saw, simple tanged chisel and mock torsion ring, together with the blade of a 'transitional' palstave. It was probably in this period that gold versions of "Ornament Horizon' twisted and ribbon torcs became most popular (Eogan, 1967). The 'central Wales' association of a twisted torc and twin-looped transitional palstave confirms the dating evidence from other regions, though conceivably examples concealed in the Penard Period were heirlooms from an earlier period. Gold torcs are widely scattered in Wales, but there is a notable concentration in the centre, with three hoards from 'central Wales', Llanwrthwl, Brecs., and Heyope, Rads. (Savory, 1958)

There were some notable inventions in the Penard period, such as socketed sickles. The primary cylinder-socket form has the same Thames Valley—Ireland axis of distribution as so much Penard material, suggesting that the blank in Wales is more apparent than real. Two other fashions were introduced from northern France, the bifid razor, rare in Wales, where none are demonstrably as early as this, and pins in a variety of late Tumulus/Early Urnfield forms. These inspired a brief but widespread manufacture of indigenous pin forms in many parts of Britain. Until recently there was little sign of this trend in Wales, but now the crook-head pin from Dinorben (Savory, 1971) and a pin with swollen shank and expanded flat-topped head from the Breiddin (Musson, 1976) can be seen as part of this sudden flowering.

Much of this metalwork was made in a distinctive, new non-leaded bronze which seems to have characterised Penard metalworking in many parts of Britain and Ireland. Penard products are widely scattered in Wales, but the concentration of Continental imports in the south and of transitional palstaves in the Marches leaves little doubt as to the main source of external influence in this period. Some survival of Cemmaes-Deansfield traditions, especially low-flanged palstaves, may be postulated but cannot be proved.

The introduction of a new range of products and a new alloy were only a small part of the changes which transformed Britain in this period. The Deverel-Rimbury tradition and its equivalents in other regions finally came to an end. The interest in formal burial which it had preserved so long finally disappeared, and henceforth, apart from a handful of eccentric inhumations with metal grave goods (Burgess, 1976e) little is known about burial customs in these islands. Deverel-Rimbury pottery was replaced by new domestic pottery, the post-Deverel-Rimbury tradition (Barrett, 1976), embodying a much greater range of shapes, thinner, finer fabrics, and more sophisticated technology. At the latest Deverel-Rimbury settlements, such as Plumpton Plain B (Hawkes, 1935) and Itford Hill (Burstow and Holleyman, 1957) in Sussex, and Eldon's Seat I (Cunliffe, 1968) in Dorset, the familiar Deverel-Rimbury wares are mixed with the new forms, especially shouldered jars. There were also widespread changes in land use. Long-established field systems as far apart as the Black Moss of Achnacree, Argyll (Barrett et al. 1976), Fengate (Pryor, 1974, 1976) and the uplands of south-west England, were abandoned in this period. The familiar pattern of stone-built upland settlement in the south-west can be traced no further and evidence here (Mercer, 1970) and at Achnacree suggests the spread of blanket peat was making life increasingly difficult. The likelihood of climatic deterioration is supported by a burst of timber trackway building in areas prone to flooding, such as the Somerset levels, (Coles et al., 1975, 1976). Furthermore, this seems to be the point at which pile dwellings appear along rivers such as the Thames and Trent. This measure against increasing dampness appears not to have been taken at the so-called Barmston 'crannog' Holderness (Varley, 1968), which was inundated in this period.

It is possible that the switch from emmer and naked barley to spelt and hulled barley may also have occurred in the Penard period (Helbaek, 1957). Profound social changes are also implicit in the widespread establishment of hill-forts. Most had timber box ramparts or used timber strengthening of some form, as can be seen at Rams Hill, Berkshire (Bradley and Ellison, 1975) and Grimthorpe, Yorkshire (Stead, 1968). They imply larger social units, while the constraints of a fixed perimeter demanded careful planning of the interior (Guilbert, 1975, 1976). They show, too, a much greater use of rectilinear structures, especially '4-posters'. These early hill-forts may have been short-lived, but they established a pattern of hill-top settlement which endured throughout the remainder of the Bronze Age.

The extent of these changes, particularly the severing of long established burial and ritual attitudes and the appearance of completely new domestic pottery traditions, would have been sufficient grounds in the past for demanding a major invasion from the Continent. Immigrations have been unfashionable in recent years, but a much better case for Penard newcomers can be made out than for Beaker or Hallstatt immigrants. It is singularly difficult to demonstrate the arrival of migrants beyond all reasonable doubt, but there was undoubtedly an increase in traffic from north-west France, reflected in the Dover and Salcombe finds, and in the French bronzes widely scattered in Britain. Comparisons with French settlements are almost impossible because so few have been published, but it is clear that prototypes for the new 'post-Deverel-Rimbury' domestic pottery can be found in northern France, as can early hill-forts (Savory, 1971). A few exotic vessels have a claim to be French imports of this period, notably the famous Plumpton Plain B vessels (Hawkes, 1935, Fig. 11) and the curious rilled vessel from Upper Chapel, Breconshire (Savory, 1971, Fig. 2).

This last find adds emphasis to the minor concentration of exotic material in south Wales. Wales clearly was profoundly influenced by the Penard changes, and even if little can be said at present about settlements and land use, hill-forts were established in the region as early as anywhere. The existence of

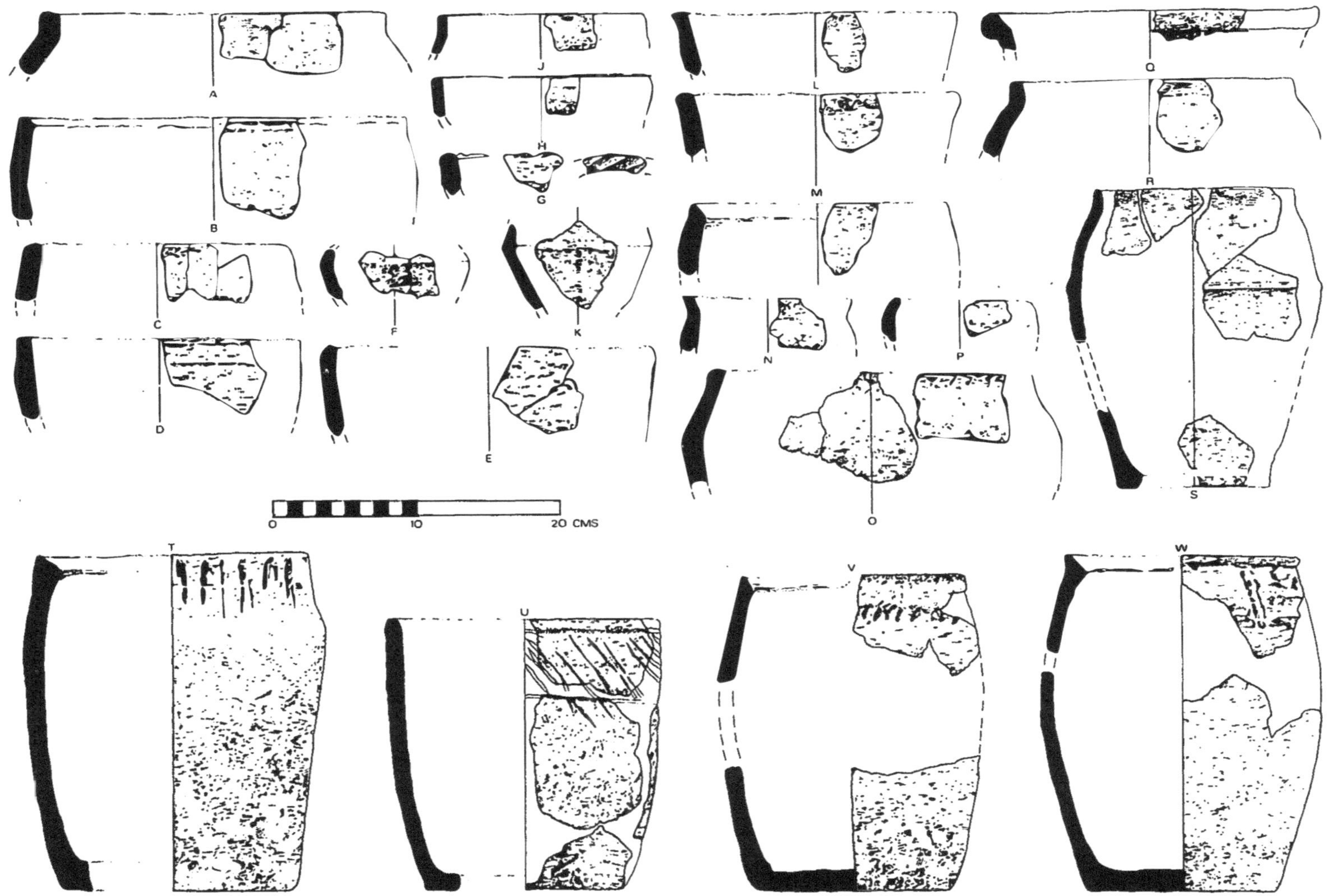

Fig. 7.5 Later Bronze Age pottery from Wales and the Marches. A-S, The Breiddin, Montgomeryshire (after Musson, 1976); T-W, Bromfield, Shropshire

an early box rampart on Dinorben has been disputed (e.g. by Alcock, 1972a) but there was clearly occupation on the hill-top in this period. The first defence at Ffridd Faldwyn, Montgomery, should also belong here (Standford, 1971), but whether it took the form of a box rampart or double palisade is not clear. Rows of 4-posters were found leading away from the entrance at this site, revealing an early example of a planned hill-fort interior. Perhaps later in the Penard period (c. 868-800 b.c.) the first rampart at the Breiddin was constructed, with a timber-reinforced front (Musson, 1976). Shouldered jars from the Breiddin show that the new pottery traditions did reach Wales (Fig. 7.5). Just how many Welsh hill-forts were constructed in the Penard period is unclear.

We can suppose that the local versions of Deverel-Rimbury burial ended in this period, as they did in England: the Sharpstones Hill, Bromfield, and Ryton-on-Dunsmore cemeteries all produced burials radiocarbon dated to the 10-8th centuries b.c. After this time as little is known about burial rites in Wales as in England.

The Isleham Period: c. 850-600 b.c./c. 1000-700 B.C.

The transformation effected during the Penard period was followed by more settled times. Knowledge of developments during the early First Millennium B.C. still rests heavily on the metalwork, and the period is named after its largest hoard, from Isleham, Cambridgeshire (Britton, 1960). Two phases can be distinguished in the metalwork, the Wilburton-Wallington phase occupying the 10th century, and the Ewart Park phase the 9th-8th centuries (Burgess, 1968a; Burgess, Coombs and Davies, 1972; for these dates, Burgess 1979). How relevant this division was to other aspects of life has yet to be established.

The Wilburton metalworking tradition, characterised by a dramatic return to lead-bronze, has long been recognised as a mainly lowland English phenomenon. Its distinctive products comprised Wilburton swords, tongue chapes, a range of pegged spearheads, including such 'fancy' forms as those with lozenge-sectioned hollow heads, lunate openings, stepped blades and fillet-defined midribs; long tubular ferrules, 'late' palstaves and slender socketed axes, notably the indented form (Burgess, 1968b). The scarcity of Wilburton material in other regions has led to a search for local alternatives. In northern England the Wallington tradition clung to the old tin-bronze and old-fashioned products, such as 'transitional' palstaves, notched-butt dirks and rapiers and looped spearheads (Burgess, 1968a). Its novelties reflect archaic traditions, for example protected-opening spearheads, and long-socketed spearheads, often with a single side-loop. Socketed axes had the square/rectangular bodies and flat or multiple moulding collars, typical of early forms, similar to, but less slender than the Wilburton equivalents. Most of these Wallington types are even more numerous in Ireland, but they are less common in Scotland and rare in Wales. With both Wilburton and Wallington metalwork poorly represented in Wales, the difficulty is to determine what metal products were in use in the region at this time.

This problem is made much easier by reducing the Wilburton/Wallington phase to a single century, and back-dating the beginning of the Ewart Park phase to _c_. 900 B.C. Its tenuous Continental connections are with Ha B1, so that the 10th century B.C. will suffice, and there is no longer justification for the traditional two century Wilburton phase. There is also a possibility that Wilburton metalwork was much more widely used in Wales and other Highland Zone areas than surviving finds indicate. Curiously the only Wilburton moulds known come not from the Wilburton heart-lands of lowland England but from Fimber, Yorkshire (Burgess, 1968a) and Rathgall, Ireland (Raftery, 1976). There is some evidence that Welsh craftsmen in the subsequent Ewart Park phase used a lot of melted-down Wilburton scrap (Northover, _pers. comm._). Thus the few Welsh Wilburton finds may be survivors of much larger quantities. They are concentrated in the upper Severn basin (Burgess, 1968b, Fig. 8), where there are two Montgomeryshire hoards, from Guilsfield (Savory, 1965) and Buttington Hall, and a few stray finds.

Wallington material is even rarer in Wales, though an unknown proportion of local 'transitional' palstaves, looped spearheads and notched-butt rapiers should belong to this phase. The socketed axes in the Guilsfield hoard are specifically not Wilburton types, but are of the broader Wallington form (Burgess, 1968a, Fig. 11). Socketed axes of Gwithian type (Burgess, 1976b), forerunners of South Welsh axes, should also belong here, but betray influence from south-west England.

In some respects the Wilburton-Wallington phase was a transitional one between Middle and Late Bronze Age metalworking. By _c_. 900 B.C., the new metallurgical fashions—lead-bronze, leaf-shaped swords, pegged spear-heads and socketed axes—had spread throughout Britain and Ireland, and clearly defined regional metalworking emerged once again. The break between Wilburton-Wallington and Ewart Park traditions has been over-exaggerated (cf. Burgess, 1968a, b). The melting down of transitional material for scrap has undoubtedly created a misleading impression of dichotomy and industrial revolution, but, increasingly, the roots of Ewart Park metalworking can be found in earlier traditions. Vital to these arguments is a group of late Wilburton hoards, containing all the usual Wilburton types but also material more appropriate to the subsequent Ewart Park phase: Isleham and Fulbourn Common, Cambridgeshire; Sturry, Kent; Blackmoor, Hampshire, Guilsford and Co. Roscommon (Burgess, 1968b, 37). The precocious interlopers include Ewart Park swords, socketed gouges, baggy socketed axes, and plate scrap. The Isleham hoard has a much longer list including, significantly, cauldron fragments. The traditions represented by these finds are followed in their respective areas by industries retaining echoes of earlier traditions. Little seems to have escaped melting down, but here we can include hoards such as those from Isle of Harty, Kent (Smith, 1955), and Marston St. Lawrence, Northamptonshire (Hawkes and Smith, 1955). In Wales Wilburton activity in the upper Severn basin was followed in the Ewart Park phase by the Broadward tradition, which retained many Wilburton elements (Burgess, Coombs and Davies, 1972). It is characterised by a profusion of spearheads, including fillet-defined and lunate opening types carried over from Wilburton, but barbed spearheads were its most eye-catching innovation. Ferrules, now with waisted base, and short tongue chapes also had a Wilburton background, but Ewart Park swords were now the standard pattern. In south-east Wales the Llantwit tradition looked more to south-west England. The Llantwit industry stands out because of the shoddiness of its main product, the 'South Welsh' ribbed socketed axe, and its use of high lead-bronze. South Welsh axes were shorter versions of the earlier Gwithian type. Faceted socketed axes and 'late palstaves' with ribbed decoration were also shared with south-west England. Weapons are not plentiful, but include Ewart Park swords and plain, pegged spearheads. The absence of South Welsh axe moulds in south Wales must be a curious quirk of fate, for they are scattered throughout southern England, not only in the south-west, where such axes are quite common, but as far west as the Middle Thames Valley, where the axes themselves are practically unknown. There are fewer South Welsh axes in south-west England than hitherto claimed, for many axes thus described are now seen to be Gwithian axes of earlier periods. The Stogursey hoard, Somerset (McNeil, 1973), remains the largest single find of South Welsh axes, but whether shipped from south Wales or made locally the evidence does not yet reveal. Similarly, a doubt hangs over the origin of the many South Welsh axes in French carp's tongue hoards (Burgess, 1968b. 21, n. 83), in view of the widespread mould distribution.

South-west Wales, as in the Middle Bronze Age, has produced puzzlingly few finds. In the north the division between west and east has already been noticed. The north-west, characterised by a high proportion of 'late' palstaves, shows a variety of external influences. Irish influence is particulary strong on Anglesey and the adjacent mainland, and is seen both in finds of gold ornaments (Lynch, 1970) and miscellaneous bronzes (Burgess, 1968b, 42). In other respects the area looked to England, but Continental contacts were few. A spectacular exception was the Late Urnfield bucket from Nannau, Merioneth, one of a small number of imported buckets which inspired the whole series of Irish-British buckets (Hawkes and Smith, 1957). That this fashion was soon adopted in Wales is shown by a lost bucket of the insular variety found near Chester (Hawkes and Smith, 1957, 148, n. 1; Briggs, 1977). Sheet metal cauldrons may well have been inspired by the buckets with which they share many technical features. British associations, notably in the Isleham hoard, make it clear that their development and chronology can no longer be tied to Greek and oriental cauldrons (pace Hawkes and Smith, 1957), which they must comfortably pre-date. The only Welsh cauldrons, part of the Llynfawr find, were old when deposited, and were probably manufactured in the Ewart Park phase.

North-east Wales has more socketed axes and far fewer 'late' palstaves than the north-west. These socketed axes were drawn from various sources for they include Irish bag-shaped, Yorkshire and South Welsh ribbed, and south-eastern forms (Burgess, 1968b, 41-2). The metalwork of this area is otherwise unremarkable, apart from the hoard of horse gear from Parc-y-meirch, Denbighshire (Sheppard, 1941), which includes Baltic rattle pendants of Montelius V.

Interest in hill-fort building seems to have died away in the Isleham period, although hill-top occupation continued and may even have been on the increase. Existing ramparts appear to have crumbled or were even pulled down (cf. Grimthorpe, Yorkshire: Stead, 1968). Many apparently open hill-top settlements are known from this period, as far apart as South Cadbury,

Somerset (Alcock, 1972b, 114-18) and Ivinghoe Beacon, Buckinghamshire (Cotton and Frere, 1968). At numerous sites the relationship between ramparts and settlements of the Ewart Park phase is unknown: Mam Tor, Derbyshire, Traprain Law, E. Lothian and Rathgall, Co. Wicklow, to name but a few. The growing list of hill-fort radiocarbon dates suggests there were two main periods of rampart building: an initial one with dates of c. 1050-800 b.c., should correspond to the Penard period, while a later group, between c. 600-450 b.c., suggests the Llynfawr period. In the intervening Isleham period hill-fort defences were not constructed, save for palisades around sites such as Fenton Hill, Northumberland (c. 690 b.c.), Navan, Co. Armagh (c. 680 b.c.) and Moel y Gaer, Flintshire (c. 620 b.c.).. Undated ramparts at such sites are likely to be of earlier or later periods.

In Wales, occupation within delapidated ramparts may be postulated at sites such as Dinorben, Ffridd Faldwyn and the Breiddin, where numerous bronzes attest intense activity in the Isleham period (Musson, 1976). There may also have been many hill-top palisade settlements like that around Moel y Gaer. Few sites have produced building plans, but rectangular structures are known in the Upper Severn area, while Moel y Gaer has post-ring buildings in this phase. Practically nothing is known of settlement outside the hill-forts.

Finds from settlements throughout southern Britain suggest burial was a casual affair in the later Bronze Age. Most occupation sites produce a scatter of human bones, suggesting deposition in all sorts of holes and corners. A handful of more formal burials, accompanied by metal grave goods, has some notable Welsh representatives (Burgess, 1976,e), such as the double inhumation with socketed axe and armlets from the Ogo, Montgomeryshire.

The work of Barrett (e.g. 1975) has established the broad outlines of pottery development in this period, but Wales has produced little evidence. The hook-rim and simple shouldered jars of the Penard period lasted into the Wilburton phase, but in the Ewart Park phase, shoulder fingertipping and neck cordons began to spread, and increasing numbers of fine bowls were used. It is to be hoped that further work on Welsh hill-forts will make clearer the local response to these developments.

The Llynfawr Period: c. 600-475 b.c./700-600 B.C.

Around 700 B.C. central European society, increasingly irradiated by Mediterranean and oriental influences, entered its Hallstatt cultural stage. Once more Europe was plunged into an unsettled period of dramatic change. For centuries iron had appeared sporadically north of the Alps, but now it became commonplace in Hallstatt society. In the first Hallstatt phase, confusingly termed Hallstatt C, bronze was still widely used, especially in sword production, but by the opening of the second, or Hallstatt D, phase, iron had largely taken over from bronze. The economic and social effects of the change-over have hardly begun to be appreciated, but were probably more serious for the suppliers than the customers, for the new iron was not superior to well-finished bronze, merely more plentiful.

The widespread distribution of Hallstatt material, and especially warrior equipment, from eastern Europe to Ireland has been taken as a sign that Hallstatt society was basically aggressive and expansionist. Hallstatt raiders have been seen in practically every corner of west and north Europe, eventually harrying the shores of Britain and Ireland (e.g. Hawkes, 1959; Burgess, 1974). Traditionally, this raiding in the 7th century gave way to Hallstatt settlement in the 6th.

In fact there is little archaeological evidence for the traditional picture of Hallstatt expansion. Indeed, the limits of the Hallstatt province were if anything smaller than the maximum extent of Urnfield traditions. Rather than Hallstatt expansion it seems logical to suppose that those territories which previously had been Urnfield, and before that Tumulus and Únětice, now became Hallstatt. Thus Eastern France lay within the Hallstatt sphere, whereas western France, Britain and Ireland lay outside it.

A more plausible explanation for the widespread distribution of Hallstatt equipment from Nice to Norway and from the Danube to Donegal, is that it went hand in hand with the spread of iron-working. The principal Hallstatt C bronze types to reach Britain were Gündlingen swords, winged chapes, razors and a variety of harness and vehicle fittings. It is unlikely that all this material could have been disseminated by trade or normal culture contact, for many of the Gündlingen swords from Britain and Ireland (Class C: Cowen, 1967) were made locally, and by alien hands (Burgess, 1979). This suggests that metalworkers were on the move in Europe, Hallstatt in the broadest sense, spreading knowledge of iron and Hallstatt bronze fashions. It is increasingly evident that iron-working spread very rapidly in the 7th century B.C. It involved a very different technology from the non-ferrous metals hitherto worked, and whether it could have spread so quickly without being carried by experts is debatable.

The arrival of iron had as shattering an effect on Irish-British as on Continental bronze-working. Hoards of the Llynfawr period are few but consistent, combining Hallstatt C bronzes with a restricted range of indigenous novelties. The mass of Ewart Park types are absent, apart from basic tool forms such as gouges and chisels, so it is clear that established industries simply collapsed. The vast numbers of hoards of Ewart Park material, especially the carp's tongue hoards of the south-east, must therefore be reinterpreted as dumped surplus scrap, never reclaimed because the rapidly shrinking role of bronze glutted the market. Thereafter, hoard deposition must soon have ceased as the new metallurgy gathered strength, and it became clear that iron rapidly became useless when buried.

It is possible that for a time both alien and indigenous craftsmen were operating in the new industrial set-up. The former, perhaps from northern France and the Low Countries, worked in iron and produced Hallstatt bronze novelties which were often modified to suit local tastes e.g. Gündlingen C swords. The indigenous craftsmen still found good markets for basic tools such as socketed gouges, tanged chisels and socketed sickles, but developed new bronze types to meet changing circumstances. Diagnostic were large socketed axes of Sompting type (Burgess, 1969), plain, ribbed or, most characteristically, bearing rib and pellet or roundel decoration. These were strongly influenced by the peculiar Armorican socketed axes (Dunning, 1959) which were being shipped into Britain in great numbers at this time. Also new was a range of axes with embellished body facets, the Blandford type (Evans, 1881,

Fig. 146; Moore and Rowlands, 1972, 30-1), which may also bear rib and pellet/roundel ornament, and which mix faceted, Sompting and Armorican influence to varying degrees. The Cardiff type of chisel, with short socket and wide, triangular blade, was also based on north-west French prototypes. Irish-British craftsmen also made an attempt to up-date the traditional Ewart Park sword, borrowing fashionable Gündlingen features such as the pommel piece, swept-back ricasso notches and broad, blunt tip. The resulting Thames-type swords (Cowen, 1967) are still unmistakeably Ewart Park weapons. Continental scholars have recently attempted to change the role of these Thames swords, claiming they are ancestral to, not derived from, Gündlingen swords (Schauer, 1972), but their absence in hoards of the Ewart Park phase, as in Late Urnfield and Carp's Tongue hoards on the Continent (pace Schauer) confirms the traditional view (Burgess, 1979).

The rapid spread of iron is shown by the frequency with which it occurs in contexts of the Llynfawr period, as far apart as Grimes Graves in Norfolk, Staple Howe in Yorkshire, Traprain Law, East Lothian, and Rathtinaun crannog 61 in Co. Sligo (Champion, 1971, Burgess, 1979). Wales, too, has produced important evidence for early iron-working, not to mention two of the few metal hoards known from this period. The Cardiff hoard (Nash-Williams, 1933) and the Llynfawr deposit (Crawford and Wheeler, 1920-1) both contain a mixture of exotic Hallstatt C types and indigenous novelties. Both too, illustrate the persistence of old tool forms, socketed gouges at Llynfawr, socketed sickles at both sites. Both contain rib and pellet-decorated Sompting axes, while Cardiff has several 'chisels' of the new triangular bladed form. On the Hallstatt side both have razors and horse gear, but Llynfawr has in addition iron objects. Its spearhead and sword are Hallstatt forms, the latter the only Hallstatt C iron sword from Britain, but a socketed iron sickle is clearly based on those found here and in the Cardiff hoard, and must have been made not too far away.

The metal used for the Cardiff and Llynfawr pieces shows a dramatic reduction in lead content and increase in tin compared with Ewart Park norms, and is in keeping with trends throughout Britain. Bronzes of specific Llynfawr types, both exotic and indigenous, have an overwhelmingly easterly distribution in Wales. No certain 'Thames' sword is known, but a scatter of Sompting axes have been found in eastern districts. The only Armorican axes come from the south-east, from Newton Nottage, Glamorgan, and in the Tintern hoard, Monmouthshire (Burgess, 1962b), and the single Gündlingen sword also has an eastern provenance, from Brogyntyn, Shropshire. These distributions point to influences spreading in from southern and eastern England, and it must be assumed that in other parts of Wales there was no time for this twilight bronze-working to develop before iron arrived and surviving Ewart Park workshops collapsed.

These economic and industrial changes must have helped to make the Llynfawr period an unsettled one. Evidence for further climatic deterioration comes from a new wave of timber trackway construction and the flooding of waterside settlements in many parts of the country. A series of tracks in the Somerset Levels and eastern England were laid down in the period c. 680-460 b.c. (Coles et al. 1975, 1976), while settlements from Ballinderry,

Co. Offaly (Hencken, 1942) to Runnymede Bridge, Surrey (Longley, 1976) were abandoned in the face of rising water levels. It is not surprising that this period saw a new burst of hill-fort construction in every corner of Britain. Old hill-forts with long decayed ramparts had their defences re-built, such as Dinorben, Ffridd Faldwyn and the Breiddin, while hill-top settlements which had hitherto managed without defences, such as South Cadbury, or had made do with palisades, as at Moel y Gaer, now received ramparts. But even more remarkable was the number of new sites: perhaps a majority of the hill-forts of Britain were built at this time. The form of the new ramparts depended very much on local circumstances. Some used the old timber box rampart, such as Moel y Gaer, but stone-faced timber-laced ramparts were much favoured, for example in Wales at Dinorben and Ffridd Faldwyn, and when these caught fire they could be vitrified by the resulting heat. Stone-faced walls of varying complexity were naturally popular in many regions, in Wales can be seen at the Breiddin and Moel Hiraddug, Flintshire, while simple dump ramparts were also popular, as built at Croft Ambrey, Herefordshire (Savory, 1976). The Marcher hill-forts of this period, such as Croft Ambrey and Credenhill, display remarkable planned interiors, full of streets and 4-poster buildings. If the latter were houses, and this is by no means certain, they can be used to arrive at a rough population estimate (Stanford, 1972b). The suggestion is that Medieval population levels had already been attained and perhaps even surpassed, with 15,000-20,000 people in Herefordshire alone, and perhaps more than 100,000 in the whole of Wales. Hill-forts in other parts of the region have rather different interior plans. Moel y Gaer combines 4-poster and round buildings (Guilbert, 1976), though its town planning is hardly less remarkable, but over much of the region only round buildings are known, though 4-posters may well await discovery. Other forms of later Bronze Age settlement have also to be located. Stanford's claim that the bulk of the population of the central Marches lived in hill-forts may or may not be true, but certainly in many parts of Britain small farmsteads or hamlets, protected at most by palisades or ditch and bank perimeter, were as much in evidence as hill-forts.

Burial in this period continued to be a casual affair, with bodies placed in holes and pits, and with stray bones a frequent find in settlements and hill-forts. If provenances are any guide then water religion was still a powerful force, for the great majority of Gündlingen and Thames swords have come from watery places. The Llynfawr deposit itself seems to have been a votive deposit thrown into a lake (Savory, 1975).

Settlement and land-use

While Bronze Age settlements are nowhere common in Britain a consistent pattern is now beginning to emerge. In the earlier Bronze Age the bulk of the population clearly lived in small farmsteads and hamlets sprinkled across an organised landscape. Some were protected by stockades, as at Gwithian, Cornwall (Megaw, 1976), Swarkeston, Derbyshire (Greenfield, 1960) and Kilellan Farm, Islay (Burgess, 1976d), while others took the form of ring ditch enclosures, such as those at Fengate (Pryor, 1974) and Playden, Sussex (Cheney, 1935). Rectilinear enclosures were also constructed, for example at Playden, and Belle Tout, Sussex (Bradley, 1970). At the other end of the period very similar stockaded and ditch and bank settlements, including such famous sites as Staple Howe in Yorkshire and West Harling, Norfolk, still housed large segments of the population, although from the Penard period onwards the settlement pattern was increasingly complicated by the growth of hill-forts, with their larger population units. In order to feed Bronze Age populations of the size envisaged, extensive field systems had to be laid out, and land divisions constructed, and these are now known from Co. Mayo to the Wessex chalk, and from the East Midlands to Dartmoor.

But Wales has yet to produce such settlements or field systems, and their location must be a primary aim of future research. Bronze Age settlements in Wales are still of the unsatisfactory patterns that have characterised much of Britain until recently: coastal sand sites, such as Newborough Warren, Anglesey (Lynch, 1970, 87-8) and Merthyr Mawr, Glamorgan, which produce great quantities of finds but little else; and cave sites, plentiful in north and south Wales, which may often have housed squatters and fugitives, but which can be of only peripheral relevance to the mainstream of settlement. Cave deposits in any case may more often reflect specialised activities, especially burial and ritual, than habitation.

A more fruitful line of enquiry at present would seem to lie in the occupation sites found underneath burial mounds, which are very common. The tendency has been to regard the juxtaposition of settlement and burial site as chance. The time has perhaps come to entertain more seriously the possibility that abodes for the living were frequently converted into abodes for the dead. Wales has produced some notable examples of such superimposition, such as those excavated by Savory in Glamorgan. The Sant-y-Nyll site (Savory, 1959-60) with its curvilinear structures deserves special mention. Another site, Brenig 51, Denbighshire (Lynch, 1974, 2-4) indicates that burial mounds may have been raised over habitation sites even more frequently than the available evidence admits. Most of the settlement layer at this site had been cleared away to make way for the cairn, and only a small patch had been accidentally left. How often were habitation sites removed completely in preparing a site for a burial mound?

Epilogue

The traditional 'Bronze Age' was therefore over by 600 B.C., and iron commonplace throughout Britain and Ireland. No mention has been made of the Celts, who inhabited these islands when the Romans arrived, and made up our Iron Age populations. It would be cowardly to avoid the thorny problem of Celtic origins on such a Welsh occasion. This account of the Bronze Age has been lukewarm towards traditional invasions, particularly those of Beaker Folk at the beginning of the period, and Hallstatt intruders at its end. Whence, then, came the Celts, particularly if we reject the Hallstatt settlers who for so long have been favoured 'first Celts'?

Prehistoric migrations are notoriously difficult to demonstrate with any degree of certainty, for too many historic examples have left little or no trace in the archaeological record. It is unlikely that any prehistoric invasion of these islands after the Neolithic settlement was on a scale which could suddenly produce an overwhelmingly Celtic population. The best evidence for a sizeable influx comes in the Penard period, but even that seems confined to Britain south of the Tees.

Instead of sporadic major incursions from the Continent we would do better to think of a constant trickle across the Channel and the North Sea, bringing European innovations and European genes. The answer to the problem of Celtic origins must lie in fact in something like Hawkes principle of 'cumulative Celticity' (Hawkes, 1972). The Celts, whether on the Continent or in Britain, did not suddenly come from anywhere, but were people who had become Celtic by accretion in the process of time. Equally, it is senseless to ask when the Celts first appeared, for the Celts known to classical writers were the lineal descendants of generations stretching back to early Neolithic farmers. When they arrived in Britain and Ireland around 4000 B.C., then these offshore islands were brought into the realms of 'cumulative Celticity'. Thereafter, incessant cross-Channel contacts ensured that out later prehistoric populations remained very closely linked to those of Europe. The cousinly cross-Channel connections described by Caesar had in fact existed not only throughout the Iron Age, but stretched back through the Bronze Age, and may even have been closer then.

ACKNOWLEDGEMENTS

I am grateful to Linda Southern and Diane Sevenoaks who patiently and skilfully typed out the various drafts of my text. Michael Lyall drew the illustrations for Figs. 7.1-7.5. John Barrett, Sheena Crawford, Frances Lynch and Adam Welfare kindly read various versions of the manuscript and made many helpful suggestions. I am also indebted to Professor Leslie Alcock for his comments on the problems of Castell Odo. Much of the metallurgical comment in this paper stems from the unpublished analyses of Dr. Peter Northover for the Board of Celtic Studies metal analysis project, and I am grateful to Dr. Northover for lengthy discussion of his results and their significance. I must also acknowledge the underlying influence of Dr. Hubert Savory, whose contribution to Welsh Bronze Age studies, and to my own work, has been immense.

BIBLIOGRAPHY

Alcock, L., 1972a. Review of Savory, H. N., 1971. Excavations at Dinorben, 1965-9, in Antiquity, XLCI, 330-1.

Alcock, L., 1972b. By South Cadbury is that Camelot. London.

Atkinson, R. J. C., 1972. 'Burial and population in the British Bronze Age', in Lynch and Burgess, 1972, 107-16.

Barrett, J., 1975. 'The later pottery: 2) Types, affinities, chronology and significance', in Bradly, R., and Ellison, A., Rams Hill = British Archaeol. Rep., 19, 101-10.

Barrett, J., 1976. 'Deverel-Rimbury: problems of chronology and interpretation, in Burgess and Miket 1976, 289-307.

Barrett, J., et al. 1976. 'Second Millennium B.C. banks in the Black Moss of Achnacree: some problems of prehistoric land use', in Burgess and Miket, 1976, 283-7.

Boon, G. C., and Lewis, J. M. (eds.), 1976. Welsh Antiquity. Cardiff.

Bradley, R., 1970. 'The excavation of a Beaker settlement at Belle Tout, East Sussex, England', Proc. Prehist. Soc., XXXVI, 312-79.

Bradley, R., 1978. The Prehistoric Settlement of Britain. London.

Bradley, R., and Ellison, A., 1975. Rams Hill: A Bronze Age Defended Enclosure and its Landscape = British Archaeol. Rep., 33.

Briard, J., et al., 1956. 'Dépôts de l'Age du Bronze De Bretagne', Travaux Laboratoire d'Anthrop. Préhist. Rennes.

Briggs, S., 1977. ' "A Roman camp kettle of copper" ', Antiq. J., LVII, 90-1.

Britton, D., 1960. 'The Isleham hoard, Cambridgeshire', Antiquity, XXXIV, 279-82.

Britton, D., 1963. 'Traditions of metal-working in the later Neolithic and Early Bronze Age of Britain: Part 1', Proc. Prehist. Soc., XXIX, 258-325.

Burgess, C. B., 1962a. 'Two grooved, ogival daggers of the Early Bronze Age from south Wales', Bull. Board Celtic Stud., XX, 75-94 esp. 85-6.

Burgess, C. B., 1962b. 'A socketed axe from central Monmouthshire and its significance for the Bronze Age in Wales and the Marches', Monmouth Antiq., I, pt. 2, 17-27.

Burgess, C. B., 1964. 'A palstave from Chepstow, with some observations on the earliest palstaves of the British Isles', Monmouth Antiq. I, pt. IV, 117-24.

Burgess, C. B., 1968a. Bronze Age Metalwork in Northern England c. 1000-700 B.C. Newcastle upon Tyne.

Burgess, C. B., 1968b. 'The later Bronze Age in the British Isles and north western France', Archaeol. J., CXXV, 1-45.

Burgess, C. B., 1969. 'Some decorated socketed axes in Canon Greenwell's collection', Yorkshire Archaeol. J., 42, 267-272.

Burgess, C. B., 1974. 'The Bronze Age', in C. Renfrew (ed.), British Prehistory—A New Outline. London. 165-232, 291-329.

Burgess, C. B., 1976a. 'Britain and Ireland in the Third and Second Millennia B.C.: a preface', in Burgess and Miket, 1976, i-iii.

Burgess, C. B., 1976b. 'The Gwithian mould and the forerunners of South Welsh axes', in Burgess and Miket, 1976, 69-75.

Burgess, C. B., 1976c. 'Meldon Bridge: a Neolithic defended promontory complex near Peebles', in Burgess and Miket, 1976, 151-79.

Burgess, C. B., 1976d. 'An early Bronze Age settlement at Kilellan Farm, Islay, Argyll', in Burgess and Miket, 1976, 181-207.

Burgess, C. B., 1976e. 'Burials with metalwork of the later Bronze Age in Wales and beyond', in Boon and Lewis, 1976, 81-104.

Burgess, C. B., 1979. 'A find from Boyton, Suffolk, and the end of the Bronze Age in Britain and Ireland', in Burgess, C., and Coombs, D. (eds.), 1979, Bronze Age Hoards: Some Finds Old and New = British Archaeol. Reports, 67, 269-83.

Burgess, C. B., Coombs, D., and Davies, D. G., 1972. 'The Broadward Complex and barbed spearheads', in Lynch and Burgess, 1972, 211-83

Burgess, C. B., and Cowen, J. D., 1972. 'The Ebnal hoard and Early Bronze Age metal-working traditions', in Lynch and Burgess (eds.), 1972, 167-81.

Burgess, C. B., and Miket, R., 1976 (eds.). Settlement and Economy in the Third and Second Millennia B.C. = British Archaeol. Reports, 33.

Burgess, C. B., and Shennan, S., 1976. 'The Beaker phenomenon: some suggestions', in Burgess and Miket, 1976, 309-31.

Burleigh, R., et al. 1972. 'Relative and absolute dating of four late Neolithic enclosures: an exercise in the interpretation of radiocarbon determinations', Proc. Prehist. Soc., XXXVIII, 389-407.

Burstow, G. P., and Holleyman, G. A., 1957. 'Late Bronze Age settlement on Itford Hill, Sussex', Proc. Prehist Soc. , XXIII, 167-212.

Butler, J. J., 1959. 'Vergeten schatvondsten uit de Bronstijd', in Honderd Eeuwen Nederland (Antiquity and Survival, II, 125-142).

Butler, J. J., 1963. 'Bronze Age connections across the North Sea', = Palaeohistoria, IX.

Case, H. J., 1977. 'The Beaker culture in Britain and Ireland', in Mercer, R. (ed.), Beakers in Britain and Europe: Four Studies = British Archaeol. Rep., S 26, 71-101.

Champion, T., 1971. 'The end of the Irish Bronze Age', North Munster Antiq. J., XIV, 17-24.

Cheney, H. J., 1935. 'An Aeneolithic occupation site at Playden, near Rye,' Antiq. J., XV, 152-64.

Chitty, L. F., 1925. 'Three bronze implements from the Edgebold Brickyard, Meole Brace', Antiq. J., V, 409-14.

Clark, J. G. D., 1966. 'The invasion hypothesis in British archaeology', Antiquity, XL, 172-89.

Clarke, D., 1970. Beaker Pottery of Great Britain and Ireland. Cambridge University Press, 2 vols.

Coles, J. M., 1962. 'European Bronze Age shields', Proc. Prehist. Soc., XXVIII, 159-90.

Coles, J. M., et al., 1975. Somerset Levels Papers, 1.

Coles, J. M., et al., 1976. Somerset Levels Papers, 2.

Coombs. D., 1975. 'The Dover harbour bronze find—a Bronze Age wreck?', Archaeol. Atlantica, 1, 193-5.

Cotton, M. A., and Frere, S. S., 1968. 'Ivinghoe Beacon excavations', Records of Bucks, XVIII, 187-260.

Cowen, J. D., 1951. 'The earliest bronze swords in Britain and their origins on the Continent of Europe', Proc. Prehist. Soc., XVII, 195-213.

Cowen, J. D., 1967. 'The Hallstatt sword of bronze: on the Continent and in Britain', Proc. Prehist. Soc., XXXIII, 377-454.

Crawford, O. G. S., and Wheeler, R. E. M., 1920-1. 'The Llynfawr and other hoards of the Bronze Age', Archaeologia, 71, 133-40.

Cunliffe, B., 1968. "Excavations at Eldon's Seat, Encombe, Dorset', Proc. Prehist. Soc., XXXIV, 191-237.

Daniel, G. E., 1950. The Prehistoric Chamber Tombs of England and Wales. Cambridge.

Davies, E., 1949. The Prehistoric and Roman Remains of Flintshire. Cardiff.

Dunning, G. C., 1959. 'The distribution of socketed axes of Breton type', Ulster J. Archaeol., XXII, 53-5.

Edeine, B., 1961. 'De quelques objets de l'age du Bronze, inédits ou peu connus, trouvés dans le departement du Calvados', Bull. Soc. Prehist. Française, LVIII, 676-92.

Ehrenberg, M. R., 1977. Bronze Age spearheads from Berkshire, Buckinghamshire and Oxfordshire = British Archaeol. Rep., 34.

Eogan, G., 1967. 'The associated finds of gold bar torcs', J. Roy. Soc. Antiq. Ire., XCVII, part 2, 129-75.

Eogan, G., 1976. "Beaker material from Knowth, Co. Meath: preliminary note', in Burgess and Miket, 1976, 251-66.

Evans, J., 1881. Ancient Bronze Implements. London

Fleming, A., 1976. 'Dartmoor reaves', Current Archaeol., V, 250-2.

Fleming, A., Collis, J., and Jones, R. L., 1973. 'A late prehistoric reave system near Cholwich Town, Dartmoor,' Proc. Devon Archaeol. Soc., 31, 1-21.

Forde-Johnston, J., 1964. 'A hoard of flat axes from Moel Arthur, Flintshire,' Flints. Hist. Soc. Pubns., 21, 99-100.

Fox, A., 1964. South West England. London.

Fox, A., and Fox, C., 1941. 'The Golden Mile Barrow, in Colwinston parish, Glamorgan', Archaeol. Cambrensis, XCVI, 185-92.

Fox, C., 1926. 'A Bronze Age barrow on Kilpaison Burrows, Rhoscrowther, Pemb.', Archaeol. Cambrensis, LXXXI, 1-32.

Fox, C., 1941. 'Stake circles in turf barrows, a record of excavation in Glamorgan, 1939-40', Antiq. J., XXI, 97-127.

Fox, C., 1943. 'A Bronze Age Barrow (Sutton 268') in Llandow Parish Glamorganshire', Archaeol. LXXXIX, 89-125.

Fox, C., 1959. Life and Death in the Bronze Age. London.

Gerloff, S., 1975. The Early Bronze Age Daggers in Great Britain and a Reconsideration of the Wessex Culture = Prähistorische Bronzefunde, VI 2. Munich.

Gimbutas, M., 1965. Bronze Age Cultures in Central and Eastern Europe. The Hague.

Green, H. S., 1974. 'Early Bronze Age burial, territory and population in Milton Keynes, Buckinghamshire, and the Great Ouse Valley', Archaeol. J., CXXXI, 75-139.

Greenfield, E., 1960. 'The excavation of Barrow 4 at Swarkeston, Derbyshire', Derbys. Archaeol. J., LXXX, 1-48.

Griffiths, W. E., 1960. 'The excavation of stone circles near Penmaenmawr, north Wales', Proc. Prehist. Soc., XXVI, 303-39.

Grimes, W. F., 1938. 'A barrow on Breach Farm, Glamorgan', Proc. Prehist. Soc., IV, 107-21.

Grimes, W. F., 1951. The Prehistory of Wales. (2 ed.), Cardiff, Nat. Mus. Wales.

Grimes, W. F., 1963. 'The stone circles and related monuments of Wales', in Foster, I. Ll., and Alcock, L., 1963, Culture and Environment. London. 93-152.

Guilbert, G., 1976. 'Planned hillfort interiors', Proc. Prehist. Soc., XLI, 203-21.

Guilbert, G., 1976. 'Moel y Gaer (Rhosesmor) 1972-1973: an area excavation in the interior', in Harding, D. (ed.), 1976, 303-17, 465-73.

Harding, D. (ed.), 1976. Hillforts: Later Prehistoric Earthworks in Britain and Ireland. London.

Hawkes, C. F. C., 1935. 'The pottery from the sites on Plumpton Plain', Proc. Prehist. Soc., I, 39-59.

Hawkes, C. F. C., 1959. 'The ABC of the British Iron Age', Antiquity, XXXIII, 170-82.

Hawkes, C. F. C., 1972. 'Europe and England: fact and fog', Helinium, XII, 105-16.

Hawkes, C. F. C., and Smith, M. A., 1955. 'Bronze Age hoards in the British Museum', Inv. Archaeol., G. B. 9-13, 2nd Set.

Hawkes, C. F. C., and Smith, M. A., 1957. 'On some buckets and cauldrons of the Bronze and Early Iron Ages', Antiq. J., XXXVII, 131-98.

Helbaek, H., 1957. 'Carbonized cereals', in Burstow and Holleyman, 1957, 206-9.

Hencken, H. O'N., 1942. 'Ballinderry Crannog No. 2', Proc. Roy. Irish Acad., 47C, 1-76.

Herity, M., 1974. Irish Passage Graves. Dublin.

Houlder, C. H., 1968. 'The henge monuments at Llandegai', Antiquity, XLII, 216-21.

Lanting, J. N., and Van Der Waals, J. D., 1972. 'British Beakers as seen from the Continent', Helinium, XII, 20- 46.

Laws, E., 1908. 'Bronze implements from the shores of Milford Haven', Archaeol. Cambrensis, 6s., VIII, 114-5 fig.

Longley, D., 1976. 'Excavations on the site of a Late Bronze Age settlement at Runnymede Bridge, Egham', London Archaeol., 3, 10-17.

Lynch, F., 1969. 'The Megalithic tombs of north Wales', in Powell, T. G. E., et al., 1969, Megalithic Enquiries in the West of Britain. Liverpool. 107-48.

Lynch, F., 1970. Prehistoric Anglesey. Llangefni, Anglesey Antiquarian Soc.

Lynch, F., 1971. 'Report on the re-excavation of two Bronze Age cairns in Anglesey: Bedd Branwen and Treiorwerth', Archaeol. Cambrensis, CXX, 11-83.

Lynch, F., 1972. 'Ring-cairns and related monuments in Wales', Scottish Archaeol. Forum., 4, 61-80.

Lynch, F., 1975. 'Brenig Valley excavations, 1974', Denbighs. Hist. Trans, 24, 1-25.

Lynch, F., 1976. 'Towards a chronology of megalithic tombs in Wales', in Boon and Lewis, 1976, 63-79.

Lynch, F., et al., 1974. 'Brenig Valley excavations 1973', Denbighs. Hist. Trans. 23, 1-56.

Lynch, F., and Burgess, C., 1972. Prehistoric Man in Wales and the West: Essays in Honour of Lily F. Chitty. Bath.

McInnes, I. J., 1968. 'The excavation of a Bronze Age cemetery at Catfoss, east Yorkshire', East Riding Archaeol., I, 1-10.

McKerrell, H., 1975. 'Correction procedures for C14 dates', in Watkins, T. (ed.), 1975. Radiocarbon, Calibration and Prehistory, Edinburgh. 47-100.

McNeil, R., 1973. 'A report on the Bronze Age hoard from Wick Park Stogursey, Somerset', Proc. Somerset. Arch. Nat. Hist. Soc. 117, 47-64.

Manby, T., 1974. Grooved Ware Sites in Yorkshire and the North of England = British Archaeol. Rep. 9

May, J., 1976. Prehistoric Lincolnshire. Lincoln.

Megaw, J. V. S., 1976. 'Gwithian, Cornwall: some notes on the evidence for Neolithic and Bronze Age settlement; in Miket and Burgess, 1976, 51-79.

Mercer, R. J., 1970. 'The excavation of a Bronze Age hut-circle settlement, Stannon Down, St. Breward, Cornwall, 1968', Cornish Archaeol., 9, 17-76.

Miles, H., 1972. 'Rhuddlan,' Current Archaeol., 3, 245-8.

Moore, C. N., and Rowlands, M., 1972. 'Bronze Age Metalwork in Salisbury Museum. Salisbury.

Morrison, A., 1968. 'Cinerary urns and pygmy vessels in south-west Scotland', Trans. Dumfriesshire Galloway Natur. Hist. Antiq. Soc., ser. 3, 80-140.

Mortimer, J., 1905. Forty Years Researches in British and Saxon Burial Mounds of East Yorkshire. London.

Musson, C .R., 1976. 'Excavations at the Breiddin 1969-1973', in Harding, D. (ed.), 1976, 293-302.

Nash-Williams, V. E., 1933. 'A Late Bronze Age hoard from Cardiff', Antiq. J., XIII, 299-300.

O'Kelly, M. J., 1973. 'Current excavations at Newgrange, Ireland', in Daniel, G., and Kjaerum, P. (eds.), 1973, Megalithic Graves and Ritual = Jutland Archaeol. Soc. Publ., 11, 137-46.

Pryor, F., 1974. 'Fengate', Current Archaeol., 4, 332-8.

Pryor, F., 1976. 'Fen-edge land management in the Bronze Age: an interim report on excavations at Fengate, Peterborough, 1971-75', in Burgess and Miket, 1976, 29-49.

Raftery, B., 1976. 'Rathgall and Irish hillfort problems', in Harding, D. (ed.), 1976, 339-57.

Renfrew, C., 1973. 'Monuments, mobilization and social organization in neolithic Wessex', in Renfrew, C. (ed.), 1973. The Explanation of Culture Change: Models in Prehistory. London, 539-57.

Renfrew, C., 1976. 'Quanterness, radiocarbon and the Orkney cairns', Antiquity, L, 194-204.

Savory, H. N., 1955. 'A corpus of Welsh Bronze Age pottery, I: Beakers', Bull. Board Celtic Stud., XVI, 215-41.

Savory, H. N., 1957. 'A corpus of Welsh Bronze Age pottery, II: Food Vessels', Bull. Board Celtic Stud., XVII, 196-233.

Savory, H. N., 1958. 'The Late Bronze Age in Wales', Archaeol. Cambrensis, CVIII, 3-63

Savory, H. N., 1959-60. 'The excavation of a Bronze Age cairn at Sant-y-Nyll, St. Brides-super-Ely (Glam.)', Trans. Cardiff Nat. Soc., LXXXIX, 9-25.

Savory, H. N., 1965. 'The Guilsfield hoard', Bull. Board Celtic Stud., XXI, part ii, 179-206.

Savory, H. N., 1971. 'A Welsh Bronze Age hill-fort', Antiquity, XLV, 251-61.

Savory, H. N., 1975. 'Some Welsh Late Bronze Age hoards—old and new', Archaeol. Atlantica, 1, 111-25.

Savory, H. N., 1976. 'Welsh hillforts: a reappraisal of recent research', in Harding, D. (ed.), 237-91.

Schauer, P., 1972. 'Zur Herkunft der Hallstatt-Schwerter', Archaeol. Korrespond., 2, 261-70.

Sheppard, T., 1941. 'The Parc-y-meirch hoard, St. George parish, Denbighshire', Archaeol. Cambrensis, XCVI, 1-10.

Smith, I. F., 1974. 'The Neolithic', in Renfrew, C. (ed.), British Prehistory: A New Outline. London, 100-36.

Smith, I. F., and Simpson, D. D. A., 1966. 'Excavation of a round barrow on Overton Hill, north Wiltshire, England', Proc. Prehist. Soc., XXXII, 122-55.

Smith, M. A., 1956 (ed.). "Grave groups and hoards of the British Bronze Age', Inv. Archaeol., G. B. 14-18, 3rd set.

Smith, M. A., 1959. 'Some Somerset hoards and their place in the Bronze Age of southern Britain', Proc. Prehist. Soc., XXV, 144-87.

Stanford, S. C., 1971. 'Invention, adoption and imposition—the evidence of the hill-forts', in Jesson, M., and Hill, D., 1971, The Iron Age and Its Hill-forts, Southampton. 41-52.

Stanford, S. C., 1972a. 'Welsh Border hill-forts', in Thomas, C. (ed.), 1972. The Iron Age in the Irish Sea Province. C.B.A., London. 25-30.

Stanford, S. C., 1972b. 'The function and population of hill-forts in the Central Marches', in Lynch, F., and Burgess, C. (eds.), 1972. Prehistoric Man in Wales and the West, 307-20. Bath.

Stead, I. M., 1968. 'An Iron Age hill-fort at Grimthorpe, Yorkshire, England', Proc. Prehist. Soc., XXIV, 148-90.

Streuver, S., and Houart, G. L., 1972. 'An analysis of the Hopewell interaction sphere', in Wilmsen, E. N. (ed.), Social Exchange and Interaction = Anthrop. Papers Mus. Anthrop. Univ. Michigan, 46, 47-79.

Thomas, N., 1972. 'An Early Bronze Age stone axe-mould from the Walleybourne below Longden Common, Shropshire', in Lynch and Burgess (eds.) 1972, 161-6.

Trump, B. A. V., 1962. 'The origin and development of British Middle Bronze Age rapiers', Proc. Prehist. Soc., XXVIII, 80-102.

Varley, W. J., 1968. 'Barmston and the Holderness crannogs', E. Riding Archaeol., 1. 12-26.

Wainwright, G. J., 1962. 'The excavation of an earthwork at Castell Bryn Gwyn, Llanidan Parish, Anglesey,' Archaeol. Cambrensis, CXI, 25-58.

Whittle, A. W. R., 1978. 'Resources and population in the British Neolithic', Antiquity, LII, 34-42.

Y. H. N., 1977. 'Important find on adventure holiday', Youth Hostelling News, No. 23,1.

CHAPTER EIGHT

THE EARLY IRON AGE IN WALES

H. N. Savory

Abstract

The distribution of human settlement in Wales during the period of 6-700 years which preceded the Roman conquest can be inferred, very roughly, from the distribution of known sites and finds. The sites now, however, consist almost entirely not of burials, as in the Neolithic and Bronze Age, but of settlements which can be recognized, even without excavation, by their defences, and are conventionally referred to, not always accurately, as "hillforts". The distribution of such sites suggests a shift of settlement away from the interior uplands of Wales towards the margins of the main river valleys and coastal plains. This can reasonably be related to the evidence for a deterioration of upland pastures during the later Bronze Age and Early Iron Age and for clearance of forest on valley slopes, particularly in the Welsh Marches, at this time.

Further analysis, however, complicates the picture, in that radiocarbon dates show that the earliest hillforts of supposed "Early Iron Age" type, in the northern Marches at least, were built long before the end of the Bronze Age, and it is clear, even without excavation, that there are regional differences between the range of hillfort types found in different parts of Wales, which are evidently related to the differing environments and histories of each area. At the same time, exploration of Early Iron Age settlements in Wales has been unevenly distributed, both as regards quality and scale, and recent large-scale excavation of large hillforts has been mostly confined to the Marches. For this reason, and because of the rarity of surviving traces of Early Iron Age field-systems outside of north-west Wales, it is not yet possible to confirm, from archaeological evidence, suspected regional differences in Wales between the relative importance of cereal and pastoral agriculture and of cattle- and sheep-rearing, although the importance of stock-raising in most parts of Wales appears to be hinted at by the layout of certain types of earthwork.

Nonetheless, study of the finds from excavations, combined with the results of field archaeology, at present seems to suggest an initial Early Iron Age phase (c. 600-300 B.C.) during which an extraneous, strongly Celtic culture; In which the population was largely concentrated in large, well-organized and fortified villages, impinged on the central and northern Marches and to some extent on the northern coasts of Wales, while a largely indigenous culture characterized by numerous scattered farmsteads, long survived further to the west in central and southern Wales—as it did, indeed, in parts of south-western England. In a later, Middle and Late La Tène phase, the resulting regional differences have been consolidated in the form of tribal entities—Cornovii and Degeangli in the North-east, Ordovices in the North-west, Demetae in the South-west and Silures in the South-east—which the Romans found in the late Ist century A.D. The relatively numerous finds of artistic metalwork in the insular La Tène tradition, especially in north Wales, suggest that, as in the Bronze Age, the exploitation of mineral resources was compensating, to a degree, in the local economy, for any deficiencies in its agricultural production.

Introduction

When the Romans, in A.D. 43, set about the task of incorporating Britain in their empire, they needed only a few years to bring under their control the whole of south-eastern Britain as far as the Severn and the Trent, but more than a quarter of a century to subjugate what is now Wales and the Marches. That is, no doubt, in large measure a reflection of the greater vulnerability of the south-eastern communities with their more highly developed political, social and economic systems, as well as the greater difficulties of the terrain in Wales. But it at least shows us that the Celtic tribes which then occupied Wales and the Marches cannot have been negligible, either in a demographic or a military sense. That, too, is what is implied by the historical and archaeological evidence for the size of the forces which were needed, not only to conquer the tribes of the Cambrian massif, but to hold them in subjection during a period of several decades after the time of Agricola (Jarrett, 1969). Until recently, however, what one could infer about the Early Iron Age inhabitants of Wales from the historical and archaeological facts of the Roman conquest did not appear to have a clear reflection in the Early Iron Age archaeology of Wales, because chance finds and dated burials and settlement sites were so few compared with those of the Neolithic and Bronze Age, and, indeed, with those of the Early Iron Age in some parts of England (Savory, 1976b).

During the last two decades, of course, the picture has changed completely. We now know that for several centuries before the Roman conquest the place of the Neolithic and Bronze Age burial sites as indicators of human settlement was being taken in Wales and the Marches by various types of walled or embanked enclosures which are conventionally, and often inaccurately, referred to as "hillforts" and which form one of the main concentrations of such sites in Britain (Fig. 8.1). It was for long supposed that in most parts of Wales such enclosures were of late date, in fact largely the result of resistance to the Roman invaders. One of the chief contributions of radiocarbon dating to Welsh archaeology has been to show that in the northern Marches, at least, as in parts of England and Scotland, hillfort-building has a long history, beginning not later than the outset of the Late Bronze Age and the first millennium B.C. In so doing it has complicated matters for the pre-historian, who can no longer

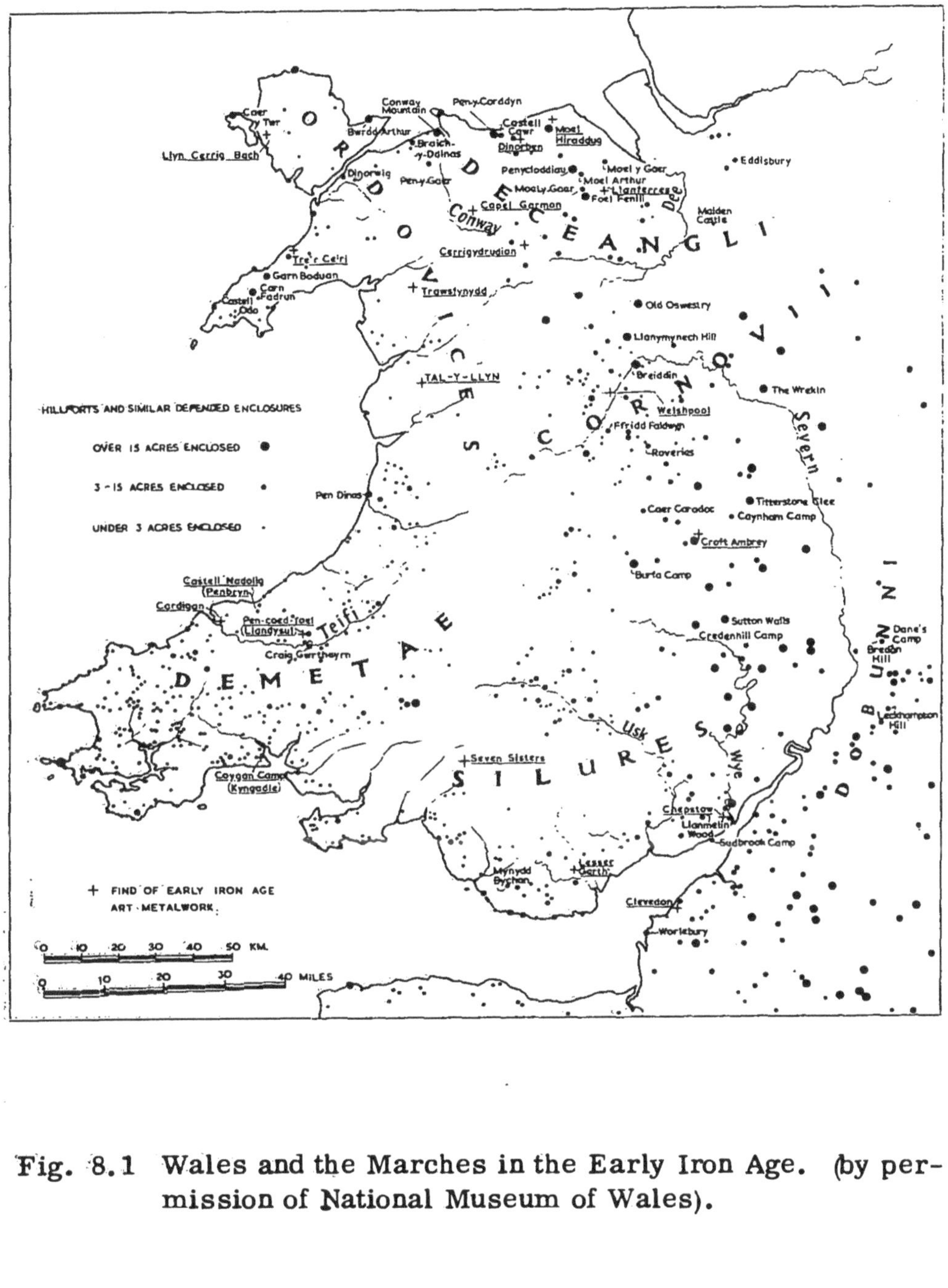

Fig. 8.1 Wales and the Marches in the Early Iron Age. (by permission of National Museum of Wales).

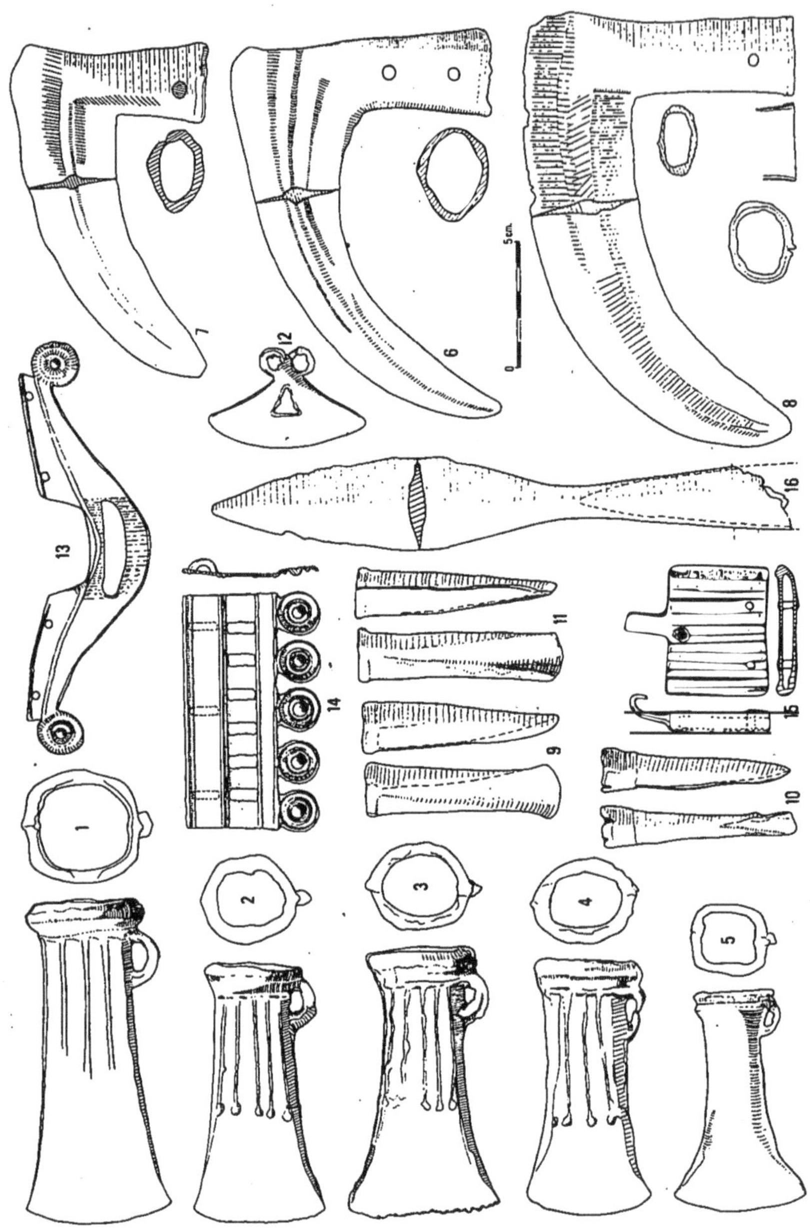

Fig. 8.2 Part of a hoard from Llyn Fawr, (Mid-Glam.). 8 and 16 of wrought iron (8 with a steel blade), rest of Bronze (by permission of National Museum of Wales).

look upon the hillfort, in the way that he has done in lowland Britain, at any rate, as an artefact serving to define the Early Iron Age archaeologically, along with the various ceramic traditions which have played such an important part in the establishment of chronological and regional groups in the same area, within the framework of the "ABC" system (Hawkes, 1931 and 1959). Evidence from pottery, too, hardly lends itself to the construction of a well-defined chronological horizon throughout Wales, dividing the Late Bronze Age from the Early Iron Age, since it is rare outside a limited number of excavated sites in the Marches and does not show any marked change at the time of the theoretical transition from Late Bronze Age to Early Iron Age, c. 600 B.C. Iron itself, which by definition should play an important part in marking that transition on excavated sites, is prevented from so doing by its poor resistance to corrosion, especially in the damp soils of Wales. Moreover, the familiar thesis that cheap and abundant iron tools would have greatly facilitated the clearance of forest for agriculture is somewhat undermined by the discovery that the use of timber in vast quantities for hillfort building, perhaps as a by-product of forest-clearance, is above all characteristic of the Late Bronze Age in the northern Marches (Savory, 1976b, 243-50). As far as we can judge at present, iron tools remained scarce and expensive in Wales as a whole during the early part of the Early Iron Age—in fact, probably until knowledge of the ore deposits which occur in various parts of Wales had been established locally in the Middle La Tène period.

Nonetheless, a few finds of early ironwork from hillforts in the Marches (p. 301 below) and the famous Llyn Fawr hoard (Fig. 8.2), in which iron objects of Hallstatt "C" date are associated with Hiberno-British bronze cauldrons and socketed sickles, do indicate a horizon, c. 600 B.C., when Britain generally was beginning to be affected by a culture, already to some degree Celtic in character, which at that time was expanding in western Europe, and has come to be accepted generally by continental archaeologists as constituting the first phase of the Iron Age there. Wales was not completely isolated from the general development of culture in western Europe at this time, and we must accept the necessity of retaining there enough of the established international terminology for the archaeology of Europe north of the Alps to make ourselves intelligible to those scholars, in Britain and on the Continent, who try to comprehend the full geographical range of archaeological forms which represent the ancient Celtic world—so important for the whole of Europe in the second half of the last millennium B.C. On this basis we can appropriately use the terms "Late Hallstatt", "Early", "Middle" (=La Tène Ic-II) and "Late La Tène", even in Wales, to refer to phases commencing c. 600 B.C., 450 B.C., 300 B.C. and 100 B.C., respectively, to which one can sometimes tentatively refer some of the archaeological phenomena now to be considered.

The Regionalism of Early Iron Age Wales and its Environmental Background

A glance at the map of Hillforts (sensu lato) (Fig. 8.1) and important chance finds of Early Iron Age metalwork in Wales and the Marches reveals a distribution which to some degree contrasts with those of Neolithic and Bronze Age chance finds and burial sites (Figs. 5.4 and 6.4) and seems to suggest that the emphasis of settlement in the Ist millennium B.C. had shifted away from the interior uplands, where, in particular, Bronze Age sites are so common, towards the main river valleys on the eastern side of the Cambrian

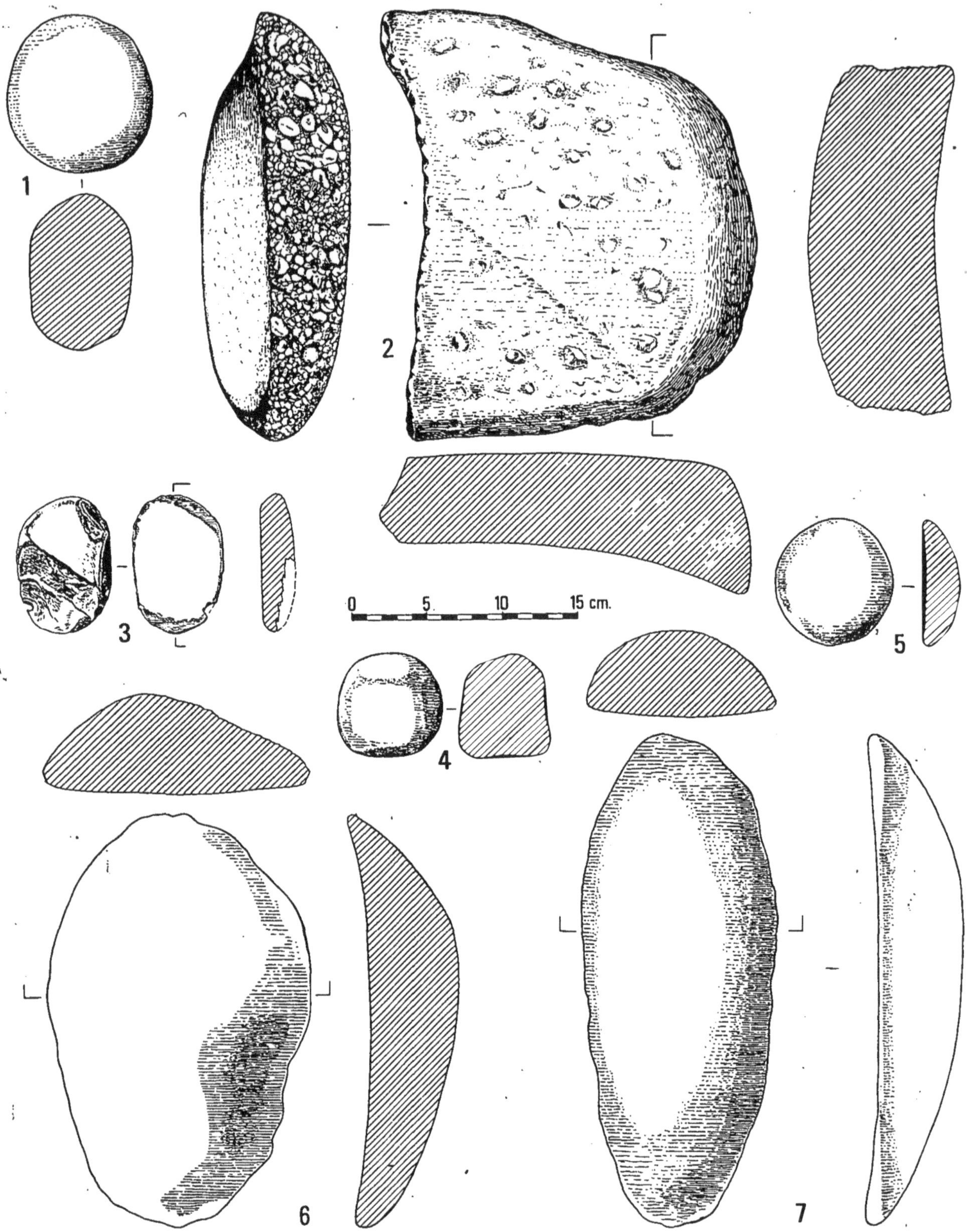

Fig. 8.3 Saddle querns and rubbers from Dinorben hillfort, Clwyd. (by permission of the National Museum of Wales)

massif and the more fertile parts of its seaboard. Such an impression would accord with the conclusions of palaeobotanists that the combined effect of deteriorating climatic conditions and human activity in tree-clearance on the uplands had brought about an increase of poor pasture there—heathland or blanket peat—by the Early Iron Age and led to increased clearance of the more fertile valley-slopes for pasture or for arable (Moore, 1973; Evans, 1975, 151; Taylor, 1975; and Turner, 1970). Particular instances of the latter would be evidence for large-scale disafforestation in the neighbourhood of Tregaron Bog in Cardiganshire c. 400 B.C. (Turner, 1970) and for extensive burning of woodland on the lower slopes of the Ystwyth valley in the same region during the 3rd century B.C.—in both cases probably for pastoral purposes in the main. Taylor (1973) has also recorded evidence of forest burning for a site at Ystwyth Forest (SH 689711), the associated charcoal layer being radiocarbon-dated to 2133 ± 110 B.P. (Q.814), i.e. in the Early Iron Age.

The truth, however, as a second glance at the map (Fig. 8.1) warns us, may be more complicated. On the environmental side, we have to take note of Webley's arguments for the role played in prehistoric Welsh communities by transhumance (Webley, 1976). As Alcock pointed out some time ago (1965) many hillforts in Wales and the Marches are sited just as conveniently for access to upland pastures as to fertile valleys, and could still have played a part in seasonal movements of cattle and sheep as well as exploitation of arable in suitable areas. Indeed, the fact that sheep seem to grow in importance in Early Iron Age Wales just as Clark suggested long ago was the case for Britain and Western Europe generally (Clark, 194) might reflect the importance for them of summer pastures on the moorlands of the interior of Wales. Until very recently it could be argued quite reasonably, as Alcock (1965, 186-8) and Bradley (1971, 74) have done, that the great size of some of the hillforts in Wales and the Marches which are sited high up, with access to moorland, is due to their use as shelters for transhuming sheep and cattle in the summer time, when inter-tribal raiding might take place. This approach has now been undermined, to some extent, by the discovery (p. 298 below) that some hillforts in the Marches—even large ones—in fact were largely covered by buildings and did not necessarily enclose a great deal of open space. Moreover, the evidence for the increase of sheep-rearing in Wales during the Late Bronze Age and Early Iron Age has not yet been satisfactorily demonstrated by statistical study of the bones from a representative series of sites spread over the whole area. This is due not only to the practical difficulties of arranging for studies of large quantities of animal bones from excavations to be carried out by experts, but to the unfortunate fact that the soil on archaeological sites in a large part of Wales is too acid to permit the preservation of bones for 2000 years or more. Accordingly, the fact that sheep seem to have been as numerous as cattle in prehistoric times at Dinorben in Denbighshire (Gardner and Savory, 1965, 223) and to have outnumbered them at Croft Ambrey in north Herefordshire (Stanford 1974, 215-22), must be received with caution in the light of the fact at Coygan Camp on the Carmarthenshire coast sheep were greatly outnumbered by cattle in the same period (Wainwright, 1967, 191). There may, therefore, have been regional differences which we cannot understand fully until bones from many more sites in different parts of Wales have been studied.

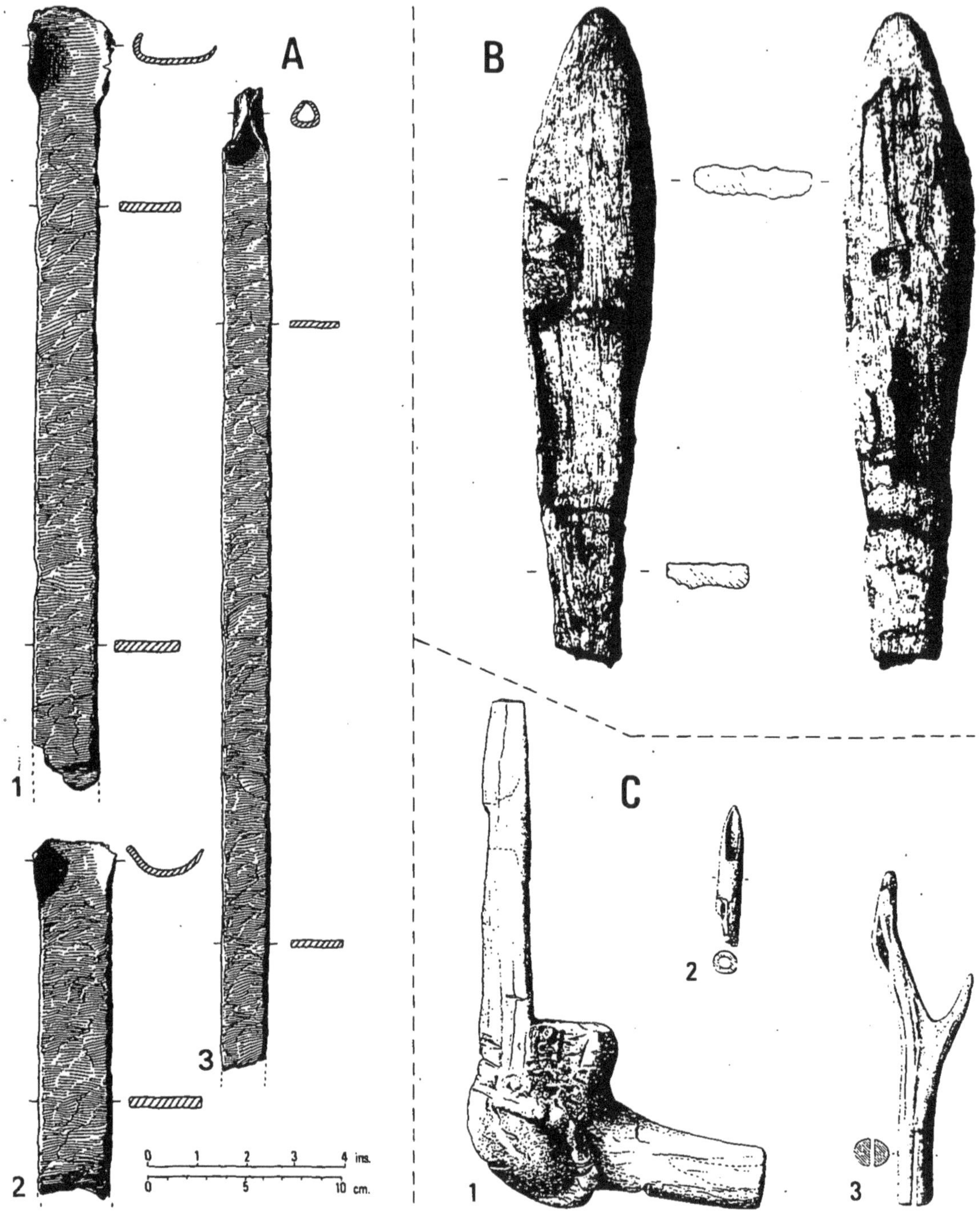

Fig. 8.4 (A) 1-2: iron bar shares; 3: currency bar: from Cerrig Bach, Gwynedd. (B) Wooden bar shares from Wales and Rath, Dyfed. (C) Antler tools from Dinorben hillfort, Clwyd. (by permission of the National Museum of Wales)

It is difficult, too, as yet to estimate the extent of arable in different parts of Wales at successive stages of the Late Bronze Age and Early Iron Age. Traces of ancient field systems like those which survived at least until recently on the chalk hills of southern England have not been identified, other than minimally, in Wales and the Marches outside of parts of north-west Wales (RCAHM (Wales), 1937, 1956, 1960, and 1964; Gresham, 1967, 187-206), and to a limited degree, Pembrokeshire (Grimes, 1950). It is not impossible that many field systems await exhumation from underneath the extensive hill peat cover of upland Wales. Even in these areas, of course, it is uncertain as to what extent the field-systems are prehistoric rather than Romano-British in date: there is, however, some agreement among those who have studied them that many of the unenclosed hut-circles which are associated with some of them in Caernarvonshire must be pre-Roman in Date (RCAHM (Wales), 1964, LXXXVII-CVI). It is highly probable that the great Caernarvonshire system of fields and associated hut-circles and homesteads formerly extended over much of the lower ground, in coastal areas of the county, where traces of them have long been removed by more recent farming activities. It is also likely that considerable areas in parts of Wales and the Marches which are at least as suitable for arable cultivation as Caernarvonshire were also laid out in field-systems which have left no trace, even in air-photographs, except in a few isolated cases, partly because of more recent cultivation and partly because methods of enclosure and cultivation were different. But because of this gap in the field archaeology of Wales we have to fall back, in most parts of the country, on the long familiar finds of artefacts connected with milling, like saddle querns (Fig. 8.3), which have been found on many settlement sites in various parts of Wales but which could conceivably sometimes have been used to mill imported grain, and the rarer finds of actual grains, as at Caynham Camp, Croft Ambrey and Midsummer Camp in the Marches (Stanford, 1972, 31) or ploughshares (Fig. 8.4) as at Walesland (Wainwright, 1971, 94-8). The fact that the latter are of wood, and were taken by the excavator to be spear-heads, helps to explain the rarity of iron ploughshares on sites in Wales and the Marches (p. 306 below). If simple wooden ards of the type known from Danish bogs were the normal type of plough in Early Iron Age Wales one cannot expect to find their traces very often in an area where soil conditions are normally unfavourable to the conservation of wooden objects. Other features of Early Iron Age agriculture lovingly brought to life by Reynolds on the slopes of Butser Hill (Hants) (Reynolds, 1976) may be suspected in Wales, without very much proof as yet; the presence of pig bones in fair quantity on some sites (Gardner and Savory, 1965, 223-5) may mean that pannage was practised in the fields, and the discovery of Iron Age potsherds apparently scattered in the fields far from any settlement in north Radnorshire (Jeffery, 1962) may mean that Iron Age people in Wales, like their Romano-British successors, were given to spreading muck from their farmyards over their arable as manure.

On the archaeological side, on the other hand, we must notice that the map (Fig. 8.1) reveals considerable differences between the types of defended enclosure which are typical of different parts of Wales. It is obvious that conspicuously large and strongly defended hillforts are much more common in the Marches than they are along the seaboard of south Wales, or in west Wales (Hogg, 1972), and that the majority of sites in the latter areas are very small. These small sites may enclose no more than an acre of two and although some are strongly fortified, others are not sited in a military way and are enclosed by walls or earthworks which seem too weak to be anything but a control for animals or an indicator of social status. There is clearly a fundamental difference between the large and strongly defended hillforts, whose construction must have involved a considerable communal effort and some of which, as we now know, must have had several hundred inhabitants, and the small and often weakly defended sites, which seem merely to have been hamlets or farmsteads inhabited by a few families. The preponderance, in the Marches and along the seaboard of north Wales, of the former type, and the preponderance, in south and west Wales, of the latter, suggests that the two areas must have had a rather different social structure during much of the thousand years or so during which these types of settlement were being built. This impression is reinforced by a detailed study of the structural features of hillforts, some of which also have regional distributions which I have analysed more fully elsewhere (Savory, 1976b). However, the distributions of the two groups overlap considerably, and it will be noted that the small sites are common enough in Montgomeryshire, north Radnorshire and the adjoining part of Shropshire, along with large hillforts which have dependent enclosures of various types, which are perhaps the counterpart of the concentric enclosures associated with some small homesteads in south and west Wales (Fig. 8.6) and can perhaps be interpreted as accommodation for cattle. While the small sites were certainly being built throughout the La Tène period, and even under Roman rule, there are cases where a large site—even as far north as Caer Drewyn near Corwen (Gresham, 1967) has replaced a small one, not necessarily at a late date, and it is possible to believe that this is the primary form of fortified settlement in much of southern and central Wales.

One hypothesis could be that we are dealing, in Wales, with an indigenous, mainly pastoral culture, rooted in the Middle Bronze Age and loosely organised in small social units liv ing in scattered homesteads. This was impinged upon, from the Late Bronze Age onwards, by an extraneous culture with a much more centralised organisation, practising mixed agriculture from nucleated villages in the northern Marches and ultimately along the northern seaboard of Wales; the interchange between these two cultures and the necessary adaptation to local conditions would in due course have produced varying local hybrid groups. This hypothesis is attractive but needs to be tested by large-scale excavations more evenly distributed over Wales than has been the case so far. At present one can only say that in each of the varying habitats which exist round the edges of the central moorland of Wales, corresponding roughly with the territories of the tribes found by the Romans—the Cornovii in the central Marches, the Degeangli in the north-east, the Ordovices in the north-west, the Demetae in the south-west, and the Silures in the south-east, the surviving "hillforts" (sensu lato) exhibit a different combination of characteristics and no doubt reflect the different vicissitudes of a particular local history extending over several centuries (Hogg, 1972a). It may, indeed, be true that the earliest hillforts (Sensu stricto) of Wales are those of the northern Marches which have yielded radiocarbon dates much earlier than any so far obtained at sites in south and west Wales (p. 307 below). That would accord with the picture now emerging of single-or double-palisaded hillforts already being built in the Middle Bronze Age in Wessex, e.g. at Ram's Hill (Berks.) (Bradley and Ellison,

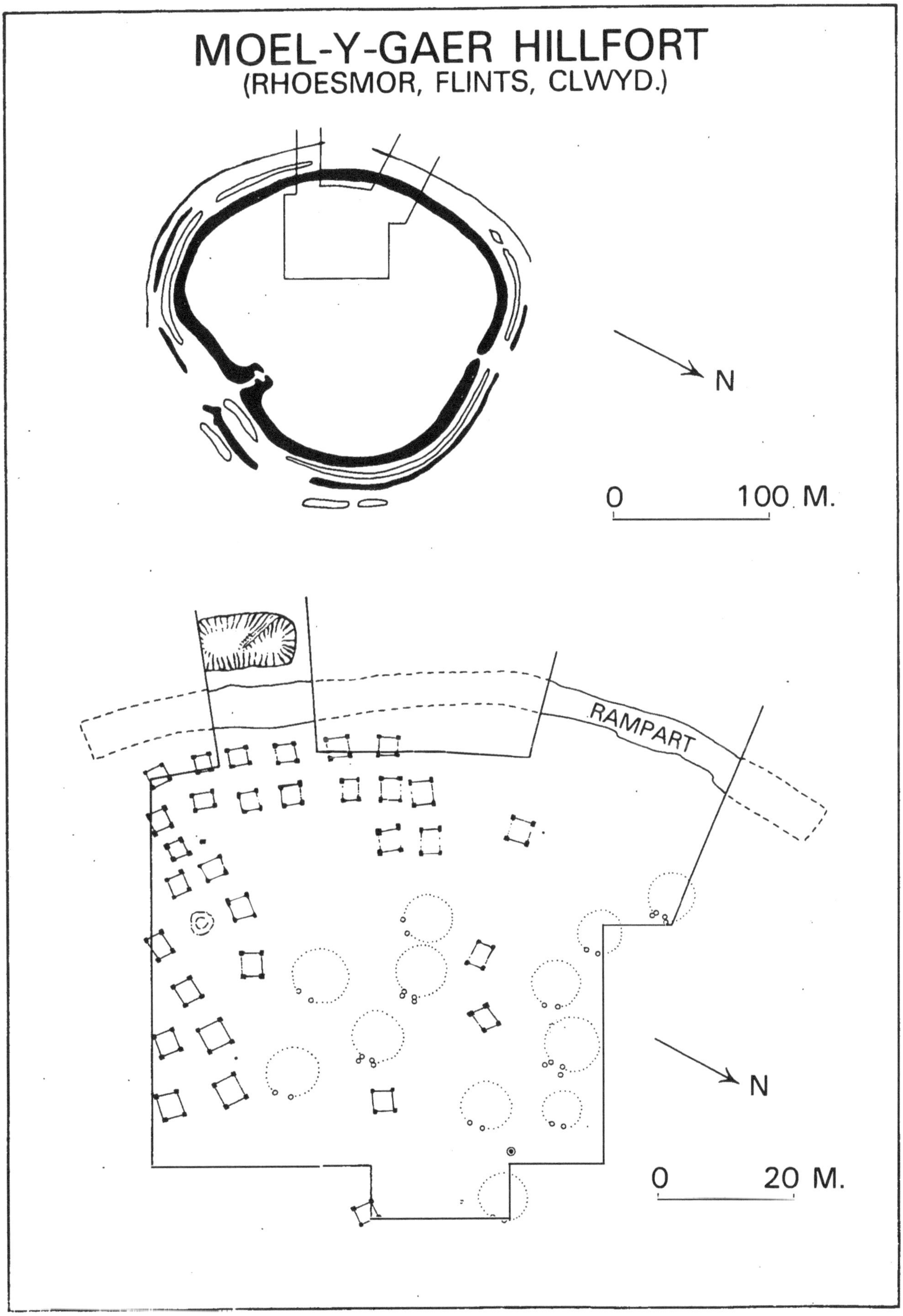

Fig. 8.5 Area plan: Moel-y-gaer hillfort, Clwyd. Planned layout of round houses and four-post structures, (by permission of the Academic Press).

1975) and possibly in the Midlands, e.g. at Mam Tor (Derbysh.) (Coombs, 1976), and the possibility that the culture responsible for these developments might impinge particularly on the Marches across the Midland plain, or by boat up the Severn from the Cotswold region. In this case one still has to explain why the southern seaboard of Wales should apparently not have been affected by such developments until the end of the Late Bronze Age (p. 270 below). It is tempting to relate this problem to the current picture of Late Bronze Age metal industries (p. 303 above), which suggests that south and west Wales belonged to a different economic and therefore perhaps political province from the northern Marches (Savory, 1976b, 250-1)—one that also embraced much of the West Country, where related settlement and earthwork types occur in the Early Iron Age, and therefore perhaps controlled traffic in the Bristol Channel. However, a new series of radiocarbon dates from a site on the south coast of Wales could still wipe away much of the apparent retardation of the area. At the same time, as we shall see (p. 301 below), the Irish Channel coasts of Wales had their own independent maritime contacts with the outside world, which may well have made their special contribution to the regional diversity which we must now consider.

Central and Northern Marches (Cornovii - Degeangli)

Our knowledge of this area if now based chiefly on the results of recent large-scale emergency excavations at the Breiddin (Mont.) and Moel y Gaer, Rhosesmor (Flints.) hillforts, supported by those of earlier, for the most part less extensive, excavations at Dinorben (Denbs.), Ffridd Faldwyn (Mont.), Moel Hiraddug (Flints.) and Old Oswestry (Shrops.) hillforts. Radiocarbon dates obtained at the first three of these sites make it clear that they were already occupied by the beginning of the Late Bronze Age and fortified or re-fortified at various times during its course, with palisades or timber-revetted or staged ramparts, and that their occupation continued thereafter far into the Early Iron Age, with various renewals of their defences. Moel y Gaer (Fig. 8.5) is the site which tells us most about the internal lay-out of a fortified village in the northern Marches at the beginning of the Early Iron Age because it occupies a flat hill-top which seems never to have been cultivated since its abandonment by its prehistoric inhabitants; stripping of about one-fifth of its internal area of 6 acres revealed well-preserved post-hole systems relating to two successive earlier settlements and the ground level of a third settlement with oblong houses which appear to have rested on sleeper-beams and to have stood in small yards. The earliest settlement consisted of a large number (17 in the area explored) of "post ring" round-houses ranging from 6.5 to 11.5 m in diameter, arranged in no particular order and dated c. 800 B.C. The second settlement had been "town-planned", c. 600 B.C., inso far as it contained, in the block area explored, two types of structure: a dozen "stake-wall" round-houses, from 5.6 m to 8 m in diameter, and over 30 "4-posters", arranged in distinct groups, as though (Fig. 8.5) the latter were complementary in function to the former. The excavator (Guilbert, 1975 and 1976) has pointed out that the posts of the "4-posters" (roughly rectangular groupings of post-holes) must have been more massive than those which framed the porches of the round-houses of phases I and II, even though they enclose a space not more than 4 m each way, and are therefore likely to have supported the raised and projecting floors of store-rooms of some sort. This interpretation appears to be supported by the results of recent large-scale area-excavations in the town-planned interior of Danebury hill-fort (Hants.) (Cunliffe, 1976), where systems of "4-posters" also occur, often arranged in rows along streets, while round-houses and storage pits were located elsewhere.

The establishment of a special storage function for "4-posters" at Moel y Gaer is important for the study of hillforts in Wales because these structures have been a regular feature at recent hillfort excavations in the Marches, apparently recurring from Hallstatt down to Middle La Tène times. In Herefordshire, Stanford has been inclined to interpret them as dwellings as well as store-houses, in the absence of any trace of round-houses (Stanford, 1973, 30f.). They are a common feature at the Breiddin (Mont.), where only a very small part of the interior of one of the largest of Welsh hillforts, enclosing 65 acres, has so far been explored. Here the phase corresponding to the second settlement at Moel y Caer saw the construction of numerous round-houses, some of which were replaced, in the 4th and 3rd centuries B.C., by large numbers of "4-posters", many of them arranged in a row behind the rampart, like some of their counterparts at Moel y Gaer (Fig. 8.5), and others at Moel Hiraddug (Flints.), where round-houses of various types also occur, as well as some rectangular houses possibly corresponding to those of the final phase at Moel y Gaer. What was stored on the "4-posters" cannot be demonstrated at present—in fact, it may have varied from time to time and from place to place—but it may often have been winter fodder. That animals were often carried through the winter, and not slaughtered until mature, has been demonstrated by Ruth and David Whithouse's study of the animal bones from Croft Ambrey (Stanford, 1974, 215-22), and the numbers of sheep at Croft Ambrey and Dinorben raise the question of the extent to which these were kept for their wool rather than their meat. To argue from the numbers of spindle-whorls found in north Wales hillforts like Dinorben (Gardner and Savory 1965, 179-84) that sheep were more important in north Wales than in Mid-Wales or south Wales, as Cunliffe has done (1974, 199), in order to account for the rarity of multiple enclosure hillforts in the former area, is somewhat precarious, for it cannot be said that spindle-whorls were proportionately fewer at Coygan Camp, for example, than at Dinorben, although in fact sheep bones were relatively poorly represented at the former site.

The earlier hillfort excavations in the central and northern Marches were generally carried out with limited resources and tended to concentrate on the sequence of defences, particularly ot gateways, so that we have difficulty in comparing their interior lay-out with that of Moel y Gaer, but it is at least clear now that Ffridd Faldwyn hillfort, near Montgomery, was well supplied with 4-posters during an early phase of its history (Guilbert, 1975, 207), but no definite trace of them was found among the recesses made for the floors of Early Iron Age round-houses at Dinorben, a hill-fort of roughly the same size at Moel y Gaer where a sloping, denuded surface, much disturbed by later occupation, 19th century tree-planting and earlier excavations, could only be partially explored. Further westwards they have not been observed, but a few have been recognised recently in south-east Wales, and they are not unknown in parts of south-west Wales (p. 302 below). The earlier excavations of defences and gate-ways however, together with the recent work at Moel Hiraddug, Moel y Gaer and Breiddin, have served to emphasise the highly organised and centralised nature of the communities which built the north-eastern hillforts, especially from the beginning of the Early Iron Age, in the 6th century B.C. It is then that a common military tradition, represented first by bastions at the entrance and later, in the 4th and 3rd centuries B.C., by guard-chambers, spreads from the Cotswolds to north-east Wales and the occasional finds of continental metal-work types, like the two swan-necked pins from Moel Hiraddug and the vase-headed pin from Ffridd Faldwyn, reflect the ultimate source of this tradition (Savory, 1976, 258-67). While this tradition was no doubt maintained by something like a corps of professional military engineers, its imposition was probably the work of powerful rulers whose authority is reflected in the guard-chambers and the roads which pass them—each renewed from time to time. That authority made possible a trade in fragile pottery, like the All Cannings Cross furrowed and hematited bowls which found their way, partly no doubt by boat up the Severn, from Wessex to Old Oswestry, to be followed in the 4th and 3rd centuries B.C. by Malvernian stamped ware, traded from Herefordshire as far as the Breiddin. Thereafter a decline may be suspected, perhaps associated with a reassertion of indigenous culture and a reversion to isolated homesteads in Mid-Wales. In its heyday this extraneous culture may have established itself here and there in north-west and south-west Wales, and the association of a possible "4-poster", a scrap of Malvernian stamped ware and a relatively large hillfort at Pen Dinas, Aberystwyth, is particularly significant (Forde, Griffiths, Hogg and Houlder, 1963 and Stanford, 1972, 34). But the Cornovii, and still more the Degeangli of the Clwyd hill-forts, do not seem to have been leaders of resistance to the Romans, and may even have been under Ordovician control in the early 1st century A.D.

North-west Wales (Ordovices)

The whole northern seaboard of Wales is linked to the Northern Marches by the possession of numerous large hillforts, some of which contain large numbers of round-houses. It is, however, difficult at present to estimate the strength of the relationship, because theer have been no modern area-excavations of their interiors, to enable us to judge whether they ever had any structures equivalent to the "4-posters" of the north-east and to what extent the obvious differences in their building techniques, whether applied to round-houses or to defences, are due to the conditions imposed by the local geology rather than different cultural origins. Certainly the practice of building dry-stone wall ramparts without ditches was not inevitable, because multiple banks and ditches were later added to some Caernarvonshire hillforts, or built on fresh sites in the same area, and the north-eastern tradition of bastions and guard-chambers is hardly represented there: indeed entrances are generally very much simpler (Savory, 1976, 267-71). Though direct dating evidence remains slight it is likely that the north-western tradition of large hillfort-building was founded by the Hallstatt period, possibly as a result of influence from the Marches at that time. But the practice of back-terracing ramparts and providing them with multiple revetment walls at front and rear is found in Ireland, Brittany and south-west Europe as well as north-west Wales and seems to denote maritime connections with the Atlantic coasts of the Continent. That traders or explorers were occasionally coming from that direction long before the Roman conquest is clear not only from ancient literary sources but from the recent dramatic discovery off Aberdaron in the Lleyn of a lead anchor-stock of

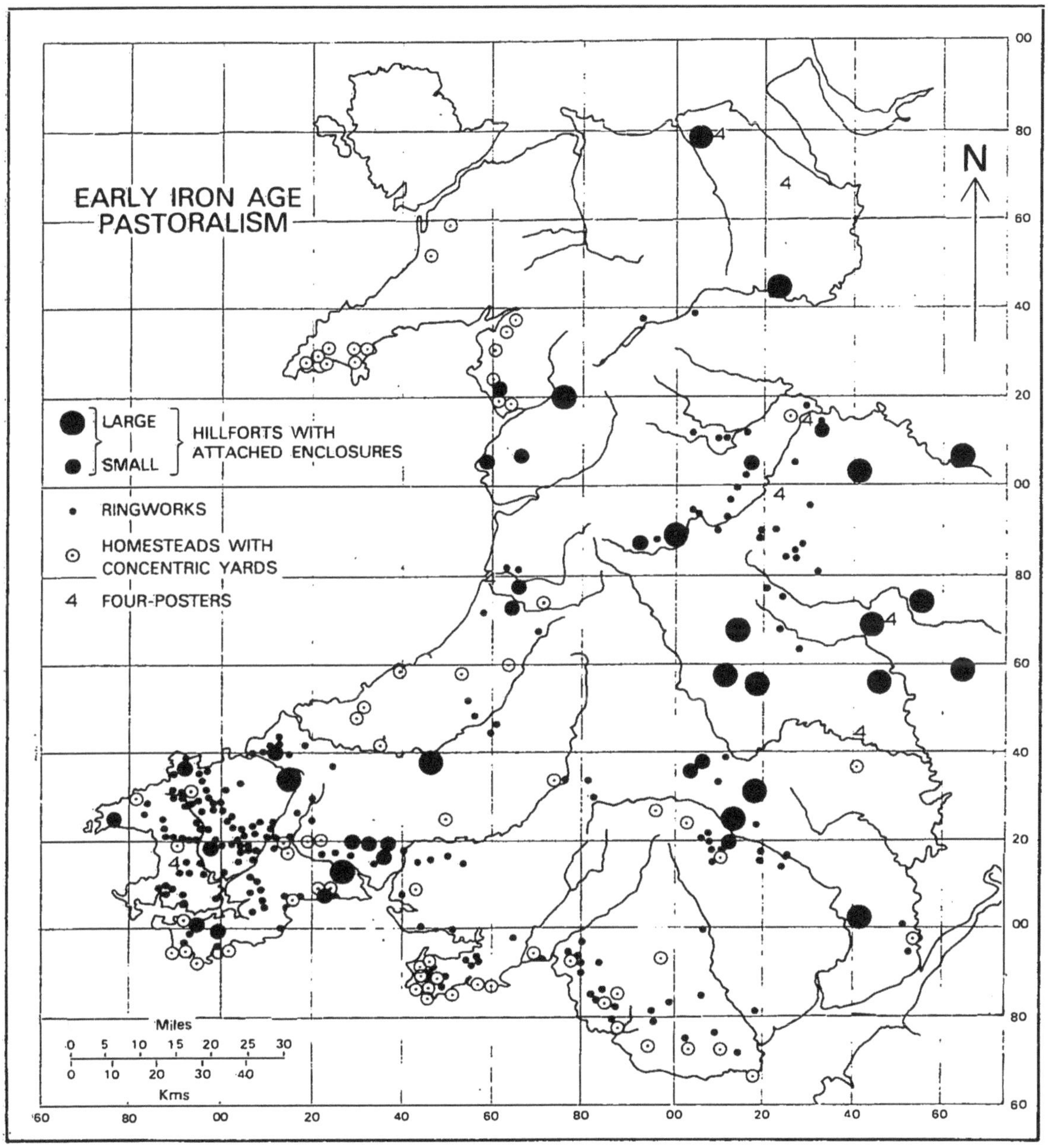

Fig. 8.6 Wales: Early Iron Age pastoralism

Hellenistic type (Boon, 1975). On the other hand, the associated practices of revetting ramparts with vertical slabs and defending the approaches to hillforts with chevaux de frise formed of stone pillars, found in south-west as well as north-west Wales, should perhaps be seen simply as an adaptation to a relatively treeless environment which is also found in parts of south-west Europe which are poor in tree-cover.

At an early stage, it seems, defended homesteads and small forts appear in parts of the north-west and a group of concentric enclosures in the Lleyn which includes Castell Odo (Alcock, 1960) recalls a type of site which is characteristic of south-west Wales and the West Country (Fig. 8.6) (p. 303 below). Unfortunately, the expected radiocarbon date from the earliest phase of occupation on this site has not been forthcoming, at the time of writing (October, 1979), but the relatively abundant pottery from this layer could be of Late Bronze Age or Hallstatt date. The palisade which had been intended at the end of the Early phase would have been characteristic for the suggested date and one must remember that something of this sort was being built at Llanstephan on the Carmarthenshire coast in the Late Bronze Age as well as in the Marches (p. 303 below). A type of homestead consisting of a single large round-house with dry-stone foundations surrounded by concentric walled yards, as at Llwyn-du Bach near Caernarvon (Bersu and Griffiths, 1949) may also reflect southern influences and open a series of enclosed homesteads which probably stretches from the Early Iron Age (aceramic in this area) far into the Roman period. It is perhaps as part of this process of reassertion of local traditions, long entrenched in south and mid Wales, that walled forts without ditches appear here and there in north-east Wales and the Shropshire Highlands, one at the Breiddin belonging to the Early La Tène phase, and round-houses with dry-stone foundations, in place of the "post ring" or "stake-wall" construction previously used in the area, appear at Caer Drewyn near Corwen and at Moel Hiraddug. Perhaps drovers or shepherds using the old Bronze Age trackways of north Wales to reach the Marches in peace time, or for other purposes in time of inter-tribal war, had a hand in this spread of north-western building traditions.

South-West Wales (Demetae)

In spite of a relatively large number of excavations at minor hillforts and embanked homesteads of "Rath", promontory and concentric type in Pembrokeshire and the Gower, most of them on a small scale or with old-fashioned technique, we still know relatively little about the evolution of Late Bronze Age and Early Iron Age settlements in south-west Wales or indeed the whole of mid Wales west of the Breiddin and Fridd Faldwyn. This is because there has been so little modern excavation in the larger hillforts which are scattered through the whole of this area, round the edge of the high moorland. It is these sites, as local seats of power, which are most likely to contain evidence of cultural change, such as the suspected expansion of the early hillfort-building culture of the Marches up the Severn valley towards Cardigan Bay and Pen Dinas, Aberystwyth, where the long sequence of development begins with a timber-revetted rampart and there is a possible "4-poster" (p. 299 above). A similar movement through mid Wales and down the Teifi and Tywi valleys might account for the paired guard-chambers in the entrance of a small hillfort near Hayscastle (Pemb.) (Crossley, 1963, 187), or the internal arrangements at Walesland Rath, presently to be considered. But in general as I have already said (p. 297 above) one might suspect that the preference for large numbers of homesteads instead of nucleated villages, fortified or not, is a primary feature of Bronze Age origin, shared with the West Country. As it happens, however, the finds of pottery likely to be of Late Bronze Age or Hallstatt date which have been made along the southern seaboard of west Wales have come from caves or what were, apparently, open settlements, and the only fortified site which has so far proved to be early is a promontary fort within which Llanstephan Castle on the Tywi estuary was built. Here three radiocarbon dates show that one of two enclosing banks had a timber spine in a bedding trench which belongs to the 8th or the 7th century B.C. (Guilbert and Schweiso, 1971a-b). Unfortunately, one can hardly hope to learn very much about the internal lay-out of this particular site, but its defences may reflect the same coastal movement as Castell Odo and possibly even Pen Dinas. Castell Cogan, not very far away, with its round-houses and timber-revetted rampart, may be nearly as early, but no radiocarbon dates are available (Grealey, Jones and Little, 1971).

The only completely excavated and well-preserved site in the whole of west Wales, Walesland Rath near Haverfordwest, belongs to a relatively late horizon, for here a weak dump rampart, enclosing a mere ½ acre, was not built, to judge by a radiocarbon date, until the late 3rd century B.C. However, it recalls Moel y Gaer and the Breiddin in so far as it contained peripheral ranges made up of rectangular structures recalling "4-posters" enclosing a small group of round-houses. Unfortunately, the acid soil on this site had destroyed animal bones so that it could not be ascertained whether the problems of winter fodder had been solved here, as at Croft Ambrey, with the help of peripheral buildings. If it had been we should understand the background to the contemporary large-scale clearances of woodlands near the Tregaron Bog and in the Ystwyth valley already referred to (p. 293 above). Nonetheless, the wooden bar-shares (Fig. 8.4) from the second phase at Walesland (1st century B.C.—1st century A.D.) make it clear that the agricultural possibilities of south Pembrokeshire were not being neglected in this late phase. On the other hand, the abundant homesteads with concentric outer yards and the numerous small or medium-sized hillforts with annexed enclosures throughout west and mid Wales (Fig. 8.1) reflect the predominantly pastoral activities of all this area. The walled forts of south-west Wales, with their chevaux de frise and vertical slab revetments, are probably of Early and Middle La Tène date: the latter feature appears at Coygan Camp, at a time when linear-incised pottery with arcaded patterns, like that found on Middle La Tène sites around the Severn estuary, was reaching the site. But the same contacts along the Bristol Channel coasts were probably responsible for the spread through the area, in later La Tène, of a different style of fortification involving the use of multiple "dump" ramparts and ditches.

South-east Wales (Silures)

So far no area-excavations in the modern manner have taken place in the interior of a sizeable hillfort in south-east Wales, on a scale which would permit comparison with Moel y Gaer or the Breiddin. Caution is therefore necessary, but it can be said that the relatively numerous sections through ramparts and explorations of entrances have as yet produced no pottery or metalwork of Late Bronze Age of Hallstatt type which is contemporary with the earliest defences, rather than "pre-rampart". The apparent retardation is confirmed by the C14 date at Twyn y Gaer hillfort, Cwmyoy, Gwent, (Probert, 1976), which suggests that the final repair of a fence enclosing an annexe attached to the eastern end of a small hillfort took place early in the 4th century B.C.: this would date a late phase of the first period of the fort followed by enlargement and several stages in the development of the eastern entrance, during which the inhabitants were clearly in touch with the Middle La Tène culture of Herefordshire. During the early phase of this site one seems to be on the Early La Tène horizon which we have already encountered in mid and south-west Wales, during which pastoral annexes were often built. At the same time early types of timber-laced or -revetted rampart have not been much in evidence so far in south-east Wales, few examples of guard-chambers or "4-posters" are known and hillforts are at best of medium size. The "dump" ramparts which are characteristic of the area may begin early, as at Cfoft Ambrey, but even a large univallate camp, Castle Ditches, Llancarfan (S. Glam.) has been shown to have been built in later La Tène, replacing a small walled fort (Hogg, 1976). There are other cases in the area where superficial study suggests that a small fort was replaced by a large one in later La Tène (Savory, 1976, 272). As in west Wales and the Marches the construction of multiple dump ramparts is a late development, applied to small homesteads as well as true hillforts. An extreme case of the former is Mynydd Bychan, Llysworney (S. Glam.) (Savory, 1954-5), where a few round-houses were enclosed in a space of $\frac{1}{2}$-acre by multiple banks and ditches with an entrance defended by a bastion and a barbican, like some large hillforts in south Wales.

The general impression at present, then, is of a conservative population, in south-east as well as west Wales, mainly pastoralist and living in numerous homesteads which began to be fortified, under threat, in the Hallstatt period, and remained, down to Roman times, the most common type of settlement, while such hillforts as were built later on were of small or of medium size. The northern frontier of the Silures seems to have rested on a belt of forested country stretching from Hereford into south Radnorshire. South of this neither the early hillfort culture of central Marches nor the Malvernian stamped ware of Middle La Tène date seems to have penetrated to any extent. The Late Bronze Age—Hallstatt hoards from Llyn Fawr and Cardiff are not an exception to this isolation from the early Celtic world of the west Midlands, because they probably represent loot carried off from the latter area by hillmen from south Wales, who otherwise continued, for some time after that, to use the products of the smiths of the final Bronze Age "Llantwit-Stogursey" group (Savory, 1976c).

The Final Phase

At the time of the Roman conquest, in the 1st century A.D., the four main natural regions of Wales, which we have just been considering, had already been occupied for some time by the tribes assigned to them by classical writers. The Cornovii and the Degeangli had perhaps been in the north-east since Hallstatt times, since there is nothing to suggest any fundamental change in this area in later La Tène times, apart from a possibility of decline and isolation. The Ordovices played a leading part in resistance to the Romans

Plate 8.IA Embossed plaque of tin-plated copper-zinc alloy, from Tal-y-Llyn, Merioneth (Gwynedd)

B Embossed crescentic plaque of bronze, from Llyn-cerrig-bach, Anglesey, first century B.C. (Gwynedd)

and the great ceremonial deposit which accumulated over a period of two centuries in a sacred lake at Llyn Cerrig Bach in Anglesey reflects the importance of their warrior class and their ability to command the services of good craftsmen or to import fine metalwork from elsewhere (Savory, 1976a, 49f., 57-60, Figs. 13-31). It may also reflect the importance of Anglesey at this time as an entrepot for trade between Britain and Ireland. If the comparatively modest local agricultural resources of their territory seem insufficient to explain the apparent military power of the Ordovices at this time one should perhaps remember that Ptolemy implies (Geographia, II, 3, 18) that they controlled the whole of north Wales and part of Cheshire and north-west Shropshire (the Degeangli presumably being their subjects). If this was so, they must have controlled the copper and lead ores of most of north Wales and it is hardly likely that they failed to exploit them, seeing what use was made of them by the Bronze Age people (p. 264 above), and the speed with which the Romans turned them to account. It is mineral wealth which is most likely to have paid for any imports into Ordovician territory, and accounts for the quality of north Wales finds of La Tène art metalwork, some of which was probably of local manufacture (Pl.8.I). There is, however, no space for me to enter into this controversial question here (Savory, 1976a, 47f.). That the Ordovices were able to keep abreast with developments in agricultural as well as military equipment is shown by the balanced sickle and the bar-share in the Llyn Cerrig Bach hoard (Fig. 8.4). The latter implies the use of dry farming techniques, with cross-ploughing, in the fields of Anglesey on the eve of the Roman conquest.

In the south, it was the Silures who led the resistance to the Romans, at first with considerable success, and Tacitus' description of them (Agricola, II.2) as a dark people resembling the inhabitants of south-west Europe confirms suspicions already entertained on archaeological grounds that they were largely aboriginal and had been Celticized late. The Demetae in the south-west cannot have been very different, but here, as with the Ordovices, there is a suspicion of a recent admixture from the Atlantic seaboard of the Continent. At the same time, the discovery of an iron bar-share, reported but not yet fully published, on the site of the early promontory fort at Llanstephan in Carmarthenshire (Guilbert, 1974), like that of a similar tool at Llyn Cerrig Bach, shows us that new technology was reaching the arable farmer on the eve of the Roman conquest,—no doubt as a result of the coastal trade of the Severn Sea which had earlier brought Dobunian types of pottery to Coygan Camp. Here, too, as in north Wales, one may now suspect that local mineral resources played a part in this coastal trade. In this case the product would have been gold, which the Romans lost no time in exploiting at Dolaucothi. Proof of this early, native exploitation is provided by the recent discovery at Carmarthen of a splendid silver brooch, decorated in relief, parcel-gilt, with insular La Tène patterns, which seems to mark the beginning of the great British series of "Trumpet" brooches, probably at an earlier date than the Roman penetration of west Wales (Boon and Savory, 1975). This, too, is a controversial question, but it is arguable that not only gold, but ornaments made from it, traded as far as East Anglia, where so many of them have been found, enabled west Wales to pay for its imports of grain or manufactured goods. There is, of course, no doubt as to the exploitation of the iron ore available in various parts of Wales during the immediate pre-Roman period:

one definite case would be the ore, derived from the southern edge of the Glamorgan coal-basin, which was found in the Late La Tène homestead at Mynydd Bychan (Savory, 1955, 48f.).

RADIOCARBON DATES

Dinorben Hillfort, Abergele, Denb.

H. N. Savory, Excavations at Dinorben, 1965-9, 76f. (dates supplied by Miss A. Bermingham, Melbourne).

V-124. Charcoal possibly derived from Period II rampart. BP 2485=535 ± 85bc.
V-176. Charcoal from timber-lacing of Period II rampart. BP 2370=420 ± 70bc.

Moel y Gaer hillfort, Rosesmor, Flints.

G. Guilbert in D. W. Harding (ed.) Hillforts, 317 and Archaeology in Wales, 15, 33.

HAR-606. Post-ring round-house beneath Phase 2 rampart. 620 ± 70bc.
HAR-1197. Post-ring round-house. 610 90bc.
HAR-604. Bedding slot for front posts of Phase II rampart. 580 ± 90bc.
HAR-1126. Post-ring round-house. 560 ± 100bc.
HAR-1196. Post-ring round-house. 470 ± 80bc.
HAR-1125. Stake-wall round-house. 480 ± 140bc.
SRR-494. 4-poster. 370 ± 45bc.
HAR-1122. Core of timber-framed and partitioned rampart A. 260 70bc.
HAR-603. Burnt twigs overlying part of 4-poster grid and covered by final rampart. 240 ± 80bc.

Breiddin Hillfort, Criggion, Mont.

C. Musson in D. W. Harding (ed.), Hillforts, 293-302.

Stake-wall round-house. 479 ± 55bc.
Other round-houses : 460 ± 100bc.; 375 ± 63bc.; 320 ± 80bc.;
4-posters: 294 ± 40bc.; 240 ± 80bc.; 238 ± 70bc.; 172 ± 45bc.

Llanstephan Promontory-fort, Carm.

G. Guilbert and Schweiso, Archaeology in Wales, II, 22.

PRN-2198. Timber spine of rampart : 500 ± 90bc.; 510 ± 70bc.; 520 ± 70bc.

Twyn y Gaer hillfort, Cwmyoy, Mon.

A. L. Probert in Boon and Lewis (eds.). Welsh Antiquity, 105-19.

BM Final repair of fence of eastern annexe. 286 ± 38bc.

BIBLIOGRAPHY

Alcock, L. (1960), "Castell Odo : An Embanked Settlement on Mynydd Ystum, Caernarvonshire", Archaeologia Cambrensis, 109, 78-135.

Alcock, L. (1965). "Hillforts in Wales and the Marches", Antiquity, 39, 184-95.

Bersu, G. and Griffiths, W. E. (1949). "Concentric Circles at Llwyn-du-bach, Penygroes, Caernarvonshire", Archaeologia Cambrensis, 100, 173-206.

Boon, G. C. (1975). "A Graeco-Roman anchor-stock from North Wales", Archaeologia Atlantica, Hamburg, I, 193-99.

Boon, G. C. and Savory, H. N. (1975). "A Silver trumpet-brooch with relief decoration, parcel-gilt, from Carmarthen, and a note on the development of the type", Antiquaries Journal, LV, 41-61.

Bowen, E. G. and Gresham, C. A. (1967). History of Merioneth, Vol. I, Dolgellau "Enclosed Homesteads", 187-206.

Clark, G. (1947). "Sheep and Swine in the Husbandry of Prehistoric Europe", Antiquity, 21, 122-36.

Coombs, D. G. (1976). "Excavations at Mam Tor, Derbyshire, 1965-9", in D. W. Harding (ed.), Hillforts, London, 147-52.

Crossley, D. W. (1963). "List of Hillforts and other Earthworks in Pembrokeshire", Bulletin of the Board of Celtic Studies, XX, 171-205.

Cunliffe, B. (1974). Iron Age Communities in Britain, London.

Cunliffe, B. (1976). "Danebury, Hampshire: a second Interim Report on the Excavations, 1971-5", Antiquaries Journal, LVI, 198-216.

Evans, J. G. (1975). The Environment of Early Man in the British Isles, London.

Forde, C. D., Griffiths, W. E., Hogg, A. H. A. and Houlder, C. H. (1963). "Excavations at Pen Dinas, Aberystwyth", Archaeologia Cambrensis, CXII, 125-53.

Gardner, W. and Savory, H. N. (1964). Dinorben, a Hillfort Occupied in Early Iron Age and Roman Times, National Museum of Wales, Cardiff.

Grealey, J., Jones, G. D. B. and Little, J. H. (1971). "Excavations at Castell Cogan, 1971: an Interim Report", Carmarthen Antiquary, 8, 17-22.

Grimes, W. F. (1950). "The Archaeology of Skomer Island", Archaeologia Cambrensis, CI, 1-20.

Guilbert, G. (1974). "Llanstephan Castle", Archaeology in Wales, 14, 11.

Guilbert, G. (1975). "Planned Hillfort Interiors", Proceedings of the Prehistoric Society, 41, 203-21.

Guilbert, G. (1976). "Moel y Gaer (Rhosesmor), 1972-3", in D. W. Harding (ed.), Hillforts, London, 303-17.

Guilbert, G. and Schweiso (1971a). "Llanstephan Castle (1971)", Carmarthen Antiquary, 8, 75-90.

Guilbert, G. and Schweiso (1971b). "Llanstephan Castle", Archaeology in Wales, II, 22.

Harding, D. W. (1976). (Ed.), Hillforts, London.

Hawkes, C. F. C. (1931). "Hillforts", Antiquity, 5, 60-97.

Hawkes, C. F. C. (1959). "The ABC of the British Iron Age", Antiquity, 33, 170-82.

Hogg, A. H. A. (1972a). "Hillforts in the Coastal Areas of Wales", in C. Thomas (ed.), The Iron Age in the Irish Sea Province (CBA Research Report, 9), London, 11-23.

Hogg, A. H. A. (1972b). "Size-distributions of Hillforts in Wales and the Marches", in Lynch and Burgess, eds., Prehistoric Man in Wales and the West (Essays in Honour of Lily F. Chitty), Bath, 293-305.

Hogg, A. H. A. (1976). "Castle Ditches, Llancarfan (Glamorgan)", Archaeologia Cambrensis, CXXV, 13-39.

Jarrett, M. G., ed. (1969), The Roman Frontier in Wales, University of Wales, Cardiff.

Jeffery, J. M. (1961). "Radnorshire", in Archaeology in Wales, 2.8.

Moore, P. D. (1973). "The Influence of Prehistoric Cultures upon the Initiation and Spread of Blanket Bog in Upland Wales", Nature, 241, No. 5388, 350-3.

Musson, C. R. (1976). "Excavations at the Breiddin, 1969-73", Harding (ed.) Hillforts, 293-302.

Probert, A. L. (1976). "Twyn y Gaer Hillfort, Gwent: an Interim Assessment", Welsh Antiquity, National Museum of Wales, Cardiff, 105-19.

Reynolds, P. T. (1976), Farming in the Iron Age, Cambridge University Press.

R.C.A.H.M. (Wales) (1964). Inventory of the Ancient Monuments in Caernarvonshire, Vol. 3, H.M.S.O., London, LXXXVII-CVI.

Savory, H. N. (1954-5). "The Excavation of an Early Iron Age Fortified Settlement on Mynydd Bychan, Llysworney (Glam.)", Archaeologia Cambrensis, CIII-IV, 85-108 and 14-41.

Savory, H. N. (1976a). Guide Catalogue of the Early Iron Age Collections, National Museum of Wales, Cardiff.

Savory, H. N. (1976b). "Welsh Hillforts: A Reappraisal of Recent Research" Harding (ed.) Hillforts, London, 237-91.

Savory, H. N. (1976c). "Wales from a Wessex Point of View" (Presidential Address), Archaeologia Cambrensis, CXXV, 1-12.

Stanford, S. C. (1972). "Welsh Border Hillforts", in C. Thomas (ed.), The Iron Age in the Irish Sea Province (CBA Research Report 9), London, 25-36.

Stanford, S. C. (1974). Croft Ambrey, Hereford.

Taylor, J. A. (1973). Chronometers and Chronicles: a Study of Palaeo-environments in west central Wales, in Progress in Geography, No. 5, Arnold, London, 247-334.

Taylor, J. A. (1975). "The Role of Climatic Factors in Environmental and Cultural Changes in Prehistoric Times", in Evans, Limbrey and Cleare (eds.), The Effect of Man on the Landscape: Highland Zone (CBA Research Report 11), London, 6-19.

Turner, J. (1970). "Post-Neolithic Disturbance of British Vegetation", in D. Walker and R. G. West (eds.), Studies in the Vegetational History of the British Isles (Essays to Godwin), Cambridge, 81-96.

Wainwright, G. J. (1967). Coygan Camp, Cardiff.

Wainwright, G. J. (1971). "The Excavation of a Fortified Settlement at Walesland Rath, Pembrokeshire", Britannia, II, 48-108.

Webley, D. P. (1976). "How the West was Won: Prehistoric Land-Use in the Southern Marches" in Boon and Lewis (eds.), Welsh Antiquity, National Museum of Wales, Cardiff, 19-35.

CHAPTER NINE

MAN-ENVIRONMENT RELATIONSHIPS

J. A. Taylor

ABSTRACT

This essay summarizes the recurrent themes and concepts developed in the eight preceding chapters on aspects of culture and environment in prehistoric Wales. The validity of established climatic and archaeological zonations is brought into question, and the advantages and limitations of new models based on ecosystems, territoriality, short-term rather than long-term cycling, both cultural and environmental, are briefly examined. Man-environment relationships are specifically discussed for each of the successive culture periods from the Upper Palaeolithic to the Iron Age. The continuing common threads in this succession are extracted and a final plea is made for more active cooperation in the field, on site and prior to, as well as during and after, excavation and sampling, between archaeologists and environmental scientists (a term which includes biogeographers, palaeobotanists, palaeozoologists, Quaternary scientists, etc.)

This final chapter takes the form of a postscript, distilled partly from the preceding eight essays and partly from alternative sources but focussing in particular on the nature and operation of the relationship between prehistoric settlements and prehistoric environments in Wales.

The man-environment equation

The man-environment equation involves relative rather than absolute terms, and variables rather than constants. It is difficult to shed the bias of modern viewpoints and it is even more difficult to achieve contemporary evaluations of behavioural attitudes of prehistoric man. But it is possible to construct a simple model based on the two major variables of the equation, i. e. environment and technology. Environment implies the general condition of plant, animal and soil resources within the stage and trend of climate change. Technology refers to the quality and range of both material culture and its expression in forms of settlement, agriculture and industry, and its resultant stability or instability and its potential for survival and expansion or construction and extinction.

Assuming that environment and technology may, on their own terms, be in the ascendant or descendant, four different combinations may theoretically

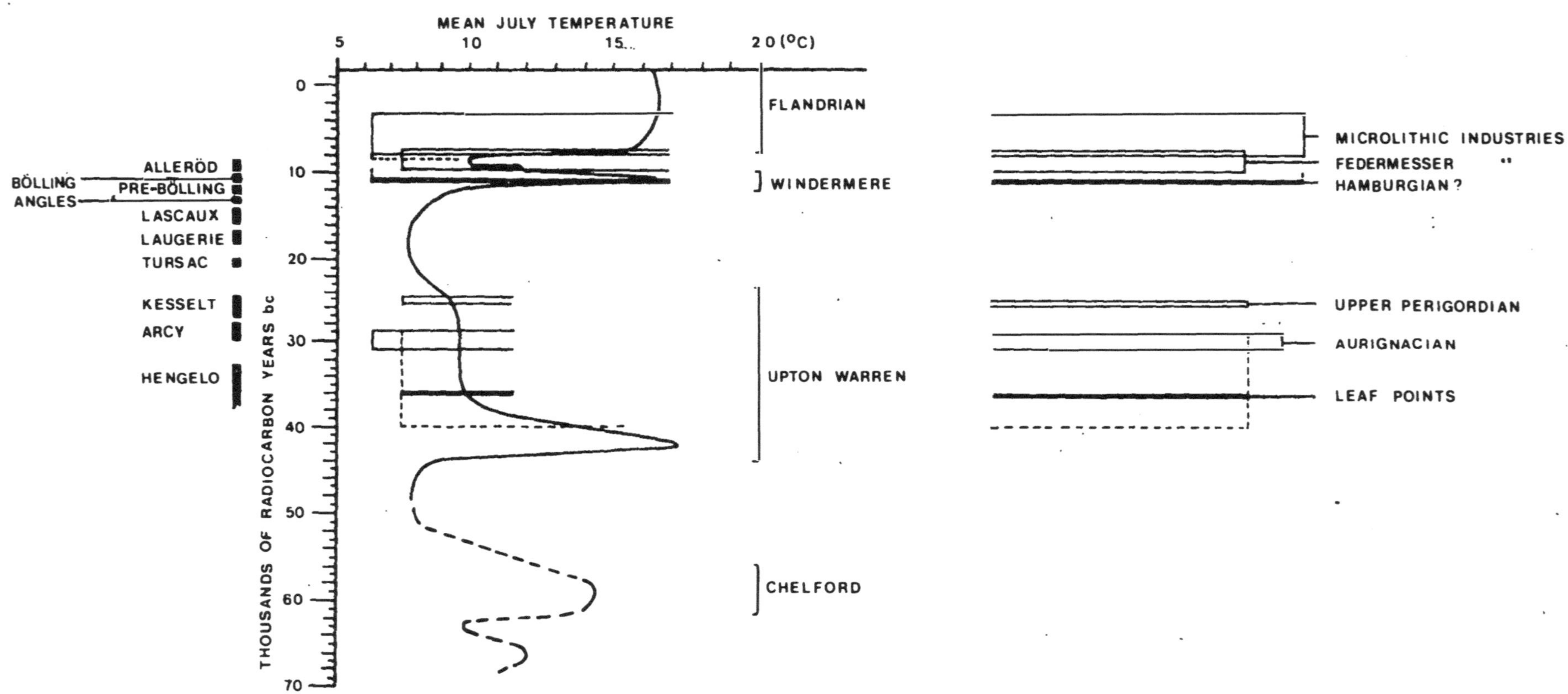

Fig. 9.1 Cultural evidence for Upper Palaeolithic activity in Britain plotted against (1) mean July temperature as derived from studies of fossil Coleoptera and (2) plotted against interstadials recognized from mainland Western Europe on the basis of pollen analyses. While a most probable date of about 36,000 bc is suggested for the manufacture of leaf points, dotted lines suggest that some might be as early as 40,000 bc when the first such spearheads appear in Central Europe, while the latest may have been associated with Aurignacian-type artifacts and hence as late as 28,000 radiocarbon years bc. Space did not allow most probable dates for the Bromme-type stemmed points found in Britain or for the Ahrensburgian assemblage from Avington VI to be indicated. This information is recoverable from the text. On the basis of the latest date from Anston Stones Cave and the earliest of the determinations from Thatcham Site IIIA an overlap between the latest occurrence(s) of a Federmesser technology and the earliest technology with microlithic equipment (but without stemmed points) is allowed for, with a dotted line indicating the oldest of the Thatcham determinations (Q 659).

occur as follows: (A) ameliorating environment; (B) deteriorating environment; (C) advancing technology; and (D) declining technology. Depending on the varying coincidences of (A) and/or (B) with (C) and/or (D) in both time and place, and the varying sequences over time of all four variables, different states of conservation or destruction of both environment and culture may be generated. For example the optimal conservational sequence would involve prolongation of (A) accompanied by either (D) or an early version of (C), implying minimal technical demand on an expanding environmental resource-base. On the other hand, a combination of (B) and (C) could imply the maximum exploitation of dwindling environmental resources unless (C) had the capacity for innovation and alternative technology and alternative conservation which could postpone the stage of natural and permanent depletion of the environment, should (B) be maintained. Again, the emergence of B and D together might preserve some stability in the man-environment relationship until either one or the other deteriorates beyond the survival limit of the other.

Thus a spectrum of tolerances and thresholds may be envisaged for although climate change is the major independent variable, biospheric change at ground is very much at the mercy of man's activities through deforestation, changes in albedo, ground hydrology, and manipulation of soil and fauna. These short-term modifications of the biosphere at ground may affect the timing and impact of climate changes operating over the longer term. In this sense regional environments or regional cultures may be differentially vulnerable to a deteriorating climate, or differentially capable of responding to an ameliorating climate. A host of interdependent variables, involving both man and 'accessible' components of the environment, would therefore militate against the application of simplistic models to interpret the broad trend of Late-glacial and Holocene interrelationships between environment and man. Nevertheless, a few broad generalisations may be attempted.

For the British Late-glacial period, i.e. the Upper Palaeolithic between about 40,000 and 10,000 B.P., Jacobi has portrayed in Figure 9.1 a repeated correlation between climatic improvements, the interstadials, and culture peaks. The major glacial maximum of the Devensian between 26,000 and 14,500 B.P. is coincident with an absence of cultural developments, and separates earlier Aurignacian settlements, contemporary with the Upton Warren interstadial, from the later Creswellian settlements identified with Windermere and associated interstadials. Such correlations are to be expected at a very cold stage with very tight margins and the absolute minimum of options available. Writing in 1973, (Taylor, 1973, p. 312) I predicted that a pattern of micro-oscillation with strictly regional interstadials would no doubt emerge, particularly for the Late-glacial period. Clearly, the application of the European-derived threefold Zonation of I, II and III based almost exclusively on pollen evidence (Godwin, 1956 and 1975) was premature. Nonetheless, a series of regional interstadials broadly analogous to the early identification of the Allerød and Bölling in Denmark, is emerging for parts of England and Wales e.g. for sites at Aberaeron (Smith, 1969; Smith and Taylor, 1969) and Clarach (Taylor, 1973), both on the coast of Cardigan Bay in west central Wales; and for sites in south-west Lancashire and north-east Cheshire (Baxter, 1980). The remarkable intensity and diversity of the periglacial environments of the Late Devensian is manifest in the work of Watson, sum-

Date	Climate	Pollen Zones	Culture Periods	(a) Central England S	(a) Central England W	(b) Welsh Borderland S	(b) Welsh Borderland W	(c) Cards. Bay Coastlands S	(c) Cards. Bay Coastlands W	(d) At Sea-Level S	(d) At Sea-Level W	(e) At 250m (820′ O.D.) S	(e) At 250m (820′ O.D.) W	(f) At 500m (1640′ O.D.) S	(f) At 500m (1640′ O.D.) W
1000 A.D.	Sub-Atlantic	VIII	Dark Ages												
A.D./B.C. 0			Romano-British	15.7 (60.3)	4.0 (39.2)	14.7 (58.5)	4.4 (40.0)	14.3 (57.7)	5.2 (41.3)	15.6 (60.1)	4.0 (39.2)	14.1 (57.4)	3.5 (38.3)	13.5 (56.3)	2.9 (37.2)
			Iron Age	15.5 (59.9)	4.1 (39.4)	15.1 (59.1)	4.6 (40.2)	14.6 (58.3)	5.3 (41.5)	16.0 (60.8)	4.1 (39.4)	13.7 (56.7)	3.6 (38.5)	13.5 (56.3)	3.0 (37.4)
				15.6 (60.1)	4.2 (39.6)	15.2 (59.3)	4.6 (40.3)	14.8 (58.6)	5.3 (41.6)	16.5 (61.7)	4.2 (39.6)	14.7 (58.5)	3.7 (38.7)	13.6 (56.5)	3.1 (37.6)
1000 B.C.	Sub-Boreal	VIIb	L.B.A. / M.B.A.	L.B.A. / IRON AGE TRANSITION **1200 B.C. Bronze Age 'Crisis Point'**											
			E.B.A.	16.4 (61.5)	3.7 (38.7)	16.0 (60.8)	4.3 (39.7)	16.1 (61.0)	5.0 (41.0)	17.3 (63.1)	3.9 (39.0)	15.9 (60.6)	3.3 (37.9)	14.6 (58.3)	2.8 (37.0)
2000				NEOLITHIC / BRONZE AGE TRANSITION **2500 B.C. Neolithic Crisis Point'**											
3000	Atlantic	VIIa	Neo-lithic	17.4 (63.3)	5.0 (41.0)	17.0 (62.6)	5.4 (41.8)	16.6 (61.8)	6.2 (43.1)	17.4 (63.3)	5.3 (41.5)	16.3 (61.3)	4.5 (40.1)	15.3 (59.5)	3.8 (38.8)
4000				MESOLITHIC / NEOLITHIC TRANSITION **4350 B.C. Mesolithic 'Crisis Point'**											
5000			Optimum	17.6 (63.7)	5.2 (41.4)	17.2 (63.0)	5.7 (42.2)	16.8 (62.2)	6.4 (43.5)	17.5 (63.5)	5.4 (41.7)	16.4 (61.5)	4.6 (40.3)	15.3 (59.5)	3.8 (38.8)
6000	Boreal	VIa	Mesolithic	17.0 (62.6)	4.3 (39.7)	16.6 (61.9)	4.7 (40.4)	16.2 (61.2)	5.4 (41.7)	17.0 (62.6)	4.6 (40.3)	15.6 (60.1)	3.2 (37.8)	14.2 (57.6)	1.7 (35.1)
7000		VIb / VIc		13.7 (56.7)	−1.0 (30.2)	13.3 (56.0)	−0.6 (31.0)	12.9 (55.2)	0.2 (32.3)	13.5 (56.3)	−1.5 (29.3)	12.5 (54.5)	−2.1 (28.2)	11.5 (52.7)	−2.7 (27.1)
8000		V / IV		12.6 (54.7)	−4.5 (23.9)	13.1 (55.5)	−4.9 (23.2)	13.4 (56.2)	−4.2 (24.5)	12.4 (54.3)	−5.1 (23.0)	11.6 (52.9)	5.4 (22.3)	10.8 (51.4)	−5.8 (21.6)
Average Mean 1901 – 1930				16.4 (61.5)	4.6 (40.3)	16.0 (60.8)	5.0 (41.0)	15.4 (59.7)	5.9 (42.6)	15.4 (59.8)	5.7 (42.2)	14.4 (58.0)	4.7 (40.4)	13.4 (56.2)	3.7 (38.6)

(d)–(f): in Wales, *ceteris paribus*

Fig. 9.2 (opposite) Chart showing estimated mean average summer and winter temperatures since circa 12,000 B.P. for (a) Central England, (b) the Welsh Borderland, (c) the Cardigan Bay Lowlands and (d) for sea-level 250 m (820' O.D.) and 500 m (1640' O.D.) ceteris paribus in Wales (after Taylor herein, Fig. 3.3). (N.B. Summer data refer to means for July/August, and winter data refer to means for December/January/February—see Fig. 3.2).

marised in Watson (1977), although the debate continues on the time-scale of origins of these periglacial phenomena (Bowen, 1977).

Surviving subsequent faunal analyses by Coope (1977a) is cold Zone III, which has been designated the 'Loch Lomond Stadial' by Sissons (1979). Its conventional dating to between 10,800 and 10,300 B. P. is obviously liable to modification at both ends and regionally over Britain as additional data accrue. Nonetheless, the survival of the base of the Holocene at about 10,000 B. P. has been even more remarkable. It is corroborated from both floristic and faunal fossil records. It has interregional consistency and may ultimately claim to have operated on a continental or even wider scale, not impossibly involving parts of both hemispheres (Kendall, 1969; van der Hammen, 1974; and Kershaw, 1978).

For the Holocene the pattern of man-environment relationships may be broadly interpreted in five successive stages with reference to the simplistic model outlined above (see Fig. 9.2 which is a replicate of Fig. 3.3). First of all, in the early Mesolithic, the rapid climatic amelioration of Zones IV and V of the early Boreal provided an increasing reservoir of environmental resources which must have been more than adequate to meet the demands of early Mesolithic man with much to spare. The superior later Mesolithic culture was located within Zone VIIa, the warm, wet Atlantic period which included the 'optimum' when the Holocene temperature curves peaked at up to 2°C higher than at present, especially in summer. The environmental amelioration was still able to meet the now increased pressure of technological advance but this was quickly followed by the transition to the Neolithic period c. 6500 and 5800 B. P. (Fig. 9.2). These centuries plus the full Neolithic, which lasted to about 5000 B. P., presented a third stage, with a combination of a revolutionary, advancing culture impinging on an environment poised for decline, slow at first but successively more rapid later in stage four. This was the long Neolithic-Bronze Age transition which lasted virtually a thousand years between about 5000 and 4000 B. P., followed by the Early Bronze Age. Stage five saw the acceleration of climatic decline in the Late Bronze Age and Iron Age and the increasing appropriation of contracting environmental resources by successively superior technologies. However, it should be stressed that the reality behind this broad application of the man/environment model lies in regional and especially altitudinal deviations in timing and coincidences, especially in areas such as Wales.

The Late-glacial

Bowen (herein Chapter One) refers to the difference of 15000 years in the timing of the climatic amelioration at the start of the Late-glacial of the Late Devensian glaciation, between the date of 13,000 B. P. on the basis of pollen evidence and 14,500 B. P. on the basis of beetle evidence (Coope, 1977b). The latter would indicate that the contemporary Zone I climate was virtually 'optimum', if only briefly (see Fig. 9.1), in stark contrast to the cold-tundra and dominantly non-arboreal pollen records for the Lower Dryas. Possible reconciliations of the two data sources lie in the difference between the timing and rates of ground micro-climatic changes as compared with macro-climatic trends normally compatible with Stevenson Screen height of 4' and more properly within the lower section of the tree-environment. Again, the angle of the sun, maintained as now, would have created periglacial topoclimates quite different from those of today in higher latitudes. The warmth of south-facing slopes would have been enhanced as would the cold, comparatively, of north-facing slopes (Crampton and Taylor, 1967; Williams, 1975). A mosaic of warm and cold habitats would have favoured early summer migrations of mobile fauna whereas tree migrations would have been restricted to, and localised within, only the most favoured of the sunnier, yet sheltered slopes. Hence, the very low counts of tree pollen. Again, opportunities for hibernation, both long or short-period, would abound for fauna in this thermally varied landscape. Indeed, by the time the birch-tundra of contemporary Zone II was established, beetle evidence already indicates a cooling climate. It may be assumed therefore that beetles also retreat substantially earlier than trees just as they colonise earlier. Their ground microclimates are more sensitive to changes in both maximum and minimum temperatures, and therefore to changes in the diurnal and seasonal ranges of temperature, i. e. temperature extremes. Climate changes begin to operate by the variation in frequency of extremes and in the shortening or lengthening of temperature ranges. Thus ground fauna of the size of beetles would be very sensitive to the early signals of climate change and would migrate promptly. Indeed, the speed with which bioclimates may change has been well demonstrated in the work of Woillard (1978 and 1979) who has proposed that the establishment of the beginning of the climatic change to Devensian glaciation in north-eastern France (i. e. the replacement of mixed deciduous forest by coniferous woodlands) took only 150 ± 75 years, and, moreover the initial "borealisation" of these forests some 115,000 years ago took less than twenty years. Presumably, reverse changes i. e. ameliorations, although relatively slower (compare the slow recovery of spring from winter with the persistence of autumn into winter in the modern British climate) was possible over somewhat longer time-scales. Another possible attendant complication might have the varying size and condition of landbridges to the European continent from the contemporary British 'peninsula', due to varying rates of ice-melt and possible réfreezing between the several cold and warm stages of the Late-glacial. The Windermere Interstadial (c. 13,000-11,000 B. P.; Coope, 1977b and Pennington, 1977) with a short-lived but impressive peak in temperatures, equivalent to those of the climatic optimum at c. 7,000 B.P., could have involved (if it was a strictly non-local climatic event embracing at least sub-continental proportions i. e. affecting N. W. Europe and the N. E. Atlantic) a short-term but rather rapid rise, then fall, in sea-level (Synge, 1977). Certainly the width and seasonal functioning of selected land-bridges could have been temporarily curtailed and then expanded. Migration rates of man, animals could have been modified as a consequence, a possibility which may have a bearing on Jacobi's radical reappraisal, promulgated herein (Chapter Two p. 76) that man did not return for certain to Britain after the Devensian glaciation till later part of Zone II, i. e. about 11,500 B. P., some 3000 years later than previously considered by Mellars (1974), Campbell (1977) and Wymer (1977).

Jacobi (herein, Chapter Two, p. 49) also explains how a Federmesser technology was innovated into Britain during the 10th millennium B. C., i. e. in late Zone II. It extended from the south-east of England as far west as Devon and extreme west Wales and as far north as the central Pennines and

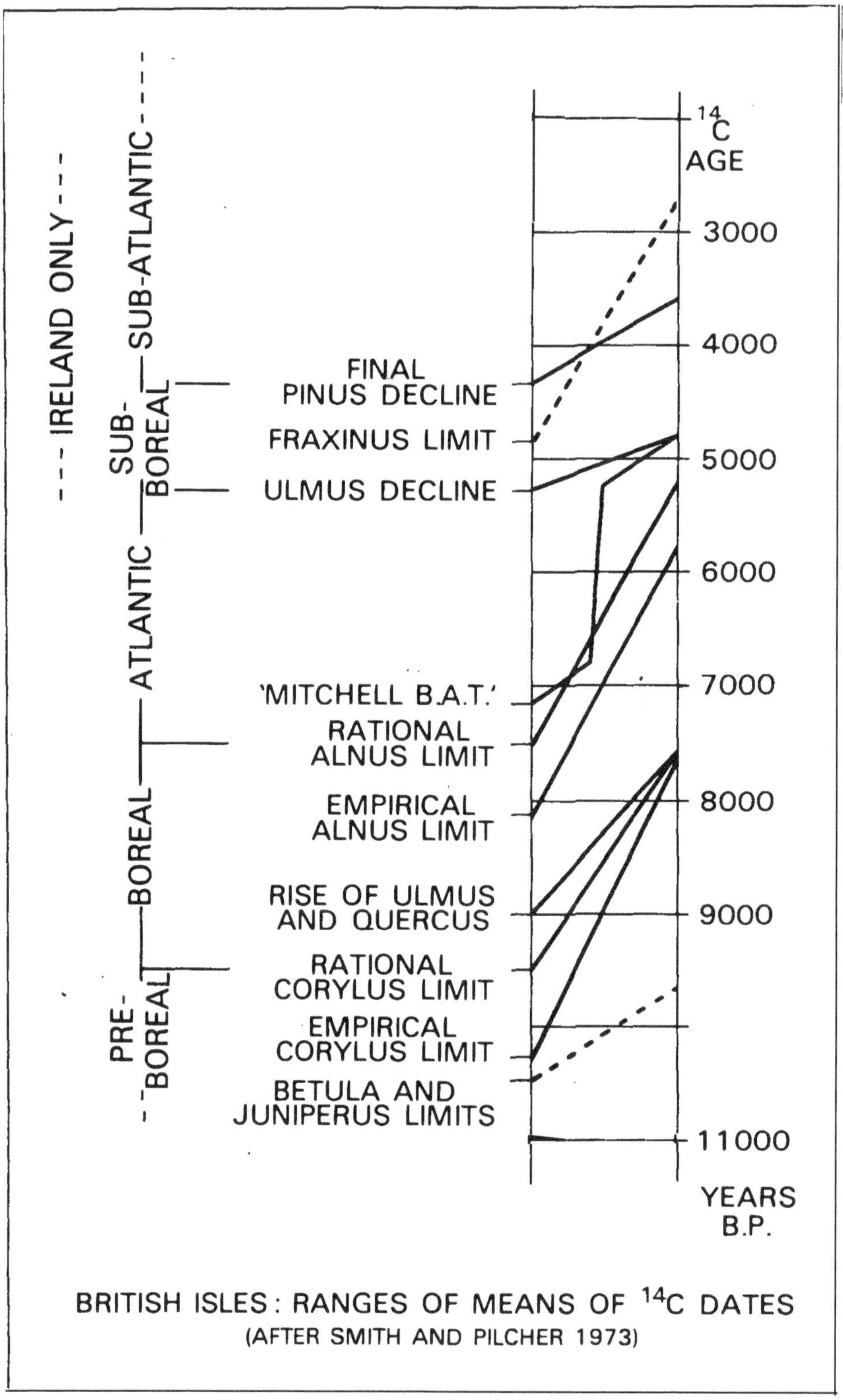

Fig. 9.3 British Isles: ranges of means for radiocarbon dates for selected pollen zone boundaries and other features (after Smith and Pilcher, 1973).

the southern Lake District (Figures 2.8 and 2.9). Presumably arriving later still to the north and west, it persisted into Zone III. This pattern of cultural retardation in north and west Britain and cultural innovation is the south and east of Britain is familiar. Federmesser types in northern England and possibly south-west Wales suggest survivals there into early Zone III.

Within the 9th millennium B. C. the correlative evidence for the Zone III recession to cold conditions of Sisson's (1979) Loch Lomond Stadial is consistent and convincing, although regional variations in onset, intensity and retreat of the combination of glacial and periglacial conditions will no doubt emerge. Sissons quotes a contrast, in July mean temperatures, between 6°C for the south-eastern Grampians and 10°C for southern England with intermediate values of 9°C for northern England and 7.5°C for the Lake District. This pattern would suggest that much of lowland England enjoyed a substantially less severe periglacial climate than the uplands to the north and west where glacier development was considerable. For example, 64 glaciers occurred in the Lake District. They were mostly cirque glaciers but some of the valley glaciers attained up to 5 km in length. In Scotland some large individual ice sheets accumulated over the Highlands e.g. one of 360 sq km surface area in the central Grampians, and another 600 m deep in the Great Glen. The surprisingly longer summers and more discontinuous permafrost of southern England in particular, but also including most of lowland England and the Welsh Borderland, must have allowed adequate survivals of tundra fauna and adequate access to the tundra-taiga ecotone (Jacobi herein, Chapter Two p.76 for cultural survivals.

In late Zone III, the Federmesser technology in southern and southeastern England was replaced by a new form involving tanged spearheads, archery equipment tipped with Ahrensburgian stemmed points and microlithic equipment identical with that found in the earliest Holocene in extreme southeast England, spreading later north and west in traditional style. Thus at about 10,000 B. P. lies a cultural discontinuity which is closely correlated in time and spatial expression with the last permanent and extensive climatic amelioration at the launching of the Holocene period.

The Holocene

As Figure 9.1 shows, the emerging overlaps between established culture periods are such as to merit independent recognition in the Holocene timescale. Similarly, Smith and Pilcher (1973) have revealed the diachroneity of pollen zonations as between upland and lowland Britain and also over relatively short distances (see Figure 9.3). Variabilities of ± 1000 years and in some instances ± 1500 years, may occur between such markers as the Boreal-Atlantic Transition. However, a major survivor of the analysis is the 'elm decline' which which, it is claimed, 'appears synchronous' at 5,200 B. P. ± 100 years. However, the distribution and elevation range of the thirty or so available sites used can hardly claim to be representative. There is not one in Wales where, without doubt, future research will help to corroborate the view propounded by Taylor (1973) that within the confines of the Principality major deviations occurred because of sharp contrast in altitude and slope, in the timing and duration of conventional pollen zones derived initially from sites at mostly lowland elevations in England and Scotland.

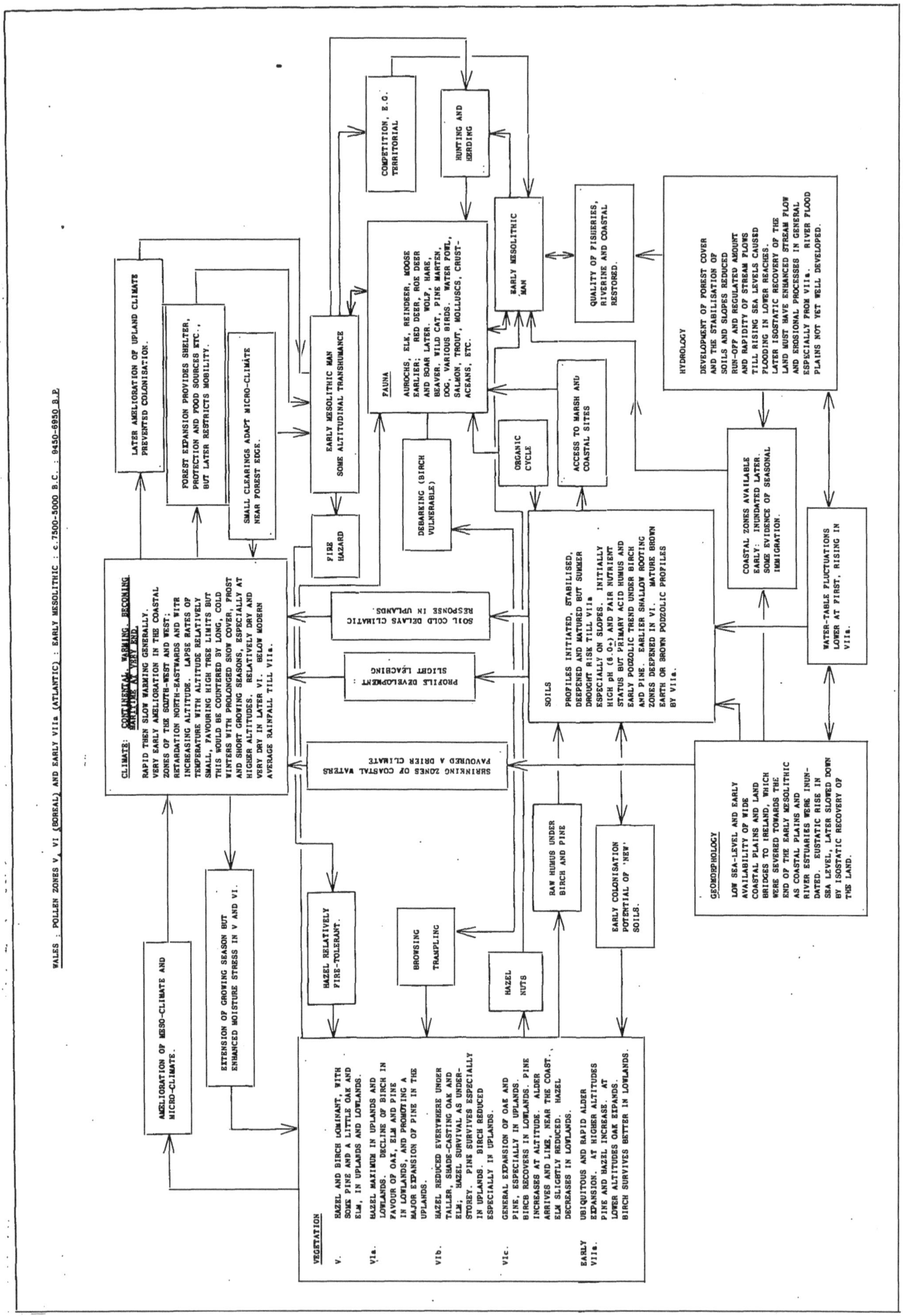

Fig. 9.4 Ecosystem model for the Early Mesolithic (after Taylor herein, Fig. 3.5).

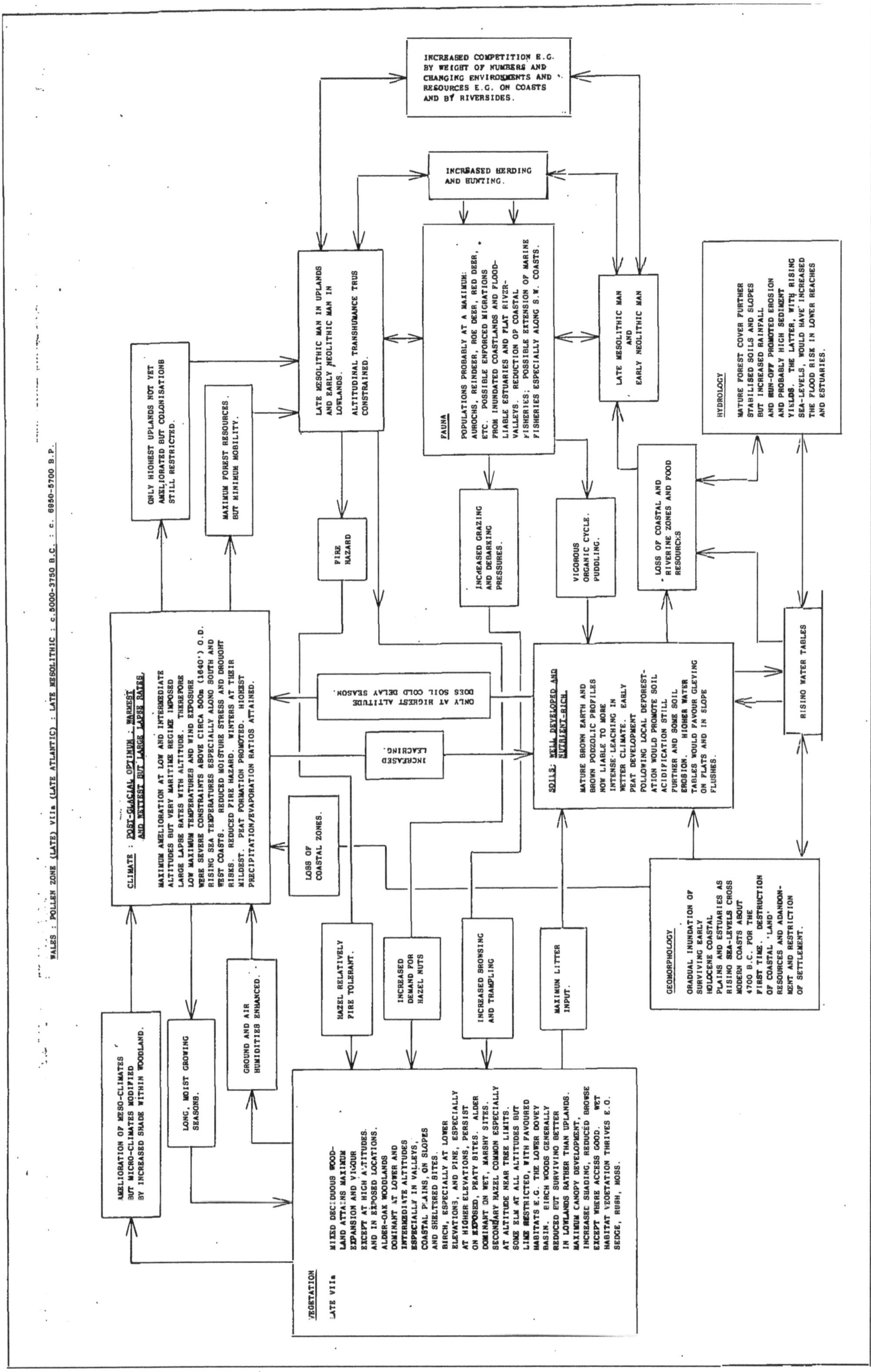

Fig. 9.5 Ecosystem model for the Late Mesolithic (after Taylor herein, Fig. 3.6).

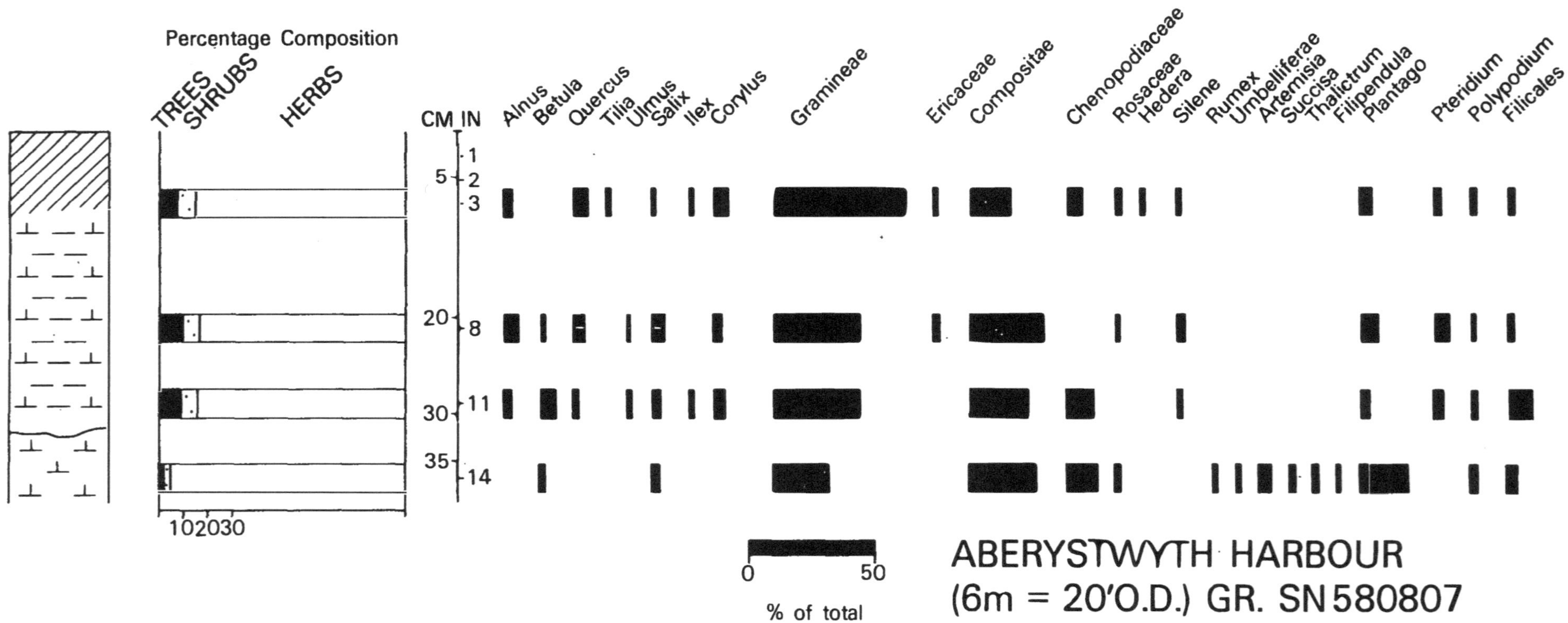

Fig. 9.6 Pollen diagram for the Mesolithic site at Aberystwyth Harbour (after Smith, R. T., 1969).

The Mesolithic

Figures 9.4 and 9.5 portray aspects of man-environment interactions for the Early and Late Mesolithic, respectively. The very early climatic release of the littorals of extreme south-west Wales at the earliest stage of the Holocene is mirrored in Figure 4.6 which reveals a concentration of Early Mesolithic sites on southwestern and western coasts with the exception of Rhuddlan in the Vale of Clwyd. Smith (1969) studied the fossil pollen record for the Aberystwyth Harbour site and his results are reproduced in Figure 9.6. Subsequent to O. T. Jones (Jones and Pugh, 1935) finding flints and charcoal in 1912, Thomas and Dudlyke (1925) collected material from a second site just below Aberystwyth hospital and, according to Jacobi herein (Chapter 4 p. 139), did not segregate the finds from each site. Whilst some of the large, obliquely blunted points and triangles exhumed are potentially very early Mesolithic, other microlith-backed bladelets and scalene micro-triangles are clearly later in date. Some continuity of settlement through the Mesolithic may be implied. The pollen diagram (Fig. 9.6) reveals a basal horizon at 37 cm which indicates a Late-glacial stage with very low tree pollen, about 2%, and dominance of Gramineae, Compositae, Chenopodiaceae and Plantago, all suggesting an open tundra environment. This basal layer was probably soliflucted from the adjacent hill slope of Pen Dinas but the overlying top-soil has been derived from hill wash at some later stage during the Holocene. At 29 cm the arboreal pollen, although still small, has increased to 9% and peaks at 11% at 21 cm. Today and at all stages of existence of the 'Plain of Cardigan' (Taylor, 1973) in Late-glacial and Holocene times, this is (was) a windy, exposed site which itself would be responsible for a reduction in site pollen. An earlier combination of birch and hazel at 21 cm is succeeded by a mixture of alder, oak, willow with some lime etc. suggesting a coincidence with the mid-Holocene climatic optimum. An uninterrupted forest cover at most here would have permitted easy access for Mesolithic man with the local availability also of both aquatic and terrestrial food sources in a congenial environment with, as now, conspicuously mild winters with very infrequent snow.

The site preferences in the late Mesolithic (see Figure 9.7) were perhaps even more specialised, mesoclimatically, and accessible to a diversity of habitats and food sources including sea, shore, estuary, river, forest and mountain. Following Jacobi (Chapter Four, herein), no less than 80% of the Late Mesolithic 'find-spots' are located in coastal Pembrokeshire or the Gower where winters were at their mildest and sea temperatures at their highest in relative contemporary terms. Hazel woods and hazel nuts also abounded in these corners of Wales. There is evidence to suggest that some form of transhumance also operated with summer settlements and transit camps inland at higher elevations which were brought into use, if only temporarily, for the first time. However, the severe lapse rates of the Atlantic climatic period precluded access to the highest plateaux where growing seasons were sharply curtailed. Again, the eustatic rise in sea-level reached the modern coastline about 6700 B. P., thereby not only severing landbridges to the continent but also inundating Early Mesolithic sites and constraining coastal settlements throughout Britain. There is a greater representation of inland sites on Figure 9.7 compared with Figure 9.8. Jacobi concludes (Chapter Four herein p. 149) that there is no irrefutable evidence either way

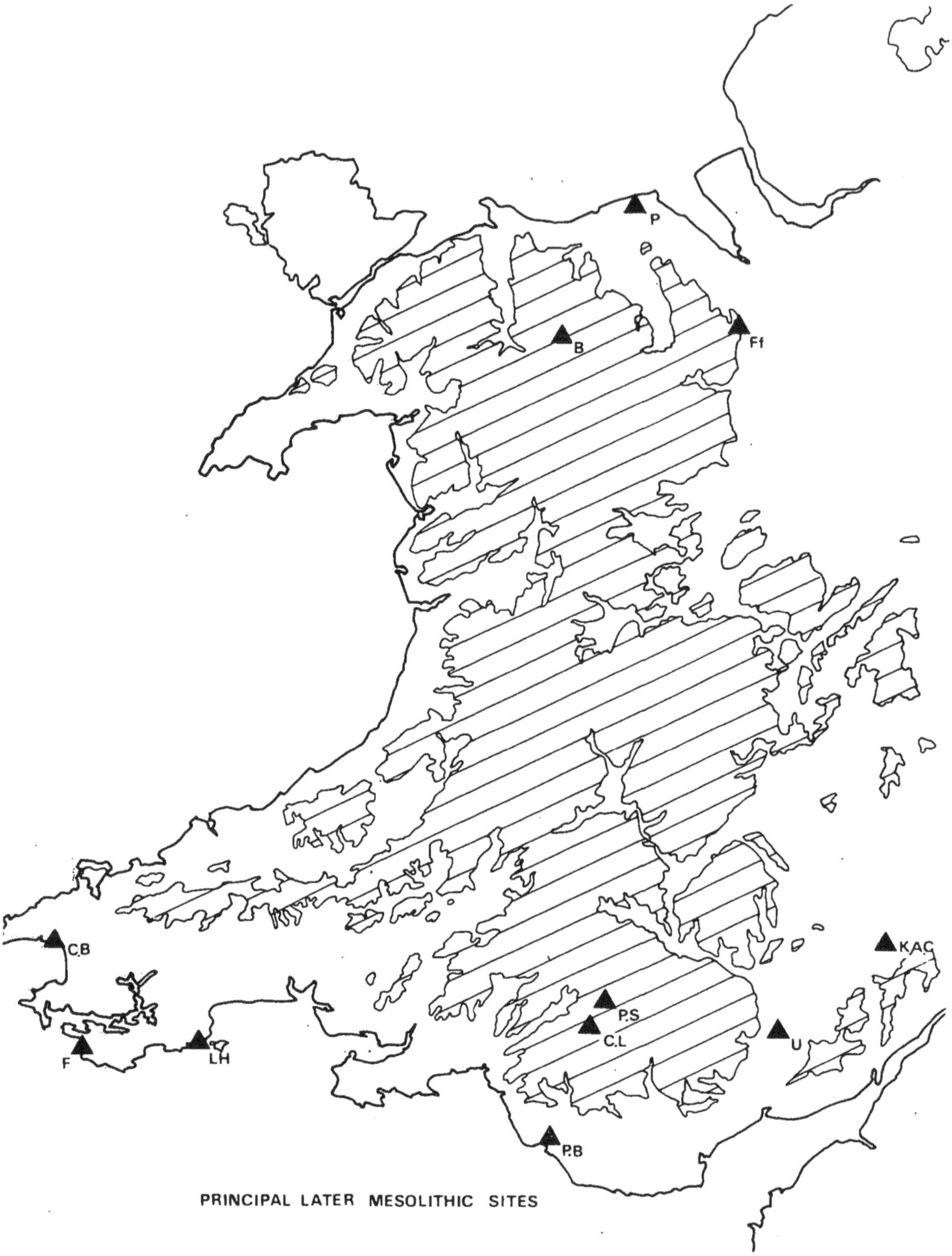

PRINCIPAL LATER MESOLITHIC SITES

Fig. 9.7 Principal Later Mesolithic Sites in Wales

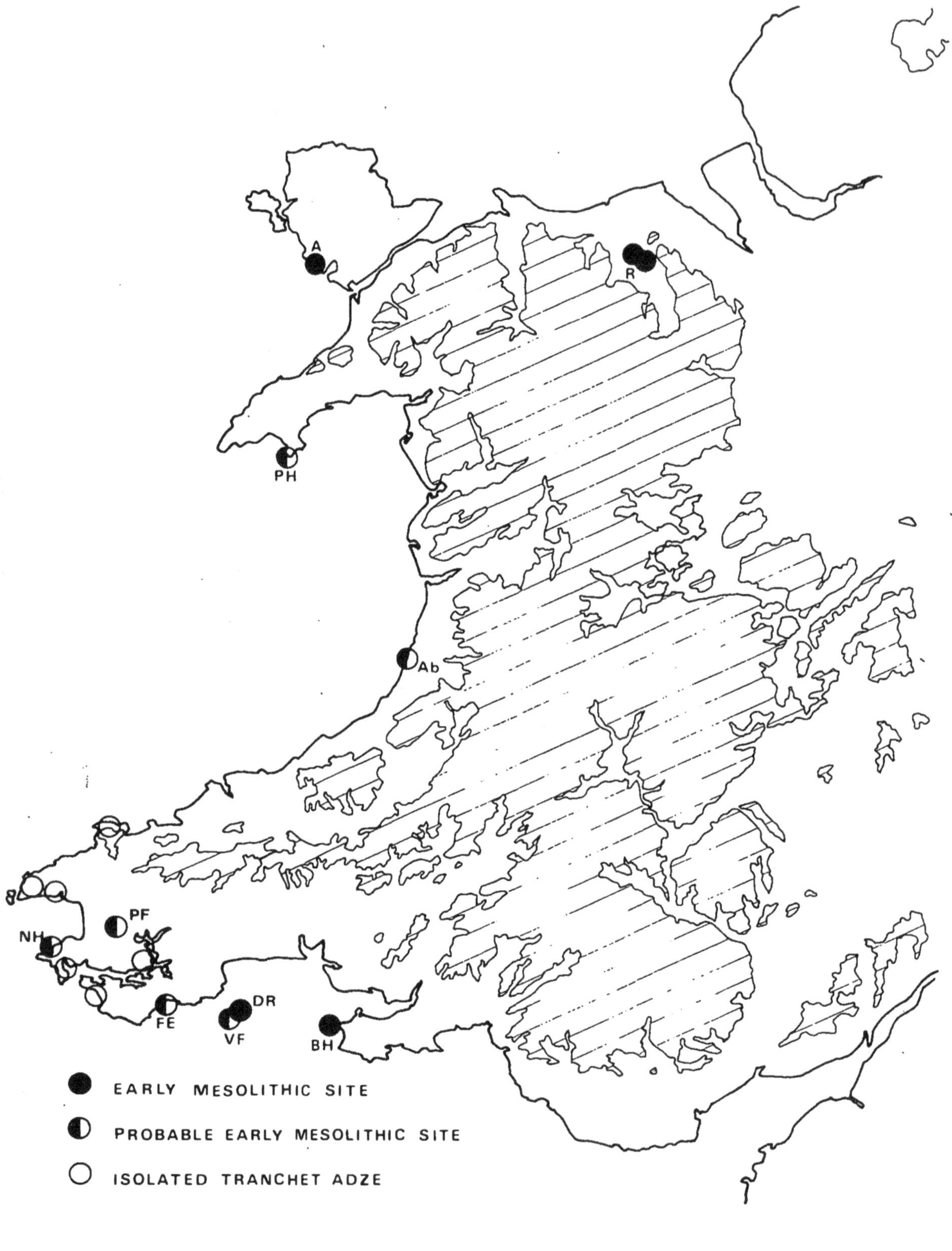
A
R
PH
Ab
PF
NH
FE
DR
VF
BH
EARLY MESOLITHIC SITE
PROBABLE EARLY MESOLITHIC SITE
ISOLATED TRANCHET ADZE

Fig. 9.8 Map of Early Mesolithic Sites in Wales (after Jacobi herein, Fig. 4.6).

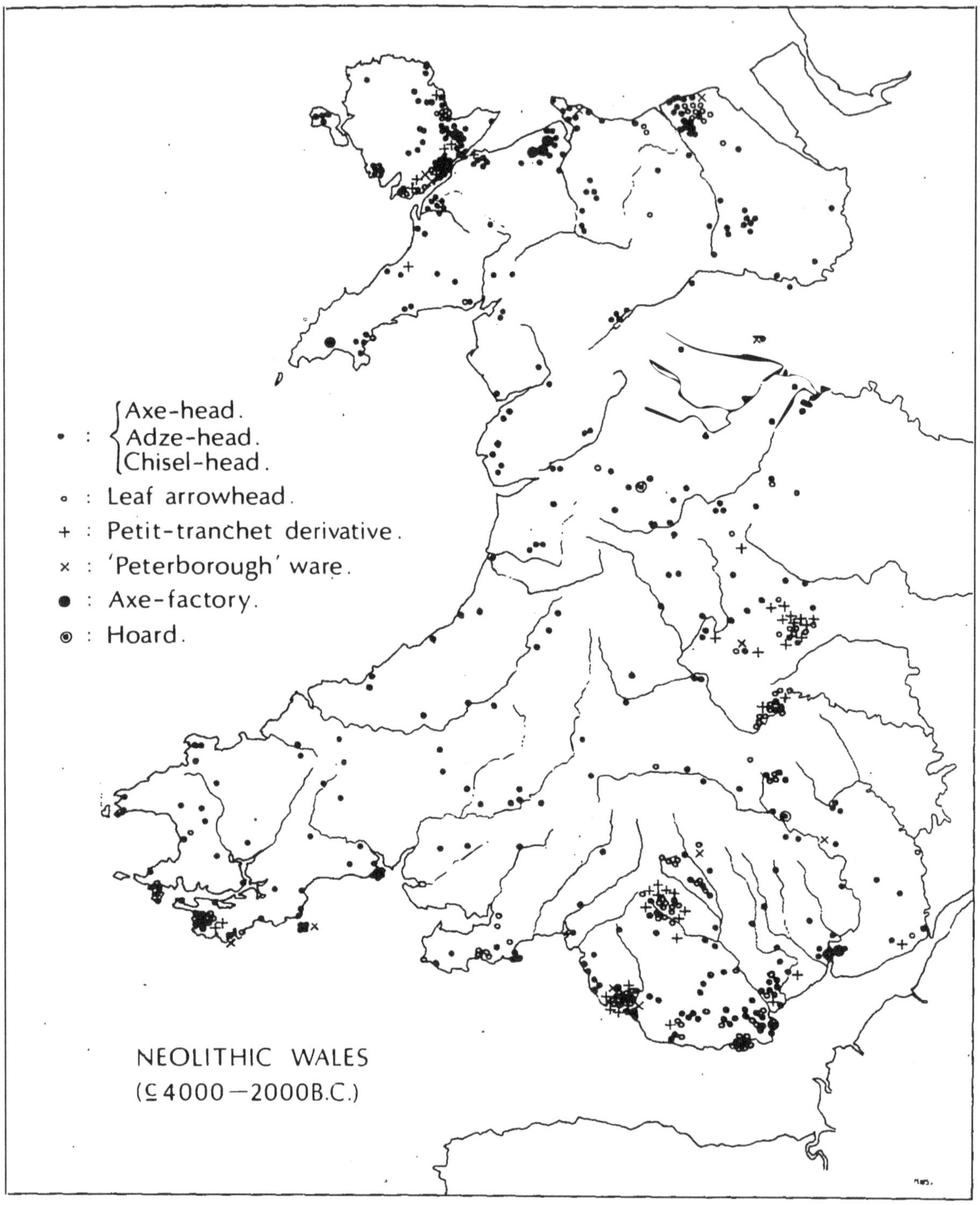

Fig. 9.9 Recorded Welsh finds of Neolithic artifacts (after Savory herein, Fig. 5.2).

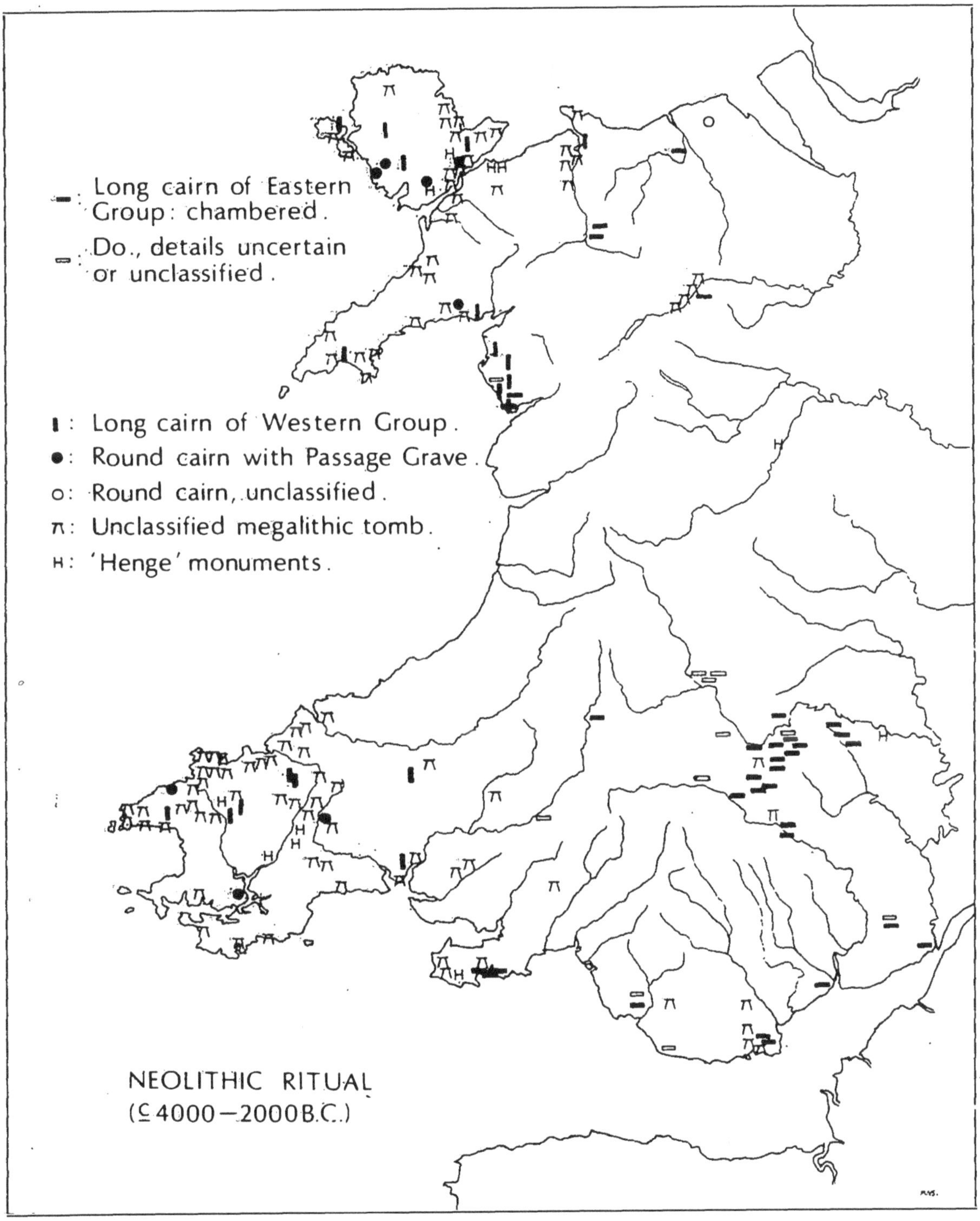

Fig. 9.10 Distribution of megalithic monuments in Wales (after Savory herein, Fig. 5.4).

for the actual presence of Mesolithic man in Wales prior to 7000 B. P. i. e. before the start of the Late Mesolithic. He is similarly agnostic about whether a Late Mesolithic culture was introduced into Wales at the same time as its parallel innovation elsewhere in Britain.

The Mesolithic-Neolithic Transition

For about 700 years or so between c. 6500 and 5800 B. P. the peripatetic Mesolithic culture was gradually replaced by the revolutionary, sedentary Neolithic settlements. A superior system and a superior technology arrived just after the peak of the climatic optimum when temperatures began to fall towards the end of the Atlantic period. As Figure 9. 9 shows, the Neolithic colonisation achieved a much more extensive penetration of the interior of Wales, often along the river valleys, but many coastal concentrations along the coasts of North and South Wales are also in evidence. Deforestations prepared the way for grazings and crops and more permanent settlements and life-style. Only the higher and remote uplands and interfluves survived this transformation of landscape which involved, for the first time in the Holocene, the beginnings of depletion of environmental resources, including soil deterioration and the onset of growth of upland peat cover—beneath which evidence of an even more extensive Neolithic occupation may be awaiting future discovery.

The Neolithic

The decline or absorption of Mesolithic influence within the new 'imported' Neolithic culture was complete by about 6500 B. P. but Savory (Chapter Five herein, p. 217) reminds us of the possibility of the survival of certain transhumant or migratory seasonal systems especially in interior Wales where the initiation of upland peat formation occurred successively in response to woodland clearances, and the late Atlantic climate deteriorated quite substantially on the higher plateaux. Climatic deteriorations affect montane zones first as Taylor (Chapter Three, p.110) explains herein. Above all, a regional development of technology became the hall-mark of the Mid-Neolithic (see Fig. 9. 10), with a marked dichotomy between easterly forms associated with adjacent England and Europe and westerly forms identified with 'Atlantic' Wales, mostly west of the major north-south watershed and displaying close affinities with Ireland. It is their house-style, burial customs, pottery types etc. , which enable the differentiation of these regional groups. It is highly probable that pastoral activities rather than arable cultivation dominated Welsh Neolithic agriculture especially on the higher remote terrain, thus constraining the population capacity of all save the favoured coastal zones and lower vales. The Later Neolithic, as with the Earlier, merged for almost one thousand years with the continuing developments of the Early Bronze Age. This transitional millennium coincided with the gentler altitudinal lapse rates of the Sub-Boreal and witnessed a large scale shift of colonisation into moorland areas e. g. Central Denbighshire and the Brecon Beacons (Lynch, 1979).

The Bronze Age

Burgess (Chapter Seven herein) emphasises the unreality of the term 'Bronze Age' and models a mosaic of territoriality which develops increasing concentration of settlement form and land use within a continuum of change

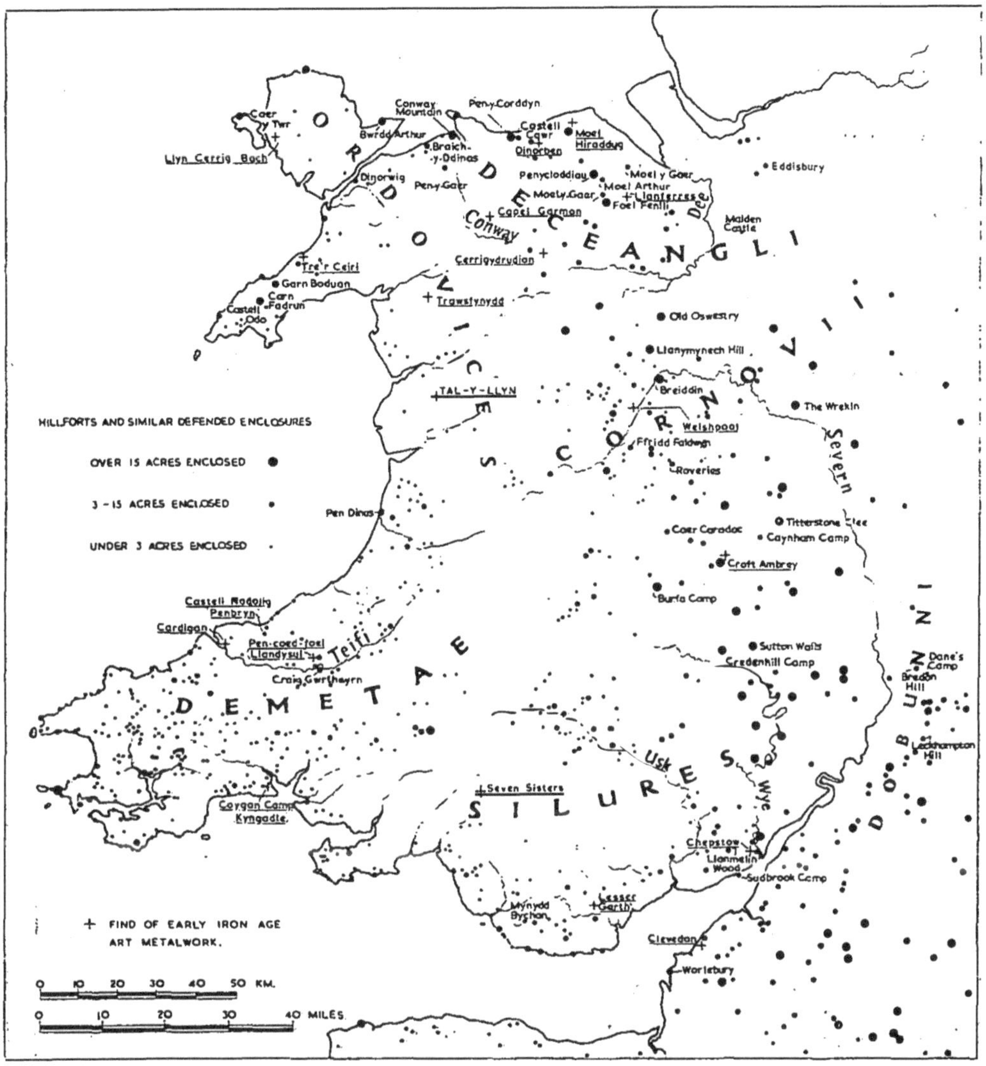

Fig. 9.11 Wales and the Marches in the Early Iron Age (after Savory herein, Fig. 8.1).

which he projects forwards into the Iron Age and possibly backwards, into the Neolithic. 'Constant trickles' and 'cumulative Celticisation' displace traditional notions of one-off invasions or colonisations. In parallel terms, the concept of an episodic and later rapid climatic deterioration, synonymous with forest depletion, peat expansion and soil acidification, reinforce the post-1200 B. C. evidence for abandonment of upland, and later some land sites, more specialised dwellings, hill forts mainly, more concentrated field-based agriculture and a more sophisticated social, economic and political organisation of both people and resources. Stanford (1972) is quoted as estimating that the population of Britain at the end of the Bronze Age could have been comparable to that at the time of the Domesday Survey.

About 700 B. C. the eventual onset of the cooler, wetter Atlantic climatic period provides an environmental watershed within the Penard period (1050-850 B. C.) which is described by Burgess as a major cultural watershed in a European, as well as in a more narrowly defined British, sense. Indeed, the natural hazards of flooding, severe winters, storms etc. especially in marginal montane zones, river floodplains, coastal reaches etc. applied to the last few centuries of Bronze Age (Fig. 3.10) as well as to the ensuing Iron Age (Fig. 3.11). Lamb (1977) reports that between 1000 B. C. and 300 B. C. temperatures decreased by up to 2°C which must represent one of the major deteriorations of the entire Holocene and historic record.

The regional character of Neolithic cultural distributions was replaced by the territories and local 'resource-catchments' which supplied the more concentrated settlements. The two-thousand-year-old Bronze Age was transformed as from c. 700 B. C. at the Hallstatt Stage and by 600 B. C. had merged with the Iron Age.

The Iron Age

The continuing superimposition of the European Hallstatt/La Tène chronology on to the British and the Welsh archaeological successions for the Iron Age will have to survive the test of future detailed field work in central Wales to resolve the problem of the location and identification of Iron Age and preceding Bronze Age field systems. Savory comments (Chapter Eight herein) on Fig. 9.11 that the peripheral emphasis of the distribution of Iron Age settlements in Wales does not necessarily imply that the interior was neglected or underused since so many hillforts were strategically placed for access to summer grazings. Indeed, transhumance survived into the historic period and into modern times in relation to seasonal movements of stock both within and between farms as part of a commercial system of management. Intriguing is the proposed increase in sheep numbers in Wales in the early Iron Age; a again future research is needed to confirm this view.

The Iron Age settlements were more intense and more fortified than those of the late Bronze Age. There is certainly evidence of deforestation of middle and lower plateaux and slopes and also valley bottoms for the first time. Witness the late Iron Age /early Roman site at Ystwyth Forest described by Taylor (1973 and 1975) where a distinctive charcoal horizon of radiocarbon age 2133 ± 110 years B.P. (Q.814) and a buried layer beneath it are coincident with severe deforestation which is well shown in the fossil pollen record.

Remarkable, however, too are the regional differences between indigenous scattered farmsteads of central and north-west Wales and the larger, better fortified, more nucleated settlements of the central and northern Welsh Marches and the north Wales coast. Obviously, a stronger arable component was possible in the agriculture of the latter. Finally, the intensification of land organisation and of territoriality is finally expressed in the four classical tribal regions of (1) the Cornovii and Degengli in the north-east, (2) the Ordovices in the north-west, (3) the Silures in the south-east and (4) the Demetae in the south-west of the Principality. As Savory argues herein (Chapter Eight), the fact that it took the Romans more than a quarter of a century to subjugate Wales and the Marches, in contrast to the very few years that were required to conquer south-east Britain, implies that the various Celtic tribes then present in Wales could not have been negligible either demographically or militarily, even allowing for the difficulty and remoteness of terrain compared with the English lowlands which the Romans must have found not only more accessible to their military and 'occupational' style but also a less forbidding physical environment as regards contour and climate.

Conclusion

Four major conclusions may be drawn as follows.

I. New Models for Old

The established Holocene climatic periods of Boreal/Atlantic/sub-Boreal/sub-Atlantic were originally proposed by Blytt (1876) over a century ago and later expanded by Sernander (1908). Not only do they reflect nineteenth century uniformitarianism but they have been applied to a quite different range of maritime climates in the British Isles which also, by virtue of geographical position, must have been vulnerable to a different sequence and timing of glacial, periglacial and Late-glacial phases and climatic recovery during the subsequent Holocene period. Again, it was in Denmark in particular that the initial three-fold division of the Late-glacial was proposed (e.g. Iversen, 1954 and 1967) eventually with at least two interstadials, the Allerød and Bølling, respectively. Already, national and regional deviations from this model for different parts of Britain are becoming established; it is very probable that more will follow when more detailed biogenic deposits are encountered in other areas. New models should reveal shorter-period oscillations in British regional palaeo-climates over the past 14,000 years or so.

A strikingly parallel but quite independently derived trend in archaeological research is towards a modification of the basis and application of the established culture periods of Palaeolithic, Neolithic, Bronze Age and Iron Age. Again, perhaps continental concepts have been superimposed on British data which, as it increases in flow, will develop its own regional chronologies. New models of territorial organisation and resource extraction and management should emphasise the over-simplification of the standard cultural breaks and the inevitable continuity of cultural processes and culture change all in terms of specific regional and often quite localised operation.

The ecosystem model as discussed and applied herein (Taylor, Chapter Three), although descriptive and strongly inferential, does provide a framework for the study of man/environment relationships and for the integration of environmental and archaeological data for specific cultural stages in specific regions.

II. The Irish Connection

Jacobi (Chapters Two and Four herein) has amply demonstrated the European connection in his interpretations of the British Upper Palaeolithic and Mesolithic. An Irish connection is also an additional and probable bridge towards a fuller interpretation of both environmental and cultural successions especially in that part of Wales west of the main north/south watershed and including Anglesey, western Gwynedd (in particular the Llŷn peninsula) western and south-western Dyfed (especially the Pembrokeshire peninsula). No doubt the Irish Sea basin, with the emphasis on those areas of it adjacent to modern shore-lines, will continue to be the best and most fruitful sources of evidence for former land-bridges and plains once vegetated and settled and now inundated.

III. Territoriality

A recurrent theme in our discussions of both culture change and environmental trends has been the changing significance and evaluation of territory, land, space, region, catchment, habitat. Behind the tribal territories of the Iron Age stretch in time the successive dispersal of settlements, disintensification of land use, desophistication of land management till we enter the very early Holocene and 'late' Late-glacial dispersed, mobile, primitive colonisations which sought out areas and sites where environmental resources were adequate and environmental hazards and risks were minimal. Prehistoric perceptions of area, distance, access and risk must clearly have varied with the culture, the technology, the land-use, the environment, the social, economic and political attitudes etc. All these factors focus on variable definition and applications of the notion of territoriality.

IV. The Field is the Laboratory

For both the archaeologist and the environmental scientist, the field represents a common laboratory. Both specialists must identify, either by objective survey or by instinct or chance, the sites and sections which will yield the evidence of prehistoric settlements and landscapes. Two major areas must, on *a priori* grounds, be expected to yield rich archaeological and palaeo-environmental evidence. These are firstly, the peat-covered uplands, not least those of Wales, where basal and sub-peat sections will eventually expose the several culture surfaces, the field systems etc., that preceded the deforestation stage (assuming it occurred extensively) and the accumulation of the peat cover (see Taylor, 1980). The second type of environment likely to yield an abundant harvest of information are the modern coastal zones, both their landward and seaward extensions, where deep cores often reveal continuous records of events over long time-scales.

It goes without saying that more active cooperation at field sites, *before* excavation and/or *before* sampling begins, could be of immeasurable mutual benefit to both archaeological and palaeoenvironmental research.

REFERENCES

Baxter, J., 1980. Private Communication.

Blytt, A., 1876. Essay on the immigration of the Norwegian flora. Christiania, Norway.

Bowen, D. Q. (ed.) 1977. Wales and Cheshire-Shropshire Lowland. INQUA Excursion Guide A8 and C8. Geo Abstracts U. E. A., Norwich, England.

Campbell, J. B., 1977. The Upper Palaeolithic of Britain (2 vols.). Oxford.

Coope, G. R., 1977a. Quaternary Coleoptera as aids in the interpretation of environmental history in Shotton, F. W. (ed) British Quaternary Studies: Recent Advances, Clarendon Press, Oxford, pp. 55-68.

Coope, G. R., 1977b. Fossil Coleopteran assemblages as sensitive indicators of climatic changes during the Devensian (Last' Cold Stage. Phil. Trans. R. Soc. Lond. B., 280, pp. 313-40.

Crampton, C. B. & Taylor, J. A., 1967. Solifluction terraces in South Wales. Biuletyn Periglacjalny 16, 15-36.

Godwin, H., 1956 and 1976. History of British Flora. C. U. P.

Hammen, T. van der, 1974. The Pleistocene changes of vegetation and climate in tropical South America. J. Biogeogr. 1, 3-26.

Iversen, J., 1954. The Late-glacial flora of Denmark and its relation to climate and soil. Danmarks Geologiske Undersøgelse, Series II, 88, 87-119.

Iversen, J., 1967. Natureus uduikling siden sidste islids maksimum. Danmarks Natur., 1.

Jones, O. T. & Pugh, W. J., 1935. The geology of the districts around Machynlleth and Aberystwyth. Proc. Geol. Assoc. 46, 247-300.

Kendall, R. L., 1969. An ecological history of the Lake Victoria Basin. Ecol. Monogr: 39, 121-76.

Kershaw, A. P., 1978. Record of last interglacial-glacial cycle from north-eastern Queensland. Nature, 272, 159-61.

Lamb, H. H., 1977. The late Quaternary history of the climate of the British Isles in Shotton, F. W. (ed.) British Quaternary Studies: Recent Advances. Clarendon Press, Oxford 20, 283-98.

Lynch, F., 1979. Private Communication.

Mellars, P. A., 1974. The Palaeolithic and Mesolithic in Renfrew, C. (ed.) British Prehistory: a new outline, 41-99, 268-79, Duckworth, London.

Pennington, W. P., 1977. The Late Devensian flora and vegetation of Britain. Phil. Trans. Roy. Soc. London, B. 280, 247-71.

Sernander, R., 1908. On the evidence of Post-glacial changes of climate furnished by the peat mosses of northern Europe. Geologiska Föreningens i Stockholm förhandlinger, 30, 465-78.

Sissons, J. B., 1979. The Loch Lomond Stadial in the British Isles. Nature, London, 280, 199-203.

Smith, A. G. & Pilcher, J. R., 1973. Radiocarbon dates and vegetational history of the British Isles. New Phyt., 72, 903-14.

Smith, R. T., 1969. Studies in the Post-glacial soil and vegetation history of the Aberystwyth area. U.C.W., Aberystwyth. Unpublished Ph.D. thesis.

Smith, R. T. & Taylor, J. A., 1969. The Post-glacial development of soils and vegetation in northern Cardiganshire. Trans. Inst. Brit. Geogrs. 48, 75-96.

Stanford, S. C., 1972. The function and population of hill forts in the Central Marches in Lynch, F. and Burgess, C. (eds.), Prehistoric Man in Wales and the West, Bath, 307-20.

Synge, F. M., 1977. Records of sea levels during the Late Devensian. Phil. Trans. Roy. Soc. London, B. 280, 211-28.

Taylor, J. A., 1973. Chronometers and chronicles: a study of palaeo-environments in west central Wales. Progress in Geography, vol. 5, Arnold, London, pp. 247-334.

Taylor, J. A., 1975. The role of climatic factors in environmental and cultural changes in prehistoric times in Evans, J. A. et al., (eds.), The Effect of Man on the landscape: the Highland Zone, C. B. A. Research Report No. 11, London, 6-19.

Taylor, J. A., 1980. The role of pedogenic factors in the initiation of peat formation and in the classification of mires. 6th Intern. Peat Congress, Duluth, U.S.A. (in press).

Thomas, R. & Dudlyke, E. R., 1925. A flint-chipping floor at Aberystwyth. J. Roy. Anthrop. Inst., 55, 73-89.

Watson, E., 1977. Mid and North Wales. INQUA Excursion Guide C.9. Geo Abstracts, U.E.A., Norwich, England.

Williams, R. G. B., 1975. The British climate during the Last Glaciation: an interpretation based on periglacial phenomena. Geol. J. Spec. Issue, 6, 95-120.

Woillard, G., 1978. Grande Pile Peat Bog: a continuous pollen record for the last 140,000 years. Quaternary Research, 9, 1-21.

Woillard, G., 1979. The Last Interglacial-glacial Cycle at Grande Pile in Northeastern France. Bull. Soc. Belge de Géologie, 88, 51-69.

Wymer, J. J., 1977. The archaeology of man in the British Quaternary in Shotton, F. W. (ed.) British Quaternary Studies: Recent Advances. Clarendon Press, Oxford.

APPENDIX I

A list of radiocarbon dates for sites in Wales: compiled by W. P. Seymour, B. Sc., in 1979.

Notes

(1) The sites are listed primarily according to COUNTY. Within each county, the letters and numbers, in brackets and underlined, refer to (i) the county (e.g. Cl for Clwyd, D for Dyfed, MG for Mid-Glamorgan, WG for West Glamorgan, GT for Gwent, GD for Gwynedd, P for Powys), (ii) the ranking according to age, oldest first, per county. For sites with more than one date, the oldest is quoted for purposes of ranking.

(2) Measurements remain as in 'Radiocarbon'.

(3) Nomenclature remains as in 'Radiocarbon' e.g. 'Würm' has been used when applicable.

(4) Where items of information (e.g. material, stratigraphy, laboratory number etc.) have been omitted in the source, referential comment has been added where possible.

(5) A list of references cited is given at the end of the Appendix.

(6) This listing is clearly liable to correction and up-dating. The Editor will be grateful to receive information in this connection, including details of new dates.

National Grid ref.	Site name/ location	Site description	Sample lab. no.	Sample material	Sample stratigraphy/comments	Date with errors	General comments with references
	CLWYD						
SJ 5138 (CL10)	Fenns' Moss, Flint	*Sphagnum* peat from various levels in peat works. Site A	Q392	*Sphagnum* peat	From just above the uppermost recurrence surface. 38-40 cm	655 ± 90	
			Q393	*Sphagnum* peat	Just below Q392 42-44 cm. Immediately below recurrence surface	745±90	Agrees with Q392. Near values for retardation layers at Tregaron
		Sphagnum peat from various levels in peat works. Site B	Q434	*Sphagnum* peat	Immediately above recurrence surface	1670±110	
			Q435	*Sphagnum* peat	Immediately below recurrence surface		Godwin, H. & Willis, E. H. (1960), Hardy, E. M. (1939)
SJ189305 (CL6)	Yogwennant, Denbigh	Burial Pit associated with beaker sherds	Birm-85	Charcoal	From within pit	3423±82	Shotton, F. W. *et al*. Day, W. (1972)
SJ 086724 (CL1)	Cae Gwyn Caves, Tremerchion	200 m E. N. E. of Ffynnon Beuno farm at entrance to Cae Gwyn Cave	Birm-146	Collagen from mammoth carpal bone	Sample underlying Upper Boulder Clay	18,000+1400 -1200	Provides limit to Irish Sea Glaciation which deposited Upper Boulder Clay. Shotton, F. W. & Williams R, E. G. (1971) Rowlands, B. M. (1971)

National Grid ref.	Site name/ location	Site description	Sample lab. no.	Sample material	Sample strati- graphy/comments	Date with errors	General comments with references
SJ 138712 (CL8)	Caerwys, Flintshire	Tufa deposit containing microlithic industry	BM-255	Collagen from post cranial of almost complete skeleton	Skeleton within tufa	2100±140	Skeleton previously considered Mesolithic. Date shows burial relatively recent and most probably intrusive. Barker, H., et al. (1971) Jackson, J. W. (1956)
SJ025779 (CL2)	Rhuddlan	Mesolithic settlement. Plentiful chert industry of Maglemosian type	BM-691	Carbonised hazel nut shells	Samples date blade industry containing non-geometric microliths of type comm common around coast of Wales	8739±86	Industry associated with BM822 more advanced than that associated with BM691: dates as expected. Burleigh, R. et al. (1976)
			BM-822	Carbonised hazel nut shells		8528±73	
SH983572 (CL4)	Brenig, Glyndwr	Principally Bronze-age site threatened by construction of a new reservoir.					
		Ring Cairn, Brenig 44, believed to have had 3 periods of major activity	HAR-500	Charcoal	Sample from central cremation	3490±70	Date later than HAR501 suggesting secondary phase of activity

National Grid ref.	Site name/ location	Site description	Sample lab. no.	Sample material	Sample strati-graphy/comments	Date with errors	General comments with references
			HAR-501	Charcoal	Sample from original pared ground surface	3630±100	Charcoal pit dug against face of ring in 1st phase of major activity
			HAR-502	Charcoal	Sample from plant incorpora-ted in outer bank	3470±70	2nd phase of activity
			HAR-503	Charcoal	Sample surround-ing urns	3230±70	
			HAR-504	Charcoal	Pit dug into top of inner bank	3290±80	Dates last burial at close of major activity at site. 3rd phase of activity
			HAR-505	Charcoal	From layer two of inner bank	3470±80	Second phase of activity on outer bank Otlet, R. L. (1977) Lynch, <u>et al.</u> (1974)
			HAR-535			3530±100	(From Burgess herein)
		Brenig 51	HAR-801			3510±70	(From Burgess herein)
			HAR-803			3500±70	(From Burgess herein)
			HAR-802			3420±70	(From Burgess herein)

National Grid ref.	Site name/ location	Site description	Sample lab. no.	Sample material	Sample strati-graphy/comments	Date with errors	General comments with references
SJ212691 (CL5)	Moely Gaer Rhosemor	Samples from Iron Age hillfort	HAR-603	Charcoal (twigs)	Sample from soil buried beneath tail of second rampart	2190±80	Terminus post quem for construction of 2nd rampart
			HAR-604	Charcoal	From packing of foundation trench for front revetment timbers of 1st rampart	2530±90	
			HAR-605	Charcoal	From occupation dirt within recti-linear area sealed beneath Iron Age rampart	3590±80	Not surprising as there is a scatter of late Neolithic to E. B. A. artifacts
			HAR-606	Charcoal	From occupation soil of Post-Ring Round House sealed under 1st rampart	2570±70	Otlet, R. L. (1977)
			SRR-497		Early layer in Iron Age hill fort	2994±40	(From Savory, H. N. herein)
SH968757	Dinorben, Denbigh	Hillfort	V123		Occupation (Pd. O) under Pd. 1 rampart	2895±95	

National Grid ref.	Site name/ location	Site description	Sample lab. no.	Sample material	Sample strati- graphy/comments	Date with errors	General comments with references
			V122	Charcoal	From collapsed beam to N. E. of Pd. 1 rampart	2845±95	
			V125	Charcoal	From ?Pd. 1 rampart	2715±85	
			V124	Charcoal	From ?Pd. II rampart	2485±85	
			V176	Charcoal	From ?Pd. II rampart under slaked lime	2370±70	
			V175	Charcoal	From within lime heap of Pd. II	2100±80	
			V177	Charcoal	From Romano-British occupation	1670±80	Lavell, C (-1970) Savory, H. N. (1971a) (1971b)
SJO832 (CL9)	Berwyn Mountain Series	Blanket bog, 8 km N. of Llaurhoeadr	UB-997	Peat	143-144 cm of section 157 cm deep, alt. 821 m Moelsych. Sample from above very compacted basal peat layers from summit-ridge non-pool-system peat	1595±40	

National Grid ref.	Site name/ location	Site description	Sample lab. no.	Sample material	Sample strati- graphy/comments	Date with errors	General comments with references
			UB-998	Peat	142–143 cm of section 310 cm depth, alt. 649 m Ygodor 2 pool system. Taken from central *Sphagnum* peat	1995±40	Cross matched to main Berwyn pollen diagram. 10 cm below increase in pollen of grazing, arable farming, open woodland indicator species, referable to Roman period
			UB-999	Peat	35–36 cm from same face as UB-998	195±60	Contamination suggested
			UB-1000	Peat	240–241.5 cm from same face as UB-998 & 999. Taken from slower forming layer		Cross matched to 250–252 cm on main Berwyn pollen diagram. 20 cm below Late Bronze Age major expansion of arable and pastoral agri- culture. General: Dates to calculate rate of peat formation in various horizons in a study of build up and erosion of pool-system peat in Berwyn blanket bog. Pearson, G. W. (1979)

National Grid ref.	Site name/ location	Site description	Sample lab. no.	Sample material	Sample strati- graphy/comments	Date with errors	General comments with references
SH9159 (CL3)	Aled Isof Reservoir	Exposures on shores of Aled Isaf Reser- voir associated with finds of flint artefacts			Oh. horizon	580 B. P.	Study by D. A. Jenkins & K. Stahr on genesis of ironpan stagno- podzols of Hiraethog series in relation to human occupation mentioned in Clayden, B. (1977)
					Wood fragments from base of Oh.	4010 B. P.	
					Iron Pan	1420 B. P.	
					Charcoal associa- ted with flints at base of Bs. horizon	5810 B. P.	
	DYFED						
SN687623 (D10)	Tregaron, Ceredigion	Dates from various layers in the S. E. raised bog with par- ticular concern for a 'retardation layer' of very highly humified peat	Q75	Very humi- fied peat	From centre of humified layer	760±70	
			Q390	18 in.-19 in. *Sphagnum* peat	Immediately above the humi- fied layer	768±90	Pollen analysis shows deforestation
			Q391	25in-26in *Sphagnum*	Immediately below humi- fied layer	1477±90	Surprisingly older than Q390
			Q388	66 in.-66.75 in. Moder- ately humi- fied *Sphagnum* peat	Immediately above main recurrence surface of bog	2669±111 & 2624±111	Corresponds with a pollen zone boundary, possibly sub-Boreal/ sub-Atlantic.
			Q389	67.5 in.- 68.25 in. highly humi- fied *Sphagnum* peat	Immediately under Q388 and just below main recurrence surface	3029±110 & 2879±110	Sufficiently older than Q388 to show bog growth had greatly slowed on ceased just before regeneration

National Grid ref.	Site name/ location	Site description	Sample lab. no.	Sample material	Sample strati- graphy/comments	Date with errors	General comments with references
							began above recur- rence surface. Godwin, H. & Willis, E. H. (1960) Turner, J. (1964)
		Pollen zone boundary determin- ations	Q930	413-416 cm detritus mud	Beginning of organic deposi- tion Betula- Pinus-Juniperus zone	10,205±220	
			Q931	404-406 cm Fen peat	At end point of detritus mud. Betula-Pinus- Corylus zone opens	9747±220	
			Q932	398-400 cm Fen peat	Middle of Corylus rise. Ulmus curve becomes continuous	9550±200	
			Q933	390-392 cm Fen peat	End of Corylus rise. Corylus- Pinus zone opens	9303±190	
			Q934	354-356 cm Fen peat with pine wood	Pollen frequen- cies of Pinus begin to exceed those of Betula. Pinus-Corylus- Ulmus zone opens	8285±150	

National Grid ref.	Site name/ location	Site description	Sample lab. no.	Sample material	Sample strati- graphy/comments	Date with errors	General comments with references
			Q935	348-350 cm Fen peat		8152±150	Check on Q934
			Q936	248-250 cm Fen peat with birch and pine wood	First of a series of three dating _Alnus_ rise	7128±180	
			Q937	242-244 cm Fen peat	Mid-point of rise of _Alnus_ curve. _Quercus-Ulmus-Alnus_ zone begins	6991±180	
			Q938	236-238 cm Fen peat	Alnus curve begins to rise	6978±140	
			Q939	228-230 cm Fen peat with wood	End of _Alnus_ rise	6526±110	
			Q940	214-216 cm Ombrogo- genous peat	Humified _Sphagnum_ and _Calluna_ peat character- ising raised bog. _Tilia_ curve becomes continuous	5979±100	

National Grid ref.	Site name/ location	Site description	Sample lab. no.	Sample material	Sample strati- graphy/comments	Date with errors	General comments with references
			Q941	172-174 Sphagnum- Eriophorum- Calluna peat	Beginning of main Ulmus decline	5110±70	
			Q942	167-169 Sphagnum- Erio- phorum- Calluna peat	Mid-point of Ulmus decline. Quercus-Alnus zone opens	4990±70	
			Q943	163-165 cm Sphagnum- Erio- phorum- Calluna peat	Last of series dating Ulnus decline	4893±70	
			Q944	152-154 cm Sphagnum- Erio- phorum- Calluna peat	Increase in frequency of Plantago pollen indicating a maximum of anthropogenic activity	4715±55	
			Q945	144-146 cm Sphagnum- Erio- phorum- Calluna peat	Pollen diagram shows rise in Ulmus and Fraxinus	4695±52	
			Q946	86-88 cm Fresh om- brogenous peat	Increase in Plantago pollen indicating a maximum anthro- pogenic effect	3336±50	

National Grid ref.	Site name/ location	Site Description	Sample lab. no.	Sample material	Sample strati- graphy/comments	Date with errors	General comments with references
			Q947	62–64 cm More humi- fied ombro- genous peat	Level shows further anthro- pogenic activity	2922±50	Switsur, V. R. & West, R. G. (1972) Hibbert, F. A. & Switzur, V. R. (1976)
		Tregaron S. E. bog	Q596	82–84 cm _Sphagnum_ peat	Sample from well below retardation layer		Beginning of phase of increased agricultural activity associated with a change from temporary clearance to more permanent pasture. Date places clearance phase in early part of the Iron-Age. Godwin, H. & Willis, E. H. (1962)
SM883347 (D1)	Aber Mawr, Pembs.	Coastal exposure N. of Mathry	NPL-98	Wood (_Pinus_)	Sample from calcareous boulder clay	>40,300	Stratigraphic evidence suggests Irish Sea till no older than shelly outwash at Mullock Bridge and Tre-llys. Callow, W. J. _et al._ (1966)
			Gr. N-5281	Wood	4–5 m depth. Below ice contact sand and gravel, _c._ 3 m calcareous Irish Sea Till con- taining the sample lies on solifluction sediments		Indicates early Würm or older. Compare with NPL-98. Vogel, J. C. & Waterbolk, H. T. (1972) John, B. S. (1965)

National Grid ref.	Site name/ location	Site description	Sample lab. no.	Sample material	Sample strati-graphy/comments	Date with errors	General comments with references
SN351101	Llanstephan Castle, Carmathens.	Stands on a promontory on the Tyri estuary, cut off by a complex series of earthworks thought to be of prehistoric origin	HAR-475	Charcoal	From shallow ditch, probably relating to orifinal timber revetted construction of bank C	2450±90	
			HAR-476	Charcoal	From primary silt in ditch X, possibly antedating timber fronted rampart	2460±70	
			HAR-477	Charcoal	From occupation soil with postholes, sealed beneath primary earthwork of 12th century A.D. Ring and Bailey Castle	2470±70	Dates show castle stands within an Iron Age hillfort. Otlet, R. L. (1977) Guilbert, G. C. (1974) Guilbert, G, C. & Schweiso, J. J. (1972)
SR8999 (D17)	Freshwater West Pembs.	Submerged peat at low tide mark on foreshore	Q530	Wood peat	4-7 cm resting on boulder clay associted with a microlithic industry	5960±120	Pollen analyses confirm post-Boreal date. Macro-remains indicate fen-wood. Date quite late for Mesolithic culture. Godwin, H. & Willis, E. H. (1964)

National Grid ref.	Site name/ location	Site description	Sample lab. no.	Sample material	Sample strati- graphy/comments	Date with errors	General comments with references
SM811080 (D4)	Mullock Bridge, Pembs.	Gravel pit in raised beach near Dale	NPL-80	Shell frag- ments (bulk sample)	From gravel pit	37,960 +1700-1400	Date suggests two possibilities. a) Several other high raised beaches in S. Britain may have been wrongly termed Monastirian. b) Pem- brokeshire has suffered isostatic uplift since the last glaciation. Callow, W. J. et al. (1965) John, B. S. (1965)
			Birm-209	Inner (a) and outer (b) fractions of marine molluscs ((Littorina)	+ 18 m altitude in gravel pit	a) 38,160 +1600-1330 b) 35,740 + 2250-1760	Closely agrees with NPL-80. Shotton, F. W. & Williams, R. E. G, (1971) John, B. S. (1967)
SM898349 (D5)	Tre Llys, N. Pembs.	Gravel pit, 200 ft. O.D. near St. Nicholas, Pembs.	NPL-97	Marine shell	In outwash gravels laid down during melting of last Irish Sea Ice sheet	37,310± 1515-1275	Date shows that gravel of the so-called S. Wales End Moraine are no younger than deposits found further S. (NPL 80). Evidence of interstadial Boreal conditions 38000 B.P.

National Grid ref.	Site name/ location	Site description	Sample lab. no.	Sample material	Sample strati-graphy/comments	Date with errors	General comments with references
							and last glaciation of W. Wales followed this. Callow, W. J. et al. (1966) John, B. S. (1965) (1967)
SN690711 (D23)	Llanafan, Ceredigion	Sloping hillside in Ystwyth valley.	Q814	Charcoal	Sample from B horizon of soil of Denbigh series. Charcoal 0.5 in. thick exposed for 3 feet laterally, overlain by 2 feet of yellow brown silt-loam. Apparently in situ above a thin burned layer	2133±110	Possibility of Pre-Roman and Iron-Age forest clearance and husbandry inducing erosion. Godwin, H. & Switsur, V. R. (1966)
SN2909 (D19) (D3)	Coygan Camp Carms.	Neolithic settlement	NPL-132	Partly charred hazlenut shells	Sample from pit C XIX	5000±95	Emphasizes early colonization of W. Seaboard. Callow, W. J. & Hassall, G. I. (1968)
	Coygan Cave Carms.	Cave 25 km S. W. of Carmarthen. Bone accumulation and Monsterian artifacts sealed by 2 calcite layers	Gr N-4399	Calcite1	Top 3 cm of Upper calcrete layer on floor of cave. Pure fresh water carbonate	3000±35	C13 content indicates 1300±500 yr must probably be subtracted from result to obtain actual age

National Grid ref.	Site name/ location	Site description	Sample lab. no.	Sample material	Sample strati-graphy/comments	Date with errors	General comments with references
			Gr. N-4423	Calcrete	Bottom 2 cm of Upper calcrete layer	5820±65	1300±500 yr should be subtracted
			Gr. N-4400	Calcrete	Top 3 cm of lower calcrete layer underlying main bone accumulation and Mousterian industry	33,200±310	1300±500 yr should be subtracted. Date apparently too young. Vogel, J. C. & Waterbolk, H. T. (1972)
			BM-499	Collagen from reindeer antler	Sample dates Mousterian of Acheulian tradition	38,684 +2713-2024	Satisfactory date compared to Gr. N-4400. Burleigh, R. et al. (1976)
SN587839 (D13)	Clarach, Ceredigion	Submerged forest bed on Clarach beach	NPL-113	Wood (*Pinus sylvestris*)	Sample from tree root *in situ*	5970±90	First evidence suggesting correlation with submerged forest at Borth. Callow, W. J. & Hassall, G. I. (1968) Taylor, J. A. (1973)
		Immediately behind storm beach at mouth of R. Clarach	Birm-229	Peat	Basal peat lens at *c.* -2 m O.D.	10,100±250	Confirms pollen analysis which suggested lateGlacial age. Shotton, F. W. & Williams, R. G. E. (1971) Smith, R. J. & Taylor, J. A. (1969)

National Grid ref.	Site name/ location	Site description	Sample lab. no.	Sample material	Sample strati-graphy/comments	Date with errors	General comments with references
		Submerged forest exposure	GU-662	Wood (Quercus)	From exposure	5592±143	Campbell, J. A. & Baxter, M. S.
SM862173 (D6)	Druidston, St. Brides Bay, Pembs.	Coastal exposure of Irish Sea Till capped by outwash deposits	I-1687	Molluscan fragments	Stratigraphy shows Irish Sea till is same age as outwash deposits	736,000	Considered to represent Main Würm age. Buckley, J. D. *et al.* (1968) Synge, F. M. (1964)
SN205479 (D2)	Banc-y-Warren, Ceredigion	Kame complex north of Cardigan	I-2559	Organic silt	Discontinuous horizon in overlying strati-fied sand, 70 ft. below top of section	31,800 +1400-1200	Date supports ex-tensive main Würm glaciation in Irish Sea Basin
			I-2802	Organic silt	In quarry section	>39,900	Buckley, J. D. & Willis, E. H. (1969) Brown, M. J. F. *et al.* (1967)
			I-2564	Wood fragments	From layer of organic lenses 17-18 ft. below top of thick out-wash	33,750 +2500-1900	Date supports theory of fluctuating inter-stadial in Irish Sea and for Main Würm glaciation. Buckley, J. D. & Willis, E. H. (1969)
SN473638 (D9)	Aberaeron Ceredigion	Exposed cliff section on beach N. of Aberavon	I-3120	Organic silt	Lenses in base of 10 ft. cliff	11,200±160	Pollen analysis sup-ports C14 date of cliff section being post-glacial. Buckley, J.D. & Willis E. H. (1969) Taylor. J. A. (1973)

National Grid ref.	Site name/ location	Site description	Sample lab. no.	Sample material	Sample strati-graphy/comments	Date with errors	General comments with references
SS147968 (D8)	Caldey Island, Tenby	Ogof-yr-Ychen cave N. E. coast of Island, c. 13.7 m altitude	Birm-340	Collagen from scapula of rhinoceros	c. 3.7 m deep in unstraified limestone rubble filling fissure on floor of cave	22,350±620	Consistent with cold fauna of late Devensian period. Magdalenian artifact found 0.5 m above dated bone. Shotton, F. W. & Williams, R. E. G. (1973)
SN467477 (D12)	Cledlyn, Lampeter	Pingo basin, 'W' at Cledlyn 12 km W, of Lampeter. Organic mud passing up into peat and over-lying 7 m of grey clayey silt	Birm-367	Peat	2.745-2.795 m below surface at centre of Pingo W.	9020±400	
			Birm-369	Peat	2.87-2.92 m below surface of bog in Pingo W	8910±290	Deviations large because of small sample size. Initial thick silt filling after collapse of Pingo ante-dates pollen zones IV and V
			Birm-368	Organic silt	Combined sample 2.90-2195 m in auger hole W, and 3.025-3.075 m in auger hole W2	9380±340	

National Grid ref.	Site name/ location	Site description	Sample lab. no.	Sample material	Sample strati-graphy/comments	Date with errors	General comments with references
		Augered samples from sediment filling collapsed pingos 'M' and 'U' a) after alkali treatment b) humic extract	Birm-485		Pingo M, W6, Bed 4 from boreholes A-D	a) 8980±450 b) 9550±290	
			Birm-486		Pingo M, W6 Bed 5 from boreholes A&B	a) 9760±230 b) 9420±190	
			Birm-487		Pingo M, W6 Bed 5 from boreholes C & D	a) 8760±310 b) 9470±190	Small samples, therefore high standard deviations. Inconsistency probably due to contamination. Dates suggest that organic sedimentation started at Pingo U at junction of pollen zones III & IV and at Pingo M slightly later.
			Birm-488		Pingo U, W5, bottom 5 cm of bed 4, boreholes a + b.	a) 9630±240 b) 10, 060±380	
			Birm-489		Pingo U, W5, bottom 5 cm of bed 4 boreholes C & D	a) 10, 160±260 b) 10, 080±320	Shotton, F. W. & Williams, R. E. G. (1973), Shotton, F. W. et al. (1975), Watson, E. (1971), Watson E, (1972), Watson, E. & S. (1972), Watson, E. (1975)

National Grid ref.	Site name/ location	Site description	Sample lab. no.	Sample material	Sample strati-graphy/comments	Date with errors	General comments with references
SN7868 (D18)	Llyn Feifi, Ceredigion	Exposure by lake shore at 1328 ft. O.D.	Q394	Pine wood	Basal layer of pine stumps on mineral soil and overlain by blanket bog	5261±120	Surrounding hills now bare of wood-land. Godwin, H. & Willis, E. H. (1960)
SN6089 (D16)	Tynslas, Ceredigion	Submerged peat bed exposed on foreshore of Cardi-gan Bay near the Dovey Estuary	Q380	Birch wood	In brushwood peat at base of submerged peat bed	6026±135	Peat overlies blue silt. Pollen zone VIIa age
			Q382	Brushwood peat	From base of peat bed -1 to -2 ft. O.D.	5898±135	The basal wood peat clearly grew very rapidly. Godwin, H. & Willis, E. H. (1961) Godwin, H. (1943)
	Borth Bog, Ceredigion	Raised bog shel-tered from Cardigan Bay on W. by big coastal dune ridge. To N. bog approaches tidal marshes on the Dovey Estuary	Q712	Juncus peat	Sample from base of clayey Juncus maritimus peat above Phrag-mites peat. Re-presents time near the culmination of a marine trans-gression	2900±110	Appears to fall in early part of pollen zone VIII. Godwin, H. & Willis, E. H. (1964)
SN614888	Borth	Exposure in drainage ditch	SSR-377 SSR-378	Cellulose from wood (Pinus)	Under 0.9 m al-luvium and 0.3 m silty peat	1100±50 1200±50	Sediments probably channel fill estuarine silt of R. Leri. Harkness, D. D. & Wilson, H. W. (1979)

National Grid ref.	Site name/ location	Site description	Sample lab. no.	Sample material	Sample strati- graphy/comments	Date with errors	General comments with references
SN3818 (D20)	Llanllwych Carmarthen	Raised bog in Towey valley 2 miles W. of Carmarthen	Q458	*Sphagnum* peat	1-2.5 cm above the one consistent and pronounced recurrence layer	3230±110	Dates in good agreement and about the same as for other British peat bogs: Godwin, H. & Willis, E. H. (1962), Thomas, K. W. (1964)
			Q459	*Sphagnum* peat	1-2.5 cm below the recurrence layer	3178±100	
SN0539 (D15)	Newport, Pembs.	Mid tide peat layer	HAR-78	Wood	Submerged forest stock from -0.6 m O.D.	6370±150	Date agrees with others from similar material from this level at other points in Cardigan Bay. Otlet, R. L. & Slade, B. S. (1974).
SN412499 (D11)	Cletwr, Ceredigion	Grey clay interbedded with peat overlying 14 cm grey clay at Cletwr pingo, 10 km S.S.E. of New Quay	Birm-389		Pingo K, W4, 3.55-3.59 m below bog surface	8260±300	Younger than expected probably due to contamination
			Birm-388		Pingo K sample W3 3.5-3.56 m below bog surface	10,170±220	Earlier than basal organic material in Cledlyn pingo. Cletwr basin shallower, therefore ice expected to melt out more quickly and organic sedimentation begin earlier. Shotton, F. W. *et al.* (1974)

National Grid ref.	Site name/ location	Site description	Sample lab. no.	Sample material	Sample strati-graphy/comments	Date with errors	General comments with references
SN3866 (D14)	Trawling Ground New Quay	Hydrocore site ZZ27, 4.8 km of New Quay	Birm-400	Peat	2.55 to 2.60 m below ocean bed, -26.55 to -26.60 m altitude	8740±110	Maximum for Flandrian transgression at site. Shotton, F. W. et al. (1974)
SM915173	Walesland Rath, Pembs.	Fortified settlement	NPL-245	Charcoal	From layer with crucible and glass beads stratified between phase 1 and 2 banks	2160±90	
			NPL-244	Charcoal	From ash layer with crucible on top of primary ditch silt	2035±90	Lavell, C. (-1970) Wainwright, G. J. (1971)
SN6979 (D24)	Rheidol, Nr. Aberystwyth	Bed of Rheidol	GU-668	Wood (*Quercus*)	From gravel on R. bed	647±136	Campbell, J. A. & Baxter, M. S. (1979)
	MID GLAMORGAN						
SN932024 (MG2) Exact G. R. unknown	Cefn Glas (Craig-y-Llyn) Blaenrhondda	Late Neolithic hut floor	HAR-744	Charcoal	Sample from hut floor make up material	4110±70	Date consistent with late Neolithic character of flints found at the site. Otlet, R. L. (1977)

National Grid ref.	Site name/ location	Site description	Sample lab. no.	Sample material	Sample strati- graphy/comments	Date with errors	General comments with references
Exact G.R. unknown (MG1)	Pen-y-Bont Ogmore-on-Sea	Late Neolithic occupation layer	BM-1112	Charcoal	Sample from near base of layer	4659±52	From Savory, H. N. (herein)
			HAR-1140	Hazel nuts	From near top of layer	4320±80	
	SOUTH GLAMORGAN						
SS957663 (SG1)	Llantwit Major	Breakwater at Col-hugh Point	I 2556	Oak (*Quercus robur*)	Remains of piles from breakwater	550±95	Piles believed to be Roman, but date indicates medieval. Indication of rates of eustatic fluctuations of sea level and rates of coastal erosion. Buckley, J. D. & Willis, E. H. (1969)
	WEST GLAMORGAN						
SS7785 (WG5)	Margam	Coastal peat beds 2ft. 9in. thick overlain by grey sand and silty clays and underlain by grey silty clays	Q265	*Phragmites* peat	0.4 cm upper surface of peat	3402±108	Dates end of marine retrogression on standstill
			Q274	Aquatic pool peat of Sphagna and Hypnaceae	Close to boundary between a lower ombrogenous *Sphagnum-Calluna* peat and an upper aquatic peat passing over into eutrophic fen peat	5605±126	Change to eutrophic conditions after this date indicative of rising ground water

National Grid ref.	Site name/ location	Site description	Sample lab. no.	Sample material	Sample strati-graphy/comments	Date with errors	General comments with references
(WG4)			Q275	Grey silty clay	Base of same peat bed where transitional from brackish to freshwater conditions. Level -10.5 ft. O.D.	6184±143	Dates indicate 3000 years of freedom from marine transgression. Godwin, G. & Willis, E. H. (1964)
SS7784 (WG3)	Raglan Burrows, Pt. Talbot	Submerged peats. Bore 19. Basal compacted peat, c. 20 cm thick at -63.1ft. O.D., over gravel in sandy silt matrix, covered by sands and silty clays to +24.4 ft. O.D.	Q660	Phragmites peat with Salix scales and nuts of Lycopus	4.5-6 cm. Taken at pollen zone boundary III/IV	10,370±170	Pollen evidence shows peat extended from pollen zone IV into zone VI
			Q661	Freshwater peat	8.5-10 cm. Sample across pollen zone IV/V boundary	9920±170	
			Q662	Fresh water peat	12.5-14 cm Sample across pollen zone V/VI boundary	8990±170	
			Q663	Fresh water peat	20-21 cm. Within pollen zone VI	8970±150	Surprisingly close to Q662. Indicates postglacial eustatic rise of sea level above -62ft. O.D. after or at this date

National Grid ref.	Site name/ location	Site description	Sample lab. no.	Sample material	Sample strati- graphy/comments	Date with errors	General comments with references
		Bore 4. Peat bed 5ft. thick over gravel at 64.6 ft. O.D.	Q664	Fine detritus mud	12 cm from peat resting on gravel	11,980±180	Pollen of general Late-glacial charac- ter
		Succeeded by sands, clays and silts to +28.4ft. O.D.	Q665	Fine	3 cm from 9 cm above Q664	11,260±170	Late-glacial char- acter of pollen but with higher _Betula_/ _Pinus_ ratio. Dates and pollen suggest Late-glacial age. Evi- dence from boreholes 19 and 14 reinforce idea that an irregular surface of drift, freshwater organic muds and peats ac- cumulated locally from zone II to zone VI when the eustatic rise of sea-level intervened. Godwin, H. & Willis, . E. H. (1964)
SS414873 (WG1)	Fall Bay	'Patella beach', regarded as interglacial	Birm-46	_Patella_ shells a) Middle CO_2 fraction b) Inner fraction	Shells from raised beach	a) 21,815±629 b) 22,796±827	Dates anomalous. Error due to contamination. Shotton, F. W. _et al._ (1968)

National Grid ref.	Site name/ location	Site description	Sample lab. no.	Sample material	Sample strati- graphy/comments	Date with errors	General comments with references
SS4377 (WG2)	Goats Hole, (Paviland Cave) Gower	Ochre burial of a young male. The so called "Red Lady of Paviland"	BM-374	Collagen from left femur and tibiae	Burial associated with upper Palaeo- lithic (Aurignacian) industry	18,460±340	Barker, H. et al. (1969), Oakley, K. P. (1968), John, B. S. (1971)
SS554868 (WG6)	Minchin Hole, Gower	Coastal Cave site Talus overlying raised beach, itself overlain by a succes- sion of cave earth, sand, sandrock and head	Birm-615	Stalagmite	From middle layer of inner talus	4500±100	Much younger than expected probably due to dissolution and reprecipitation of $CaCO_3$ during Holocene. Williams, R. E. G. & Johnson, A. S. (1976) Bowen, D. Q. (1966) Sutcliffe, A. J. & Bowen, D. Q, (1973)
	GWENT						
ST3888 (GT2)	Llanwern, Newport	Peat from contact with an overlying marine clay at +96ft. 6in. O.D. to 10ft. O.D.	Q691	Peat	Peat band 1ft. thick overlain by 5ft. silty marine clay and 4ft. made ground. Underlain by about 12ft grey blue clay and sands and gravels	2660±110	Godwin, H. & Willis, E. H. (1964), Locke, S. (1974)

National Grid ref.	Site name/ location	Site description	Sample lab. no.	Sample material	Sample strati- graphy/comments	Date with errors	General comments with references
ST444853 (GT1)	Undy Wharf, Monmouths.	Submerged forest exposed on foreshore	IGS-C14/32 (St 3279)	Wood	Sample from peat band overlain by recent clay	3485±100	Welin, E. et al. (1972)
SO365068 (GT3)	Coed-y-Bwyndd Bettws Newydd	Hillfort excavations	HAR-546	Charcoal	From carbonized remains of wattle wall framework, of 2nd phase of round-house	2390±70	Otlet, R. L. (1977)
			HAR-567	Charcoal	As HAR-546	2350±90	Babbidge, A. V. (1970-71)
	GWYNNEDD						
SH2329 (GD16)	Mynydd Rhiw, Lleyn	Neolithic axe-factory site	Q387	Charcoal	From quarry pits	3094±90	Too young for Neolithic date. Either contamination or charcoals secondary but no evidence to support latter. Godwin, H. & Willis, E. H. (1961), Houlder, C. (1961)
SH632633 (GD9)	Nant Ffrancon, Bethesda Caerns	Lacustrine sediments in glacial basin. Pollen zone boundary determinations	Q890	Silty lake mud	608-620 cm. First organic deposition. Juniperus pollen curve rising. In Betula-Pinus-	10,080±220	

National Grid ref.	Site name/ location	Site description	Sample lab. no.	Sample material	Sample strati- graphy/comments	Date with errors	General comments with references
					Juniperus pollen assemblage zone 597-608 cm		
			Q890(bis)	Silty lake mud	In Betula-Pinus-Juniperus zone at peak of Juniperus curve	9915±220	
			Q891	Lake mud	587-591 cm. At decline of Juniperus pollen curve in Betula-Pinus-Juniperus zone	9820±200	
			Q892	Lake mud	584-587cm near end of fall of Juniperus pollen curve in Betula-Pinus-Juniperus zone	9745±200	
			Q893	Lake mud	570-573 cm. Near end of fall of Juniperus pollen curve. In Betula-Pinus-Juniperus pollen assemblage zone.	9631±200	

National Grid ref.	Site name/ location	Site description	Sample lab. no.	Sample material	Sample strati-graphy/comments	Date with errors	General comments with references
			Q894	Lake mud	562-565 cm. At point when Corylus expansion begins. Opening of Betula-Pinus-Corylus zone	9098±180	
			Q895	Lake mud	557-560 cm. Beginning of Ulmus curve. In Betula-Pinus-Corylus zone	8932±170	
			Q896	Lake mud	551-554 cm. Dating rise of Corylus. In Betula-Pinus-Corylus zone	8809±170	
			Q897	Lake mud	546-549 cm. Beginning of Corylus-Pinus pollen assemblage zone	8642±150	
			Q898	Lake mud	528-531 cm. At opening of Pinus-Corylus-Ulmus pollen assemblage zone. Alnus curve begins here	8455±150	

National Grid ref.	Site name/ location	Site description	Sample lab. no.	Sample material	Sample strati- graphy/comments	Date with errors	General comments with references
			Q899	Lake mud	509-512 cm. In Pinus-Corylus-Ulmus zone at level of Corylus peak	8118±120	
			Q899(bis)	Lake mud	502-508 cm	8162±120	
			Q900	Lake mud. Slightly more organ-ic	420-423 cm. On-set of rise of Alnus pollen curve and beginning of Quercus-Ulmus-Alnus pollen assemblage zone.	6884±110	
			Q901	Organic lake mud	410-413 cm. Alnus pollen rising rapidly in Quercus-Ulmus zone	6790±100	
			Q902	Organic lake mud	400-403 cm. At completion of Alnus rise in Quercus-Ullmus-Alnus zone	6726±100	

National Grid ref.	Site name/ location	Site description	Sample lab. no.	Sample material	Sample stratigraphy/comments	Date with errors	General comments with references
			Q903	Organic lake mud	215-218 cm. Immediately before Ulmus decline	5160±70	
			Q904	Coarse wood peat	207-210 cm. Ulmus pollen declining and weed pollen increasing in frequency.	5054±70	
			Q905	Organic lake mud with wood fragments	190-193 cm. Last of 3 samples dating Ulmus decline	4870±60	
			Q906	Wood fragments with organic lake mud	170-172 cm. Frequency of Ulmus pollen shows a slight recovery	4422 ±60	
			Q907	Wood peat	164-166 cm. Ulmus pollen lower with weeds of cultivation reaching high frequencies	4256±50	Onset of major deforestation. Results compare with dates from Tregaron Switsur, V. R. & West, R. G. (1973) Hibbert, F. A. & Switsur, V. R. (1976)

National Grid ref.	Site name/ location	Site description	Sample lab. no.	Sample material	Sample strati- graphy/comments	Date with errors	General comments -with references
SH7277 (GD15)	Penmaenmawr, Caerns.	Stone circle 278	N. P. L. -10	Carbonised wood	From old surface beneath stoney bank	3355±155	Associated with scraps of pottery
			N. P. L. -11	Carbonised wood	From old surface against main upright	3470±145	Dates erection or at least use of circle
		Monument 280	N. P. L. -12	Carbonised wood	From contents of fire pit dug into old surface of circle	3080±145	Dates a time during use of circle. Callow, W. J. et al. (1963) Griffiths, W. E. (1962)
SH5256 (GD3)	Moel Tryfan, Caerns.	Exposure in quarry at 396 m	I 2303	Mixed shell fauna	From gravel over-lying 6 to 8 ft. of till and under-lying 40ft. similar till	33,740 +2100-1800	Glaciation postdates this. Indicates con-temporaneity of Welsh and Irish Sea glacia-tions during main Würm. Buckley, J. D. & Willis, E. H. (1969)
SH4845 (GD6)	Brincir, Caerns.	Fluvioglacial deposit exposed at Clynnog Hawr, Brincir Moraine, 11.5 miles S. of Caernarfon	I 2801	Organic silts	Isolated nodules of silts scattered throughout 35 ft. sand and gravel sequence. Sample 15-20ft. below top of section	16,830± 970-860	Date supports re-advance of Irish Sea and Welsh ice post-this date, depositing moraine. Buckley, J. D. & Willis, E. H. (1969)

National Grid ref.	Site name/ location	Site description	Sample lab. no.	Sample material	Sample strati- graphy/comments	Date with errors	General comments with references
SH553259 (GD1)	Mochras, Merioneth.	U. C. W. Aberystwyth, Dept. of Geology, Borehole, W. of Llanbedr	Birm-77	Lignitic plant re- mains	Remains in silty clay at 100.4 m depth under 53 m of till	>47,500	Originally claimed as Pleistocene, now re- garded as Cainozoic. Shotton, F. W. et al. (1969), Wood, A. & Woodland, A. W. (1969)
SH449373	Glannllynau, Caernarfon	Kettle Hole with well documented pollen and Coleop- teran spectra	Birm-212	Moss	Washed from silty clay 1.34 m below base of zone I detritus mud	14,468±300	Dates start of Late- glacial infilling of kettle hole. Arctic/ subarctic Coleopteran assemblage
			Birm-232	Plant debris	From silty clay 22.5 to 25 cm below base of zone I detritus mud	11,714±255	
			Birm-233	Plant debris	From silty clay 15 to 17.5 cm below base of zone I detritus mud	11,617±270	Shotton, F. W, & Williams, R. E. G. (1971), Boulton, G. S. (1977)
			Birm-276	Terrestrial seeds	From Late-Glacial detritus mud	12,560±230	Deposit contained rich thermophilous insect assemblage associated with low arboreal pollen frequency. Shotton, F. W. & Williams, R. E. G. (1973)

National Grid ref.	Site name/ location	Site description	Sample lab. no.	Sample material	Sample strati- graphy/comments	Date with errors	General comments with references
			GaK-1603	Bulk sample		12,850±250	Coope, G. R. & Brophy, J. A. (1972) Saunders, G, E. (1968) Simpkins, K. (1974) Coope, G. R. (1977)
SH238408 (GD2)	Morfa Nefyn Caernarfon	Exposure 25 m W. of beach road	Birm-302	Wood a) after alkali pre-treatment b) Humose extract	Washed from organic debris from thick succession of fluvio-glacial sand and gravel overlying till	a) 37,300 +1630–1360 b) 43,000	Sample must be derived. Shotton, F. W. & Williams, R. E. G. (1973)
SH775819 (GD11)	Llandudno	Borehole near railway station, 3.99 m O.D.	SSR-61	Peat	8.15–9.20 m depth overlain by estuarine clay, beach deposits and blown sand. Underlain by estuarine clay on Irish Sea till	7635±52	Helps date last major rise in level of Irish Sea. Harkness, D. D. & Wilson, H. W. (1974)
SN665987 (GD14)	Cefn Cynhafel Merioneths.	Peat bog 10.5 km E of Towyn	SSR-74	Wood	From peat	4225±55	Harkness, D. D. & Wilson, H. W. (1974)
SH362850 (GD13)	Bedd Branwen, Llanbabo, Anglesey	Samples associated with Bronze Age urns and other pottery from an excavation of a barrow	BM-452	Charcoal	From base of hollow close to central stone	4923±75	Predates barrow construction

National Grid ref.	Site name/ location	Site description	Sample lab. no.	Sample material	Sample strati- graphy/comments	Date with errors	General comments with references
			BM-453	Charcoal	Contents of Pigmy cup over- lying bones	3224±81	Dates primary burial
			BM-455	Charcoal	From cremation deposit on old ground surface	3257±80	Secondary burial inserted after barrow complete
			BM-456	Charcoal	From old ground surface	3353±60	Dates barrow construction. Burleigh, R. et al. (1976), Lynch, F. (1971)
SH606492 (GD18)	Dinas Emrys, Beddgelert Caernarfons.	Defended chieftain's stronghold 1.6 km N. E. of Beddgelert	Birm-659	Wood	From earliest posthole, 1.2 m deep	650±100	Date unexpectedly late. Sample originally thought to be Iron-Age with site extending into 12th century. Williams, R. E. G. & Johnson, A. S. (1976), Savory, H. N. (1960)
SH275415 (GD4)	Porth Dinllaen, Llyn	Drift exposure	?	Marine shells	In drift	31,800±?	Referred to in Lewis, C. A. (1970) p. 346
SH289247 (GD5)	Porth Neigwl, Llyn	Drift exposure	?	Marine shells	In drift	29,000±?	Referred to in Lewis, C. A. (1970) p. 346
SH594710 (GD12)	Llandegai, Caerns.	Henge Monuments	NPL-222		From cremations outside entrance of Henge II	3740±145	
			NPL-221		End of primary silting in ditch, Henge I	4420±140	

National Grid ref.	Site name/ location	Site description	Sample lab. no.	Sample material	Sample strati- graphy/comments	Date with errors	General comments with references
			NPL-224		Cremation circle outside entrance of Henge I	4480±140	
			NPL-220		Primary fire pit Henge I	4740±150	
			NPL-223		From postholes preceeding henges	5240±150	Renfrew, C. (ed.) (1974), Houlder, C. (1968)
SH823182 (GD10)	Bwlch-y-fign, Dinas Mawddwy	Valley bog 610 m above sea level	Birm-902	Peat	Onset of organic accumulation at the site	8330±120	Date at transition between pollen zones IV and V when, despite a continental climate, the wetness of the site initiated peat formation, aided by high rainfall and low evaporation at this altitude. (Taylor, J. A., _in press_)
SH6017 (GD17)	Llanaber	Submerged forest exposure	GU-665	Wood (_Larix_)	From exposure	734±52	Campbell, J. A. & Baxter, M. S. (1979)

National Grid ref.	Site name/ location	Site description	Sample lab. no.	Sample material	Sample strati- graphy/comments	Date with errors	General comments with references
	POWYS						
SO 214981 (p. 9)	Hen Domen, Montgomery	Castle built c. 1070 A.D.	Birm-133	Charcoal	From soil layer buried by ram- part and above pebble floor of pre-rampart building	970±290	Shotton, F. W., et al. (1970)
SN 800916 (p. 3)	Uwchygaireg, Machynlleth	Hilltop peat 2.44 m deep, weathered into vertical face and resting on weathered mudstones	I. G. S. C-14/103 (St 3834)	Peat	From basal 0.08 m	5570±100	
			I. G. S. C-14/104 (st 3827)	Peat	0.84 m above base	4065±100	
			I. G. S. C-14/105 (St 3818)	Peat	1.22 m above base	3640±100	Welin, E., et al. (1974)
SN 735980 (p. 1)	Glaspwll, Machynlleth	Ditch cut in hill peat	I. G. S. -C14/131 (St 3873)	Wood	From bottom of ditch, 0.91 m below top of peat	5345±100	
			I. G. S. -C14/132 (St 3867)	Plant remains	From diatomaceous silt beneath 1.22 to 1.52 m of dark peat	11,430±160	
			I. G. S. -C14/133 (St 3820)	Peat	0.08 m layer under- lying 0.3 m-0.6 m fawn silty gravel over- lying 0.6 m fine solifluction gravel	7515±100	Welin, E., et al. (1974)

(contd.)

National Grid ref.	Site name/ location	Site description	Sample lab. no.	Sample material	Sample strati-graphy/comments	Date with errors	General comments with references
SN800916 (P3)	Uwchygarreg, Machynlleth	Hilltop peat 2.44 m deep, weathered into vertical face and resting on weathered mudstones	IGS C-14/103 (St3834)	Peat	From basal 0.08 m	5570±100	
			IGS C14/104 (St3827)	Peat	0.84 m above base	4065±100	
			IGS C14/105 (St3818)	Peat	122 m above base	3640±100	Welin, E. et al. (1974)
SN735980 (P1)	Glaspwll, Machynlleth	Ditch cut in hill peat	IGS C14/131 (St3873)	Wood	From bottom of ditch, 0.91 m below top of peat	5345±100	
			IGS C14/132 (St3867)	Plant remains	From diatomaceous silt beneath 1.22 to 1.52 m of dark peat	11,430±160	
			IGS C14/133 (St3820)	Peat	0.08 m layer underlying 0.3m-0.6 m fawn silty gravel overlying 0.6 m fine solifluction gravel	7515±100	Welin, E. et al. (1974)

National Grid ref.	Site name/ location	Site description	Sample lab. no.	Sample material	Sample strati- graphy/comments	Date with errors	General comments with references
SN 735980 (P. 2)	Garthgwinion Machynlleth	Ditch	SSR-147	Plant debris	Lenses from c. 0.9 m depth in stoneless clay/ silt	2937±55	
			SSR-148	Silty peat	From 0.9 m depth in 1.06 to 1.22 m thick band	3270±45	
			SSR-149	Silty peat	From base of 0.6 m thick layer overlying 0.6 m solifluction gravel	4159±55	
			SSR-150	same as SSR-149	Rootlets removed	4104±55	
			SSR-152	Silty peat	From base of 0.44 m thick seam overlying c. 0.7 m silty solitfluction gravel	5473±40	
			SSR-153	same as SSR-152	Rootlets <u>not</u> removed	5395±40	
			SSR-151	Peat	From base of c. 0.9 m thick layer overlying c. 0.15 m seam of soliflucted gravel	6894±60	Harkness, D. D. & Wilson, H. W. (1974)

(contd.)

National Grid ref.	Site name/ location	Site description	Sample lab. no.	Sample material	Sample strati- graphy/comments	Date with errors	General comments with references
			SSR-148	Silty peat	From 0.9 m depth in 1.06 to 1.22 m thick band	3270±45	
			SSR-149	Silty peat	From base of 0.6 m thick layer overlying 0.6 m solifluction gravel	4159±55	
			SSR-150	Same as SSR-149	Rootlets removed	4104-55	
			SSR-152	Silty peat	From base of 0.44 m thick seam overlying c. 0.7 m silty solifluction gravel	5473±40	
			SSR-153	Same as SSR-152	Rootlets not removed	5395±40	
			SSR-130	Gytta	310-314 cm from base of sediment core in organic lake mud. Pollen as SSR-129	2410±60	
SO139261		Collected from 6 m of water c. 200 m off-shore and 500 m N. of stream entry	SSR-131	Gytta	255-259 cm. From transition zone immediately below red-brown silty clay. Pollen analysis not yet complete	1681±60	Dates beginning of accelerated soil erosion

National Grid ref.	Site name/ location	Site description	Sample lab. no.	Sample material	Sample strati- graphy/comments	Date with errors	General comments with references
SO 139265	LG 10	Collected in 7.6 m of water 200 m off- shore and 900 m N. of major stream entering lake from S	SSR-129	Lacustrine gytta	240-244 cm from transition zone immediately below red brown silty clay. Presence of pollen of Fagus, Fraxinus, Tilia, Ulnus, Quercus and other trees, and 30% non arboreal pollen including cereals and weeds	1790±60	Dates beginning of accelerated soil erosion in catch- ment
			SSR-130	Gytta	310-314 cm from base of sediment core in organic lake mud. Pollen as SSR 129	2410±60	
SO 139261		Collected from 6 m of water c. 200 m offshore and 500 m N. of stream entry	SSR-131	Gytta	255-259 cm. From transition zone im- mediately below red-brown silty clay. Pollen analysis not yet complete	1681±60	Dates beginning of accelerated soil erosion

National Grid ref.	Site name/ location	Site description	Sample lab. no.	Sample material	Sample strati-graphy/comments	Date with errors	General comments with references
			SSR-132	Gytta	302-306 cm. From base of sediment core in organic lake mud. Pollen as for SSR 129	2508±70	With SSR-130 dates a stable period in land use. Harkness, D. D. & Wilson, H. W. (1974)
SO 151316 (P. 4)	Pen-y-Wyrlod, Talgarth, Brecon	Long cairn	HAR-674	Human bone	From long cairn in charnel deposit in Chamber NE II	4970±80	1st date published for Welsh laterally chambered long cairn. Otlet, R. L. (1977)
SJ 292144 (P. 6)	Breiddin	Hillfort. Late Bronze Age and Iron Age occupations with earlier late Neolithic/Early Bronze Age activity	BM-798	Charcoal (whittled willow or poplar)	From *in situ* remains of haft of late Bronze Age socketed axe.	2704±50	Descriptions and comments updated by C. R. Musson, November 1979.
			BM-878	Charcoal (species not identified)	From paired postholes of early hillfort defence	2750±41	
			BM-879	Charcoal (oak; socket shows timber c. 150 mm diam)	From posthole in continuous foundation trench of early hillfort defence	2778±71	

(contd.)

National Grid ref.	Site name/ location	Site description	Sample lab. no.	Sample material	Sample strati- graphy/comments	Date with errors	General comments with references
			BM-880	Charcoal (species not identified)	From occupation deposit behind early hillfort defence	2818±64	
			BM-881	Charcoal (species not identified)	From burning on line of round- house wall, behind later hill- fort defence	2429±55	
			BM-882	Charcoal (species not identified)	From pit or bowl-hearth in interior of hillfort	3826±106	Apparently belongs with Late Neolithic/ Early Bronze Age phase of occupation
			BM-883	Charcoal (young oak, hazel and sloe)	From intrusion into back of late hillfort defence	2059±105	
			BM-884	Charcoal (ash, and perhaps other species)	From throughout a posthole of 4 post structure in interior	2188±70	Not demonstrably structural timber *in situ*
			BM-885	Charcoal (species not identi- fied)	From basal deposits below hillfort defence	3024±62	Expected date Late Neolithic/Early Bronze Age. Possible contamination by later charcoal ?

National Grid ref.	Site name/ location	Site description	Sample lab. no.	Sample material	Sample strati-graphy/comments	Date with errors	General comments with references
			BM-963	Charcoal (hazel and perhaps other species)	From pit tentatively associated with roundhouse in interior	2325±63	
			BM-964	Charcoal (hazel)	From posthole of 4 post structure in interior	2244±40	From post-socket 400-500 mm diam.
			BM-965	Charcoal (oak and perhaps other species)	From 3 postholes of 4-post structure behind later hill-fort defence	2122±45	Small amounts of charcoal from post-sockets
							Dates indicate complex sequence of occupations. First hillfort defence (BM-798, 878-80) is preceded by earlier activity on hilltop (BM-882 & 885). Roundhouses (BM-881 & 963) follow after construction of later defence, followed by a 4 post structure (BM 884 & 964) Burleigh, R. <u>et al.</u> (1976)

National Grid ref.	Site name/ location	Site description	Sample lab. no.	Sample material	Sample strati- graphy/comments	Date with errors	General comments with references
			HAR-467	Charcoal (mixed species)	From postholes of roundhouse porch in interior	2410±100	Not demonstrably structural timber in situ
			HAR-468	Charcoal (various species)	From combination of small samples from 13 postholes of 4 posted buildings in interior	2190±80	Not demonstrably structural timber in situ
			HAR-469	Charcoal (oak)	From posthole of 4-post structure in interior	2120±70	From post-socket 400-500 mm across. Otlet, R. L. (1977) Musson, C. R. (1970) Musson, C. R. (1972)
SN 827933 (P5)	Glaslyn, Nr. Machynlleth	Eroding peaty inter-fluve sloping to the headwater of the Afon Clywedog 1 km S. of Lake Glaslyn	I-4928	Peat	Base of 1.2 m section overlying fossilised silty gravel represent-ing compressed remains of a bured soil of pre-peat age	3870±100	Pollen at this point indicates substantial woodland cover dominated by Corylus, Quercus & Alnus but including Betula, Pinus and Tilia. High Ericaceae pollen Taylor, J. A. (1973)
SO 1091 (P10)	Newton	River Severn Valley	GU-667	Wood (Quercus)	From gravel river bed	885±83	Campbell, J. A. & Baxter, M. S. (1974)

National Grid ref.	Site name/ location	Site description	Sample lab. no.	Sample material	Sample strati- graphy/comments	Date with errors	General comments with references
SN 792975 (P7)	Pantybedw Machynlleth	Freshly cut drainage ditch	SSR-137	Silty peat	1.28-1.38 m below surface from base of layer 0.46 m thick overlying clay and beneath 0.92 m surface peat	3609±40	
			SSR-138	Wood	from base of surface peat 0.92 m below surface	3345±40	
			SSR-139	Wood (Quercus)	Embedded in upper level of surface peat, depth 0 to 0.61 m below surface	1835±40	Sample dated represented initial 40 yr of tree growth
SN 797971	Melin Dulas Machynlleth		D.D.R.-140	Wood (Quercus)	From peat accumulated on surface of soliflucted till	3330±45	Harkness, D. D. & Wilson, H. W. (1974)
SO 1326 (P8)	Llangorselake Brecon	2 sites along north-south axis of a trough where water depth greater than elsewhere in the lake			2 stratigraphic members distinguished in both cores: an upper red-brown silty clay overlying very dark brown organic mud		

APPENDIX I

REFERENCES

Babbidge, A. V. (1971). Excavations at Coed-y-Bwnydd, Bettws Newydd. Monm. Antiq. 3 (1) pp. 59-60.

Barker, H., Burleigh, R., Meeks, N. (1969). British Museum natural radiocarbon measurements VI Radiocarbon 11 pp. 272-294.

Barker, H., Burleigh, R., Meeks, N. (1971). British Museum natural radiocarbon measurements VII Radiocarbon 13 pp.157-188.

Boulton, G. S. (1977). A multiple till sequence formed by a Late Devensian Welsh ice cap, Glanllynnau, Gwynedd. Cambria 4 (1) pp. 10-31.

Bowen, D. Q. (1966). Dating Pleistocene events in south west Wales. Nature 211 pp. 475-6.

Brown, M.J. F., Ellis-Gruffydd, I. D., Foster, H. D., Unwin, D. J. (1967). A new radiocarbon date for Wales, Nature 213 pp.1220-1.

Buckley, J. D., Trautman, M. A., Willis, E. H. (1968). Isotopes radiocarbon measurements VI Radiocarbon 10 pp. 246-94.

Buckley, J. D., & Willis, E. H. (1969). Isotopes radiocarbon measurements VII Radiocarbon 11 pp. 53-105.

Burleigh, R., Hewson, R., & Meeks, N. (1976). British Museum natural radiocarbon measurements VIII Radiocarbon 18 pp. 16-42.

Burrows, C. J. (1976). Radiocarbon dates from Late Devensian deposits at Nant Ffrancon, Caernarfon. New Phyt. 75 pp.167-71.

Callow, W. J., Baker, M. J., & Pritchard, D. H. (1963). National Physical Laboratory radiocarbon measurements I Radiocarbon 5 pp. 34-38.

Callow, W. J., Baker, M. J., & Hassall, G. I. (1965). National Physical Laboratory radiocarbon measurements III Radiocarbon 7 pp. 156-61.

Callow, W. J., Baker, M. J., & Hassall, G. I., (1966). 'National Physical Laboratory radiocarbon measurements Radiocarbon 8 pp. 340-7.

Callow, W. J., & Hassall, G. I. (1968). National Physical Laboratory radiocarbon measurements V Radiocarbon 10 pp. 115-118.

Campbell, J. A., & Baxter, M. S. (1979). Glasgow University radiocarbon measurements IX. Radiocarbon 21, pp. 171-179.

Clayden, B. (1977). Palaeosols. Cambria 4 (1) pp. 84-97.

Coope, G. R. (1977). Coleoptera as clues to the understanding of climatic changes in north Wales towards the end of the last (Devensian) Glaciation. Cambria 4 (1) pp. 65-72.

Coope, G. R., & Brophy, J. A. (1972). Late-Glacial environmental changes indicated by a Coleopteran succession from north Wales. Boreas 1 pp. 97-142.

Day, W. (1972). The excavation of a Bronze-age burial mound at Ysgwennant, Llansilin, Denbighshire. Arch. Camb. 121 pp. 17-50.

Godwin, H. (1943). Coastal peat beds of the British Isles and North Sea. J. Ecol. 31:2 pp. 199-247.

Godwin, H, & Switsur, V. R. (1966). Cambridge University natural radiocarbon measurements VIII. Radiocarbon 8 pp. 390-406.

Godwin, H., & Willis, E. H. (1960). Cambridge University natural radiocarbon measurements II. Radiocarbon 2 pp. 62-72.

Godwin, H., & Willis, E. H. (1961). Cambridge University natural radiocarbon measurements III. Radiocarbon 3 p. 60-76.

Godwin, H., & Willis, E. H. (1962). Cambridge University natural radiocarbon measurements V. Radiocarbon 4 pp. 57-70.

Godwin, H., & Willis, E. H. (1964). Cambridge University natural radiocarbon measurements VI. Radiocarbon 6 pp. 116-137.

Griffiths, W. E., (1962). Dating the Bronze age in north Wales. Proc. Prehist. soc. 28 p. 387.

Guilbert, G. C. (1974). Llanstephan Castle: 1973 interim report. Carm. Antiq. 10 pp. 37-44.

Guilbert, G. C., & Schweiso, J. J. (1972). Llanstephan Castle—an interim report of the 1971 excavation. Carm. Antiq. 8 pp. 75-90.

Hardy, E. M. (1939). Studies of the Postglacial history of British Vegetation V: The Shropshire and Flint Maelor Mosses. New Phyt. 38 pp. 364-96.

Harkness, D. D., & Wilson, H. W. (1974). Scottish Universities Research and Reactor Centre radiocarbon measurements II. Radiocarbon 16 pp. 238-51.

Harkness, D. D..& Wilson, H. W. (1979). Scottish Universities Research and Reactor Centre radiocarbon measurements III. Radiocarbon 21 pp. 203-256.

Hibbert, F. A., & Switsur, V. R. (1976). Carbon-14 dating of Flandrian pollen zones in Wales and northern England. New Phyt. 77 pp. 783-807.

Houlder, C. (1961). The excavation of a Neolithic stone implement factory on Mynydd Rhiw, Caernarfon. Proc. Prehist. Soc. 28, p. 108-43.

Houlder, C. (1968). The Henge monuments at Llandegai. Antiquity 42 pp. 216-21.

Jackson, J. W. (1956). The Caerwys Tufa. Liverpool and Manchester Geological Journal i (4) p. xxiv-xxviii.

John, B. S. (1965). A possible Main Würm glaciation in west Pembrokeshire. Nature 207 pp. 622-3.

John, B. S. (1967). Further evidence for a middle Würm interstadial of S. W. Wales Geol. Mag. 104 pp. 630-33.

John, B. S. (1971). The 'Red Lady' of Paviland: a comment. Antiquity 45 pp. 141-44.

Lavell, C. (1971). Archaeological Site Index To Radiocarbon Dates for Great Britain and Ireland C. B. A.

Lewis, C. A. (1970). The Glaciations of Wales and Adjoining Regions. Longman.

Locke, S. (1974). The Postglacial deposits of the Caldicot Level and some associated archaeological discoveries. Monm. Antiq. 3 pp. 1-16.

Lynch, F. (1971). Report on the re-excavation of two Bronze-age Cairns in Anglesey: Bedd Branwen and Treionwerth. Arch. Camb. 120 pp. 11-83.

Lynch, F., Waddell, J., Allen, D., & Grealey, S. (1974). Brenig Valley excavations 1973 interim report. Denbs. Hist. Trans 23 pp. 9-64.

Musson, C. R. (1970). The Breiddin 1969. Current Archaeol. 19 pp. 215-18.

Musson, C. R. (1972). Two winters at The Breiddin. Current Archaeol. 33 pp. 263-67.

Oakley, K. P., (1968). The date of the 'Red Lady' of Paviland. Antiquity 45 pp. 141-14.

Otlet, R. L. (1977). Harwell radiocarbon measurements II Radiocarbon 19 pp. 400-423.

Otlet, R. L., & Slade, B. S. (1974). Harwell radiocarbon measurements I Radiocarbon 16 pp. 178-91.

Pearson, G. W. (1979). Belfast radiocarbon dates IX Radiocarbon 21 pp. 274-290.

Renfrew, C. (Ed.) (1974). British Prehistory: A New Outline. Duckworth.

Rowlands, B. M. (1971). Radiocarbon evidence of the age of an Irish Sea Glaciation in the Vale of Clwyd. Nature 230 p. 112.

Saunders, G. E. (1968). A Fabric analysis of the ground moraine deposits of the Lleyn Peninsula of S. W. Caernarvonshire. Geol. Jnl. 6 pp.105-18.

Savory, H. N. (1960). Excavations at Dinas Emrys, Beddgelert, Caernarvonshire. 1954-56. Arch. Camb. 109 pp. 13-77.

Savory, H. N. (1971a). A Welsh Bronze-age hillfort. Antiquity 45 pp. 251-61.

Savory, H. N. (1971b). Dinorben 1965-69. Cardiff.

Shotton, F. W., Blundell, D. J., & Williams, R. E. G. (1968). Birmingham University radiocarbon dates II Radiocarbon 10 pp. 200-206.

Shotton, F. W., Blundell, D. J., & Williams, R. E. G. (1969). Birmingham University radiocarbon dates III Radiocarbon 11 pp. 268-70.

Shotton, F. W., Blundell, D. J., & Williams, R. E. G. (1970). Birmingham University radiocarbon dates IV Radiocarbon 12 pp. 385-399.

Shotton, F.W., & Williams, R. E. G. (1971). Birmingham University radiocarbon dates V Radiocarbon 13 pp. 141-56.

Shotton, F. W., & Williams , R. E. G. (1973). Birmingham University radiocarbon dates VII Radiocarbon 15 pp. 451-68.

Shotton, F. W., Williams, R. E. G., & Johnson, A. S. (1974). Birmingham University radiocarbon dates VIII Radiocarbon 16 pp. 285-303.

Shotton, F. W., Williams, R. E. G., & Johnson, A. S. (1975). Birmingham University radiocarbon dates IX Radiocarbon 17, p. 255-75.

Simpkins, K. (1974). The Late glacial deposits at Glanllynnau, Caernarvonshire. New Phyt 73 pp. 605-18.

Smith, R. T. & Taylor, J. A. (1969). The Postglacial development of soils and vegetation in N. Cardiganshire. Trans. Inst. Brit. Geogr. 48 pp. 75-96.

Sutcliffe, A. J., & Bowen, D. Q. (1973). Preliminary report on excavations in Minchin Hole, April-May 1973. William Pengelly Cave Studies Trust Newsletter 21 pp. 12-25

Switsur, V. R. , & West, R. G. (1972). University of Cambridge natural radiocarbon measurements X Radiocarbon 14. pp. 239-246.

Switsur, V. R., & West, R. G. (1973). University of Cambridge natural radiocarbon measurements XI Radiocarbon 15. pp. 156-64.

Synge, F. M. (1964). The age of Irish Sea till at west Pembrokeshire. Geol. Assoc. Proc. 75 p. 431.

Taylor, J. A. (1973). Chronometers and chronicles; a study of palaeoenvironments in west-central Wales. Prog. Geogr. 5 p. 247-334.

Thomas, K. W. (1964). The stratigraphy and pollen analysis of a raised bog at Llanllwch near Carmarthen. New Phyt. 63 pp. 101-17.

Turner, J. (1964). The anthropogenic factor in vegetational history 1: Tregaron and Whixall Mosses. New Phyt 63 pp. 73-90.

Vogel, J. C., & Waterbolk, H. T. (1972). Groningen radiocarbon dates X Radiocarbon 14. pp. 6-110.

Wainwright, G. J. (1971). Britannia 2, 48 pp. 48-108.

Watson, E. (1971). Remains of Pingos in Wales and the Isle of Man. Geol. Jnl. 7. pp. 381-92.

Watson, E. (1971). Pingos of Cardiganshire and the latest ice limits. Nature 236 pp. 343-4.

Watson, E. (1975). Guide to Field Excursions. Co-ordinating Committee for Periglacial Research symposium pp.1-10.

Watson, E. & S. (1972). Investigations of some Pingo basins near Aberystwyth, Wales. 24th Int. Geol. Cong., Montreal. Sec. 12 pp221-23.

Welin, E., Engstrand, L., & Vaczy, S. (1972). Institute of Geological Sciences radiocarbon dates II Radiocarbon 14 p. 140-44.

Welin, E., Engstrand, L., & Vaczy, S. (1974). Institute of Geological Sciences radiocarbon dates V Radiocarbon 16 pp. 95-104.

Williams, R. E. G., & Johnson, A. S. (1976). Birmingham University radiocarbon dates X. Radiocarbon 18 pp. 249-267.

Wood, A, & Woodland, A. W. (1969). Borehole at Mochras, west of Llanbedr, Merionethshire. Nature 219 pp. 1352-4.

www.ingramcontent.com/pod-product-compliance
Ingram Content Group UK Ltd.
Pitfield, Milton Keynes, MK11 3LW, UK
UKHW061843041026
15591UKWH00015B/895

9 780860 540793